Biogenic Trace Gases: Measuring Emissions from Soil and Water

METHODS IN ECOLOGY

Series Editors

J.H. LAWTON FRS
Imperial College at Silwood Park
Ascot, UK

G.E. LIKENS
Institute of Ecosystem Studies
Millbrook, USA

Biogenic Trace Gases: Measuring Emissions from Soil and Water

EDITED BY

P.A. MATSON

Department of Environmental Science, Policy and Management
University of California, Ecosystem Sciences Division
Berkeley, California, USA

R.C. HARRISS

National Aeronautics and Space Administration,
Mission to Plant Earth, Washington, DC 20546, USA

Blackwell
Publishing

© 1995 by
Blackwell Science Ltd
Editorial Offices:
Osney Mead, Oxford OX2 0EL
25 John Street, London WC1N 2BL
23 Ainslie Place, Edinburgh EH3 6AJ
238 Main Street, Cambridge
  Massachusetts 02142, USA
54 University Street, Carlton
  Victoria 3053, Australia

Other Editorial Offices:
Arnette Blackwell SA
1, rue de Lille, 75007 Paris
France

Blackwell Wissenschafts-Verlag GmbH
Kurfürstendamm 57
10707 Berlin, Germany

Blackwell MZV
Feldgasse 13, A-1238 Wien
Austria

First published 1995
Transferred to digital print 2002

Set by Best-set Typesetter Ltd., Hong Kong

Printed and bound in Great Britain by
Marston Lindsay Ross International Ltd, Oxford

DISTRIBUTORS

Marston Book Services Ltd
PO Box 87
Oxford OX2 0DT
(*Orders*: Tel: 01865 791155
        Fax: 01865 791927
        Telex: 837515)

USA
Blackwell Science, Inc.
238 Main Street
Cambridge, MA 02142
(*Orders*: Tel: 800 215-1000
        617 876-7000
        Fax: 617 492-5263)

Canada
Oxford University Press
70 Wynford Drive
Don Mills
Ontario M3C 1J9
(*Orders*: Tel: 416 441 2941)

Australia
Blackwell Science Pty Ltd
54 University Street
Carlton, Victoria 3053
(*Orders*: Tel: 03 347-5552)

A catalogue record for this title
is available from the British Library

ISBN 0–632–03641–9

Library of Congress
Cataloging-in-Publication Data

Biogenic trace gases: measuring emissions from
  soil and water edited by P.A. Matson and
  R.C. Harriss.
      p.   cm.—(Methods in ecology)
  Includes bibliographical references and index.
  ISBN 0-632-03641-9
  1. Atmospheric chemistry—Technique.
  2. Bioclimatology—Technique.
  3. Biogeochemistry—Technique.
  4. Agricultural ecology—Technique.
  I. Matson, P.A. (Pamela A.) II. Harriss,
  Robert C. III. Series.
  QC879.6.B566 1995
  574.5′222—dc20   94-38066

Contents

List of contributors

J.H. BUTLER *National Oceanic and Atmospheric Administration, R/E/CG1, 325 Broadway, Boulder, CO 80303—3328, USA*

J.P. CHANTON *Department of Oceanography, Florida State University, Tallahassee, FL 32306—3048, USA*

D.J. COOPER *Plymouth Marine Laboratory, Prospect Place, Plymouth PL1 3DH*

P.M. CRILL *Institute for the Study of Earth, Oceans and Space, University of New Hampshire, Morse Hall, Durham, NH 03824—3525, USA*

E.A. DAVIDSON *Woods Hole Research Center, PO Box 296 Woods Hole, MA 02543, USA*

F. FEHSENFELD *NOAA Environmental Research Laboratories, 325 Broadway, Boulder, CO 80303—3328, USA*

R.C. HARRISS *National Aeronautics and Space Administration, Mission to Planet Earth, Washington, DC 20546, USA*

G.L. HUTCHINSON *USDA-ARS-NPA—Natural Resources Research Centre, Soil-Plant-Nutrient Research Unit, PO Box E, Fort Collins, CO 80522—0470, USA*

C.E. KOLB *Centre for Chemical and Environmental Physics, Aerodyne Research, Inc., Billerica, MA 01821—3976, USA*

D.H. LENSCHOW *National Center for Atmospheric Research, PO Box 3000, Boulder, CO 80307, USA*

G.P. LIVINGSTON *NASA Ames Research Center, Earth Sciences Division—SGE: 239-20, Moffett Field, CA 94035—1000, USA*

S. MacINTYRE *Marine Science Institute, University of California, Santa Barbara, CA 93106, USA*

P.A. MATSON *Department of Environmental Science, Policy and Management, University of California, Berkeley, 108 Hilgard Hall, Berkeley, CA 94720—3110, USA*

P.C. NOVELLI *Cooperative Institute for Research in Environmental Science, R/E/CG1, University of Colorado, Boulder, CO 80309—0449, USA*

C.S. POTTER *Jonson Controls World Services, NASA-Ames Research Center MS 239-20, Moffett Field, CA 94035, USA*

D.S. SCHIMEL *National Center for Atmospheric Research, PO Box 3000, Boulder, CO 80307—3000, USA*

J.P. SCHIMEL *Institute of Arctic Biology, University of Alaska, Fairbanks, AK 99775, USA*

S.E. TRUMBORE *Department of Geosciences, University of California, Irvine, CA 92717—3100, USA*

R. WANNINKHOF *NOAA, Atlantic Oceanographic and Meterological Laboratory, Ocean Chemistry Division, 4301 Rickenbacker Causeway, Miami, FL 33149, USA*

G.J. WHITING *NASA Langley Research Center, MS 483, Hampton, VA 23681, USA*

J.C. WORMHOUDT *Center for Chemical and Environmental Physics, Aerodyne Research, Inc., Billerica, MA 01821—3976, USA*

M.S. ZAHNISER *Center for Chemical and Environmental Physics, Aerodyne Research, Inc., Billerica, MA 01821—3976, USA*

The Methods in Ecology Series

The explosion of new technologies has created the need for a set of concise and authoritative books to guide researchers through the wide range of methods and approaches that are available to ecologists. The aim of this series is to help graduate students and established scientists choose and employ a methodology suited to a particular problem. Each volume is not simply a recipe book, but takes a critical look at different approaches to the solution of a problem, whether in the laboratory or in the field, and whether involving the collection or the analysis of data.

Rather than reiterate established methods, authors have been encouraged to feature new technologies, often borrowed from other disciplines, that ecologists can apply to their work. Innovative techniques, properly used, can offer particularly exciting opportunities for the advancement of ecology.

Each book guides the reader through the range of methods available, letting ecologists know what they could, and could not, hope to learn by using particular methods or approaches. The underlying principles are discussed, as well as the assumptions made in using the methodology, and the potential pitfalls that could occur—the type of information usually passed on by word of mouth or learned by experience. The books also provide a source of reference to further detailed information in the literature. There can be no substitute for working in the laboratory of a real expert on a subject, but we envisage this Methods in Ecology Series as being the 'next best thing'. We hope that, by consulting these books, ecologists will learn what technologies and techniques are available, what their main advantages and disadvantages are, when and where not to use a particular method, and how to interpret the results.

Much is now expected of the science of ecology, as humankind struggles with a growing environmental crisis. Good methodology alone never solved any problem, but bad or inappropriate methodology can only make matters worse. Ecologists now have a powerful and rapidly growing set of methods and tools with which to confront fundamental problems of a theoretical and applied nature. We hope that this series will be a major contribution towards making these techniques known to a much wider audience.

John H. Lawton, Gene E. Likens

Preface

Change in the composition of the atmosphere is one of the best documented and most important of global changes, yet many of the reasons for atmospheric changes are poorly known. An understanding of the role of the biosphere as source or sink of important atmospheric trace gases, and of why sources or sinks are changing, is critical to predicting the future and controlling change.

In the past decade, research directed toward the understanding of biosphere–atmosphere gas exchange has evolved from a disciplinary effort of atmospheric scientists to an interdisciplinary effort that cuts across the fields of atmospheric chemistry, analytical chemistry, biogeochemistry, ecology, geography, microbiology, micrometeorology, and soil physics, to name a few. Today, interdisciplinary scientists and multidisciplinary teams often attempt to evaluate not just gas fluxes but the biological and physical processes that control them. Moreover, information on controlling processes is being incorporated into mechanistic models that estimate processes and gas fluxes at scales ranging from microscales to the global system.

Because the study of biosphere–atmosphere trace gas exchange is relatively new, and because the multiple approaches used in measuring gas fluxes as well as in analysing the underlying controlling processes have evolved from a diversity of scientific disciplines, no compendium of methods has been available for use by established scientists or newcomers to the field. While this book cannot cover every topic of interest, we hope that it will provide information about the range of currently available methods and approaches for measurement of trace gas exchange; analytical methods that will be available in the near future; and process studies and mathematical modelling and extrapolation approaches for describing processes and fluxes across a range of spatial and temporal scales. We present this array of methods and approaches in recognition of the fact that selection of specific approaches must be based on the scientific objectives of each particular study as well as the environmental and logistical constraints of the area being studied.

CHAPTER 1

Trace gas exchange in an ecosystem context: multiple approaches to measurement and analysis

P.A. MATSON & R.C. HARRISS

1.1 Introduction

The changing composition of the atmosphere is the best-documented and one of the most important of the on-going human-caused global changes. Rapid increases in the concentration of CO_2 (0.5%/y), CH_4 (1%/y), N_2O (0.25%/y) and a number of other radiatively and chemically important trace gases have occurred over the last several centuries (Blake & Rowland, 1988; Prinn *et al.*, 1990; Watson *et al.*, 1990). The increases in these gases are documented by a 20 year or longer record of atmospheric measurements at a range of sampling stations around the world (e.g. Rasmussen & Khalil, 1986; Prinn *et al.*, 1990; Boden *et al.*, 1992; Steele *et al.*, 1992) and by the long-term record provided through the analysis of gases trapped in air bubbles in glacial ice (e.g. Raynaud & Barnola, 1985; Raynaud *et al.*, 1988; Oeschger & Arquit, 1989). These long-term records make it clear that the rate and magnitude of change for CO_2, CH_4 and N_2O is greater than any atmospheric pertubation experienced on Earth over the last several hundreds of thousands of years.

The importance of global changes in atmospheric composition is also indisputable. Many gases contribute to the radiative balance of Earth (Houghton *et al.*, 1990, 1992); other gases have critical roles in regional and global atmospheric chemistry (Thompson, 1992; Graedel & Crutzen, 1993). Some gases, notably CO_2, also have direct effects on the plants, animals and ecosystems of Earth (Mooney *et al.*, 1987; Field *et al.*, 1992). Numerous research programmes are focused on the interactions between biological and physical components of the global system, including the International Global Atmospheric-Biospheric Chemistry Project (IGBP Report 13, see Matson & Ojima, 1990) and other projects of the International Geosphere-Biosphere Programme (IGBP Report 12, 1990) addressing these issues. The state of knowledge concerning these interactions have been detailed in a number of books and articles, including Andreae & Schimel (1989), Houghton *et al.* (1990, 1992), Turner *et al.* (1990), Schlesinger (1991), Moore & Schimel (1992), Tolba & El-Kholy (1992) and Graedel & Crutzen (1993).

Although the existence and some of the consequences of a changing atmospheric composition are well known, the reasons for the increases

1

are not always understood. For some gases (e.g. CO_2), the sources of the increase are quite clearly defined, but there is uncertainty for some of the sinks. For others (e.g. CH_4), the identity of the sources are reasonably well delineated, but their importance relative to each other is less sure. For still others (e.g. N_2O), global budgets are out of balance, suggesting that sources or sinks are not all accounted for, or that current estimates of those sources are substantially inaccurate. For many other gases, especially the short-lived gases (e.g. NO, CO, non-methane hydrocarbons (NMHC)) whose atmospheric budgets cannot easily be constrained by global atmospheric measurements, knowledge of the background sources and sinks and causes of their changes are rudimentary (Chameides et al., 1988; Jacob & Wofsy, 1990; Khalil & Rasmussen, 1990; Altshuller, 1991; Erickson & Taylor, 1992; Parrish et al., 1993).

Knowledge of the causes of change in atmospheric composition is critical as we look to the future. Predictions of global climate change, for example, are dependent on our ability both to predict future emissions of the greenhouse gases from industrial or biotic sources (Houghton et al., 1990, 1992), and to understand how the sources and sinks themselves will respond to climate change (e.g. Nisbet, 1989; Harriss & Frolking, 1992). More importantly, if the sources of the increases are not well understood, possibilities for limiting or preventing emissions are compromised (e.g. Hogan et al., 1991; Pearman, 1992). Understanding the characteristics of biogenic gas fluxes between the biosphere and atmosphere, and the processes that control them, are essential to this understanding.

In this book, we describe a range of methods that are currently used for measurement or estimation of biogenic trace gas fluxes and their controlling processes. We have limited our discussion to those gases produced primarily by microbial processes in sediments, soils and surface waters (e.g. CH_4, CO_2, N_2O, NO, NH_3, some sulphur gases), although several chapters extend their discussions to other gases as the authors see fit. We do not discuss fluxes of CO_2 between plants and atmosphere nor do we discuss NMHC emissions from plants. For discussions of methods related to these gases, we suggest Baldocchi et al. (1988), Field et al. (1989), Pearcy et al. (1989), Field & Mooney (1990), Zimmerman et al. (1990), Sharkey et al. (1991) and Enders et al. (1992).

We have sought to provide information on different approaches for measuring fluxes (e.g. enclosure and micrometeorological), and standard analytical methods for use with each flux measurement approach. We also include information on analytical methods that are likely to be available in the near future; the use of isotopes and tracers for estimating fluxes; approaches for measuring microbial processes that control gas fluxes; and application of mathematical modelling and other extrapolation approaches for describing fluxes at broader spatial and temporal scales.

Our reason for providing this combination of information is that it is rarely useful only to measure trace gas flux; for flux data to be useful, they must be placed in the context of the ecosystem processes and properties that regulate gas fluxes. Moreover, we provide this range of methods and approaches in recognition that different objectives and different environments require different approaches, and often multiple approaches.

1.2 Fluxes in the context of ecosystem processes

Exchanges of biogenic trace gas between surfaces and the atmosphere depend on the production and consumption of the gases by microbial processes, on physical transport through soils, sediments or water, and on flux across surface–air boundaries. These biological and physical processes in turn depend on other biotic and abiotic properties of ecosystems. As our goals are to:

1 develop an understanding of the factors that control flux;
2 organize our measurements so that they are useful for regional to global scale estimates; and
3 use our knowledge of controls and spatial variation to predict how anthropogenic and natural disturbances will affect fluxes in the future,

we believe it is necessary to examine trace gas fluxes in the context of these ecosystem processes and properties.

Numerous ground-based and aircraft-based studies have examined trace gas fluxes from a variety of ecosystems. From these studies it is clear that there is great variability across time, among ecosystems of a given type (e.g. tropical forests), and across ecosystems of different types (temperate vs. tropical forests, forests vs. wetlands (Keller et al., 1983, 1986; Goodroad & Keeney, 1984; Harriss et al., 1985; Sebacher et al., 1986; Matson & Vitousek, 1987; Johansson et al., 1988; Schmidt et al., 1988; Vitousek et al., 1989; Eichner, 1990; Matson et al., 1990; Sanhueza et al., 1990; Davidson et al., 1991; Williams & Fehsenfeld, 1991; Morrissey & Livingston, 1992; Bartlett & Harriss, 1993, to name a few). This spatial and temporal variability in fluxes, in fact, has been seen as one of the major impediments to estimating and predicting gas fluxes at regional and global scales. However, it can also be argued that rather than being ignored or averaged, this variability in fluxes within and between ecosystems can be used to improve estimates and understanding (Matson et al., 1989).

The basis for this argument rests on the knowledge that biogenic trace gas fluxes arise from relatively well-defined biological and physical processes, which in turn are controlled by factors that vary in space and time. Robertson (1989) viewed these factors in a conceptual sequence from the proximal regulators of a process (e.g. for denitrification: availability of

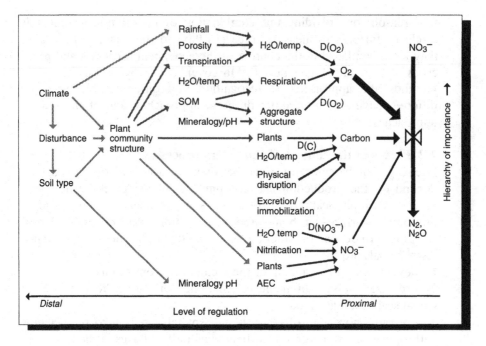

Fig. 1.1 The major factors regulating denitrification in soil (example for rain forest soils). (From Robertson, 1989.)

oxygen, nitrate and carbon; see Fig. 1.1) through a series of intermediate levels to the most distal factors (e.g. climate, soil type and disturbance). The proximal regulators often vary on the scale of small enclosures or below; they are related mechanistically to the particular processes that produce and transport gases. The distal factors, on the other hand, vary at much coarser scales (Matson *et al.*, 1989). Some, such as soil types, geological parent material and disturbance type, vary more or less discontinuously in space and time, and can be defined in discrete units. Others, such as climate, vary continuously. The matrix of these factors results in a range of ecosystems having different biogeochemical characteristics (e.g. Vitousek & Sanford, 1986), and consequently different trace gas fluxes.

Knowledge of the more proximal regulating factors provides a key both to understanding why fluxes are what they are, and to understanding the conditions under which they might change. Many trace gas studies have not attempted to measure controlling processes, in part because such measurements increase the time and resources required to study any given site. Equally importantly, until recently most of the site-based measurements of trace gas fluxes were carried out by atmospheric chemists, whereas most of the evaluations of soil and microbial processes were carried out by ecologists, soil microbiologists and soil physicists. Today, interdisciplinary researchers and research teams often attempt to evaluate not just gas fluxes but some of the more important controlling mechanisms (e.g. Whalen *et al.*, 1990; Whiting *et al.*, 1991; Davidson *et al.*, 1993; Keller *et al.*, 1993; also see Chapter 10, this volume), and information on the controlling mechanisms is being incorporated in simulation models that estimate fluxes at a range of spatial scales (e.g. Mosier and Parton, 1985; Li *et al.*, 1992; Potter *et al.*, 1993; also see Chapter 11, this volume).

Placing both flux and process measurements in the context of the coarser scale factors that regulate gas flux (e.g. climate, soil type, parent material, disturbance type, etc.) is key to extrapolation and to the development of regional and global models. Several estimates of sources of trace gas at regional and global scales have used this approach. For example, Matson & Vitousek (1987) showed that N_2O fluxes in tropical forests are positively related to indices of nitrogen turnover in the soil, and that both vary as a function of soil type; based on this information, they stratified tropical forests on the basis of soil type, climate and disturbance to organize previous N_2O flux measurements and to develop an estimate of the tropical N_2O source (Matson & Vitousek, 1990; Matson & Vitousek, in press). Matthews & Fung (1987) and Bartlett & Harriss (1993) used global data bases of vegetation, soil and inundation to stratify wetlands into major groups, and used average methane fluxes for the five groups to calculate global emissions distributions.

Not every trace gas study can afford to measure fluxes and at the same time evaluate the relative importance of the controlling mechanisms. There is a need for studies that focus primarily on the spatial and/or temporal distribution of gas fluxes in a given region or under a given set of circumstances, even if they cannot at the same time carry out mechanistic studies across the range of sites. Nevertheless, these studies should define the more distal regulating factors like soil type, climate, disturbance history, soil moisture and temperature characteristics, and basic soil characterization (see Chapter 10, this volume), in order that the gas flux data be placed in a biogeochemical context that contributes to the understanding, extrapolation and prediction of fluxes. Moreover, we suggest that the matrix of more distal factors provide a framework in which to develop hypotheses and to select sites that will further the understanding of global sources and sinks of trace gases.

1.3 Multiple approaches for estimation of flux
The chapters in this book discuss a number of different approaches for measurement of trace gas fluxes between soils, sediments and water

surfaces and the atmosphere. The authors outline the advantages and limitations of the different flux measurement approaches and analytical approaches, and they indicate the conditions and range of questions under which the various approaches are most useful. It is clear from these discussions that no single measurement approach will work for all gases in all situations. Choice of the approach or combination of approaches that should be used depends in part on the scientific question being asked, but the physical characteristics of the sites being studied, the analytical capabilities with regard to the gases being measured, and the funding and facilities available to the investigator must also be considered.

In the earlier days of trace gas studies, researchers commonly looked at only one gas per study and used a single measurement approach. As an interdisciplinary science of biosphere–atmosphere interactions continues to mature, an increasing number of investigations are addressing fluxes of several different gases per study, using a range of analytical approaches and sometimes a variety of flux measurement techniques. Moreover, many more of these studies include measurements of critical ecosystem properties. The use of multiple approaches can result in a much broader, more synthetic understanding of fluxes and the factors that control them, at scales ranging from square metres to hundreds of square kilometres.

One good example of a multidisciplinary, multiple-approach study of trace gas exchange between the biosphere and atmosphere is the Arctic Boundary Layer Experiment (ABLE-3A). In ABLE-3A, an integrated set of ground-based, airborne and satellite measurements were used to characterize the influence of Alaskan tundra ecosystems on the chemistry and dynamics of the lower atmosphere (Harriss et al., 1992). Enclosure measurements and ecosystem characterization were combined with satellite data to determine regional fluxes of methane (Bartlett et al., 1992; Morrissey & Livingston, 1992). Micrometeorological eddy correlation techniques were used on ground-based towers and on research aircraft to measure independently methane fluxes at local to regional spatial scales (Fan et al., 1992; Ritter et al., 1992).

In the ABLE-3A study, each method had specific advantages and limitations. The enclosure techniques were relatively inexpensive, and because they were portable, were deployable to remote sites. They were useful in identifying the sources of spatial variation in fluxes, and in developing relationships between specific ecosystem properties and processes and flux. For regional flux estimates, however, the enclosure-based point measurements of flux had to be extrapolated in space and time to landscape and regional scales using land cover maps. (See Chapters 2 and 4 for information on enclosure methods and Chapter 6 for analytical approaches; see Chapters 10 and 11 for ecosystem process measurements and extrapolation approaches.)

The eddy correlation flux estimates using tower-based micrometeorological techniques were more costly than the enclosure approaches and required electrical power and other logistical support. Their great advantage, however, was that measurements were nearly continuous, with data gaps only during periods of no wind or during maintenance periods. Because the tower was located at the intersection of lake, dry tundra and wet tundra habitats, fluxes from each ecosystem could be determined depending on wind direction. On the other hand, because of cost and logistical support, the tower was established only at one site; therefore the variability among similar ecosystem types could not be evaluated. (See Chapters 5, 7 and 8 for micrometeorological measurement and analytical approaches.)

Airborne eddy correlation flux measurements over tundra ecosystems were also included. These measurements provided a rapid, quantitative survey of gas exchange, integrated over all the different habitats of the tundra. Because aircraft-based measurements are very expensive and require substantial logistical support and infrastructure (including access to appropriate airports), they cannot be carried out with high temporal frequency; in ABLE-3A, aircraft measurements occurred during 21 flights over a 44-day period. Because they integrate all the fluxes over a large area (from hectares to square kilometres), spatial resolution of both the ecosystem properties and processes that control fluxes is difficult.

The ABLE-3A study provided the first intercomparison of these different flux measurement techniques. Estimates of the tundra methane source based on enclosure extrapolations and on tower and aircraft eddy correlation flux measurements were in close agreement (Bartlett & Harriss, 1993), indicating that the different techniques can be used in a complementary fashion and in ways that take advantage of their different capabilities. The enclosures are most useful for identifying inter- and intrasite variability and for developing an understanding of controlling processes. The tower-based measurements provide excellent temporal resolution of fluxes, help identify broad relationships between flux and weather and other environmental controls, and, to a limited extent, identify the source strengths of different ecosystems. Finally, aircraft-based measurements provide integrated flux estimates for a region, allowing tests of regional extrapolations and model estimates (Matson & Harriss, 1988).

Studies such as ABLE represent massive efforts that include numerous investigators from a number of different disciplines, using many different measurement platforms and approaches. There are cases, however, in which multiple approaches are possible and advisable even for single investigators or small research groups. For example, studies of nitrogen cycling and nitrogen trace gas fluxes at the ecosystem to regional scale most logically include measurements of both N_2O and NO. Because these gases have very different roles in the atmosphere (N_2O is a long-lived greenhouse gas; NO is a short-lived, chemically reactive gas), they were usually studied separately in the past, by different research groups for different reasons. However, both gases are produced by the same microbial processes (although not always in the same proportion), and variations in the fluxes of both reflect differences in nitrogen cycling among sites (Firestone & Davidson, 1989; Davidson et al., 1993; Vitousek & Matson, 1993). They appear to respond similarly to disturbance and fertilization (Hutchinson & Brams, 1992; Williams et al., 1992; Davidson et al., 1993; Keller & Matson, 1993), and the relative flux between the two gases reflects soil texture and moisture differences (Davidson et al., 1991; Garcia-Mendez et al., 1991; Davidson, 1992, 1993; Williams et al., 1992; see also Chapter 10, this volume).

Different measurement and analytical approaches are required for NO and N_2O. N_2O flux measurements require use of enclosures (Chapter 2), with analysis most often by gas chromatography with electron capture detection (Chapter 6); because of analytical limitations, micrometeorological approaches for measurement of N_2O flux are not presently feasible. NO flux can be measured using enclosures (Chapter 2) with chemiluminescence detection or laser-induced fluorescence (Chapters 6 and 7), but because NO can be taken up in plant canopies, flux at the soil–atmosphere interface cannot always be interpreted as flux to the atmosphere. Micromet approaches are available for measurement of flux through the canopy (Chapters 5 and 7; also see Williams et al., 1992), and in many cases it is useful to use both enclosure and micromet approaches together (Johannson, 1989; Bakwin et al., 1990). Ideally, such a multiple-approach study will use mechanistic studies to address the proximal and distal regulation of soil–air flux of both N_2O and NO measured using enclosures (Chapters 2, 6 and 10); at the same time, micromet techniques will allow measurement of NO flux from the surface through the canopy to the atmosphere (Chapters 5 and 7).

There are numerous other examples where multiple approaches, for both measurement of flux and analysis of the ecosystem processes that control flux are useful. The study of trace gas exchange has evolved beyond the simple measurement of a single gas moving out of or into soils or sediments. We hope that this book will provide insight and information about the range of methods and approaches for measurement of trace gas exchange; the choice of methods and approaches that are most appropriate to any given site or scientific question ultimately lies with the people doing the research.

Acknowledgements

We thank P. Vitousek and C. Potter for comments of earlier drafts, and NASA–Ames Research Center Earth System Science Division and the University of New Hampshire Institute for the Study of Earth, Oceans and Space.

References

Altshuller A.P. (1991) The production of CO by the homogeneous NO_x-induced photo-oxidation of volatile organic compounds in the troposphere. *Journal of Atmospheric Chemistry*, **13**, 155–182.

Andreae M.O. & Schimel D.S. (1989) *Exchange of Trace Gases between Terrestrial Ecosystems and the Atmosphere.* John Wiley & Sons, Chichester.

Bakwin P.S., Wofsy S.C., Fan S.-M., Keller M., Trumbore S.E. & da Costa J.M. (1990) Emission of nitric oxide from tropical forest soils and exchange of NO between the forest canopy and atmospheric boundary layers. *Journal of Geophysical Research*, **95**, 755–16764.

Baldocchi D.D., Hicks B.B. & Meyers T.P. (1988) Measuring biosphere–atmosphere exchanges of biologically related gases with micrometeorological methods. *Ecology*, **69**, 1331–1340.

Bartlett K.B. & Harriss R.C. (1993) Review and assessment of methane emissions from wetlands. *Chemosphere*, **26**, 261–320.

Bartlett K.B., Crill P.M., Sass R., Harriss R. & Dise N. (1992) Methane emissions from tundra environments in the Yukon–Kuskokwim delta, Alaska. *Journal of Geophysical Research*, **97**, 16645–16660.

Blake D.R. & Rowland F.S. (1988) Continuing worldwide increase in tropospheric methane, 1978–1987. *Science*, **239**, 1129–1131.

Boden T.A., Sepanski R.J. & Stoss F.W. (1992) *Trends '91: A Compendium of Data on Global Change. Carbon Dioxide Information Analysis Center.* Publication No. ORNL/CDIAC-49, Oak Ridge, TN, 60 pp.

Chameides W.L., Lindsay R.W., Richardson J. & Kiang C.S. (1988) The role of biogenic hydrocarbons in urban photochemical smog: Atlanta as a case study. *Science*, **241**, 1473–1475.

Davidson E.A. (1992) Sources of nitric oxide and nitrous oxide following wetting of dry soil. *Soil Science Society of America Journal*, **56**, 95–102.

Davidson E.A. (1993) Soil water content and the ratio of nitrous oxide to nitric oxide emitted from soil. In: Oremland R.S. (ed) *The Biogeochemistry of Global Change: Radiatively Active Trace Gases.* Chapman & Hall, New York.

Davidson E.A., Matson P.A., Vitousek P.M. *et al.* (1993) Processes regulating soil emissions of NO and N_2O in a seasonally dry tropical forest. *Ecology*, **74**(1), 130–139.

Davidson E.A., Vitousek P.M., Matson P.A., Riley R., Garcia-Mendez G. & Maass J.M. (1991) Soil emissions of nitric oxide in a seasonally dry tropical forest of Mexico. *Journal of Geophysical Research*, **96**, 15439–15445.

Eichner M.J. (1990) Nitrous oxide emissions from fertilized soils: summary of the available data. *Journal of Environmental Quality*, **19**, 272–280.

Enders G., Dlugi R., Steinbrecher R. *et al.* (1992) Biosphere/atmosphere interactions: integrated research in a European coniferous forest ecosystem. *Atmospheric Environment*, **26A**, 171–189.

Erickson D.J. & Taylor J.A. (1992) 3-D tropospheric CO modeling: The possible influence of the ocean. *Geophysical Research Letters*, **19**, 1955–1958.

Fan S.M., Wofsy S.C., Bakwin P. *et al.* (1992) Micrometeorological measurements of CH_4 and CO_2 exchange between the atmosphere and subarctic tundra. *Journal of Geophysical Research*, **97**, 16627–16644.

Field C.B. & Mooney H.A. (1990) Measuring photosynthesis under field conditions: Past and present approaches. In: Hashimoto Y., Kramer P.J., Nonami H. & Strain B.R. (eds) *Measurement Techniques in Plant Science*, pp. 185–205. Academic Press, San Diego.

Field C.B., Ball J.T. & Berry J.A. (1989) Photosynthesis: Principles and field techniques. In: Pearcy R.W., Mooney H.A., Ehleringer J.R. & Rundel P.W. (eds) *Plant Physiological Ecology: Field Methods and Instrumentation*, pp. 209–253. Chapman & Hall, London.

Field C.B., Chapin F.S. III, Matson P.A. & Mooney H.A. (1992) Responses of terrestrial ecosystems to the changing atmosphere. *Annual Review of Ecology and Systematics*, **23**, 210–235.

Firestone M.K. & Davidson E.A. (1989) Microbiological basis of NO and N_2O production and consumption in soil. In: Andreae M.O. & Schimel D.S. (eds) *Exchange of Trace Gases between Terrestrial Ecosystems and the Atmosphere*, pp. 7–21. John Wiley, New York.

Garcia-Mendez G., Maass J.M., Matson P.A. & Vitousek P.M. (1991) Nitrogen transformations and nitrous oxide flux in a tropical deciduous forest in Mexico. *Oecologia*, **88**, 362–366.

Goodroad L.L. & Keeney D.R. (1984) Nitrous oxide emission from forest, marsh, and prairie ecosystems. *Journal of Environmental Quality*, **13**, 448–452.

Graedel T.E. & Crutzen P.J. (1993) *Atmospheric Change: An Earth System Perspective*. Freeman & Co, New York.

Harriss R.C. & Frolking S.E. (1992) The sensitivity of methane emissions from northern freshwater wetlands to global warming. In: Firth P. & Fisher S. (eds) *Global Climate Change and Freshwater Ecosystems*, pp. 48–67. Springer Verlag, New York.

Harriss R.C., Gorham E., Sebacher D.I., Bartlett K.B. & Flebbe P.A. (1985) Methane flux from northern peatlands. *Nature*, **315**, 652–654.

Harriss R.C., Wofsy S.C., Bartlett D.S. *et al.* (1992) The Arctic boundary layer expedition (ABLE-3A): July–August 1988. *Journal of Geophysical Research*, **97**, 16383–16394.

Hogan K.B., Hoffman J.S. & Thompson A. (1991) Methane on the greenhouse agenda. *Nature*, **354**, 181–182.

Houghton J.T., Callander B.A. & Varney S.K. (1992) *Climate Change 1992: The Supplementary Report to the IPCC Scientific Assessment*. Cambridge University Press, New York.

Houghton J.T., Jenkins G.J. & Ephraums J.J. (1990) *Climate Change: The IPCC Scientific Assessment*. Intergovernmental Panel on Climate Change. Cambridge Press, New York.

Hutchinson G.L. & Brams E.A. (1992) NO versus N_2O emissions from an NH4+-amended Bermuda grass pasture. *Journal of Geophysical Research*, **97**, 9889–9896.

International Geosphere – Biosphere Programme: A study of Global change (IGBP) (1990) The Initial Core Projects. Report No. 12. Stockholm, Sweden.

Jacob D.J. & Wofsy S.C. (1990) Budgets of reactive nitrogen, hydrocarbons, and ozone over the Amazon forest during the wet season. *Journal of Geophysical Research*, **95**, 16737–16754.

Johansson C. (1989) Fluxes of NO_x above soil and vegetation. In: Andreae M.O. & Schimel D.S. (eds) *Exchange of Trace Gases between Terrestrial Ecosystems and the Atmosphere*, pp. 229–246. John Wiley, New York.

Johansson C., Rodhe H. & Sanhueza E. (1988) Emission of NO in a tropical savanna and cloud forest during the dry season. *Journal of Geophysical Research*, **93**, 7180–7192.

Keller M. & Matson P.A. (1993) Biosphere–atmosphere exchange of trace gases in the tropics: Evaluating the effects of land use changes. In: Prinn R. (ed) *Global Atmospheric–Biospheric Chemistry. Proceedings of the first International Global Atmospheric Chemistry Scientific Conference*, Eilat, Israel 18–22 April.

Keller M., Kaplan W.A. & Wofsy S.C. (1986) Emissions of N_2O, CH_4 and CO_2 from tropical soils. *Journal of Geophysical Research*, **91**, 11791–11802.

Keller M., Goreau T.J., Wofsy S.C., Kaplan W.A. & McElroy M.B. (1983) Production of nitrous oxide and consumption of methane by forest soils. *Geophysical Research Letters*, **10**, 1156–1159.

Keller M., Veldkamp E., Weitz A. & Reiners W.A. (1993) Pasture age effects on soil–atmosphere gas exchange in a deforested area of Costa Rica. *Nature*, **365**, 244–246.

Khalil M.A.K. & Rasmussen R.A. (1990) Atmospheric carbon monoxide: Latitudinal distribution of sources. *Geophysical Research Letters*, **17**, 1913–1916.

Li C., Frolking S.E. & Frolking T. (1992) A model of nitrous oxide evolution from soil driven by rainfall events: 1. Model structure and sensitivity. *Journal of Geophysical Research*, **97**, 9759–9776.

Matson P.A. & Harriss R.C. (1988) Prospects for aircraft-based gas exchange measurements in ecosystem studies. *Ecology*, **69**(5), 1318–1325.

Matson P.A. & Ojima D.S. (eds) (1990) *The Terrestrial Biosphere Exchange with Global Atmospheric Chemistry*. Terrestrial biosphere perspective of the IGAC project: Companion to the Dookie Report. IGBP Report No. 13. Stockholm, Sweden.

Matson P.A. & Vitousek P.M. (1987) Cross-system comparisons of soil nitrogen transformations and nitrous oxide flux in tropical forest ecosystems. *Global Biogeochemical Cycles*, **1**, 163–170.

Matson P.A. & Vitousek P.M. (1990) Ecosystem approach to a global nitrous oxide budget. *Bioscience*, **40**(9), 667–672.

Matson P.A. & Vitousek P.M. Biosphere–atmosphere interactions in a tropical deciduous forest ecosystem. In: Mooney H.A., Medina E. & Bullock S.H. (eds) *The Tropical Deciduous Forest Ecosystem*. Springer-Verlag, in press.

Matson P.A., Vitousek P.M. & Schimel D.S. (1989) Regional extrapolation of trace gas flux based on soils and ecosystems. In: Andreae M.O. & Schimel D.S. (eds) *Exchange of Trace Gases between Terrestrial Ecosystems and the Atmosphere*, pp. 97–108. John Wiley, New York.

Matson P.A., Vitousek P.M., Livingston G.P. & Swanberg N.A. (1990) Sources of variation in nitrous oxide flux from Amazonian ecosystems. *Journal of Geophysical Research*, **95**, 16789–16798.

Matthews E. & Fung I. (1987) Methane emission from natural wetlands: global distribution, source area and environmental characteristics of sources. *Global Biogeochemical Cycles*, **1**, 61–86.

Mooney H.A., Vitousek P.M. & Matson P.A. (1987) Exchange of material between terrestrial ecosystems and the atmosphere. *Science*, **238**, 926–932.

Moore B. III & Schimel D.S. (1992) *Trace Gases and the Biosphere*. UCAR Office for Interdisciplinary Studies, Boulder, Colorado.

Morrissey L.A. & Livingston G.P. (1992) Methane emissions from Alaska tundra: An assessment of local spatial variability. *Journal of Geophysical Research*, **97**, 16661–16670.

Mosier A.R. & Parton W.J. (1985) Denitrification in a shortgrass prairie: A modeling approach. In: Caldwell D.E., Brierly J.A. & Brierly C.L. (eds) *Planetary Ecology. Selected Papers from the Sixth International Symposium on Environmental Biogeochemistry*, pp. 441–452. Van Nostrand Reinhold, New York.

Nisbet E.G. (1989) Some northern sources of atmospheric methane: production, history, and future implications. *Canadian Journal of Earth Science*, **26**, 1603–1611.

Oeschger H. & Arquit A. (1989) Resolving abrupt and high-frequency global changes in the ice-core record. In: Bradley R.S. (ed) *Global Changes of the Past*. UCAR Office for Interdisciplinary Earth Studies. Boulder, Colorado.

Parrish D.D., Holloway J.S., Trainer M., Murphy P.C., Forbes G.L. & Fehsenfeld F.C. (1993) Export of North American ozone pollution to the North Atlantic Ocean. *Science*, **259**, 1436–1439.

Pearcy R.W., Ehleringer J., Mooney H.A. & Rundel P.W. (eds) (1989) *Plant Physiological Ecology: Field Methods and Instrumentation*. Chapman and Hall, London.

Pearman G.I. (ed) (1992) *Limiting Greenhouse Gases: Controlling Carbon Dioxide*

Emissions. Wiley & Sons, New York.

Potter C.S., Randerson J., Field C.B. *et al.* (1993) Terrestrial ecosystem production: A process model based on global satellite and surface data. *Global Biogeochemical Cycles*, **7**, 811–842.

Prinn R.G., Cunnold D., Rasmussen R. *et al.* (1990) Atmospheric emissions and trends of nitrous oxide deduced from ten years of ALE-GAGE data. *Journal of Geophysical Research*, **95**, 18369–18385.

Rasmussen R.A. & Khalil M.A.K. (1986) Atmospheric trace gases: Trends and distributions over the last decade. *Science*, **232**, 1623–1624.

Raynaud D. & Barnola J.M. (1985) An Antarctic ice core reveals atmospheric CO_2 variations over the past few centuries. *Nature*, **315**, 309–311.

Raynaud D., Chappellaz J., Barnola J.-M., Korotkevitch Y.S. & Lorius C. (1988) Climate and CH_4 cycle implications of glacial-interglacial CH_4 change in the Vostok ice core. *Nature*, **333**, 655–657.

Ritter J.A., Barrick J., Sachse G. *et al.* (1992) Airborne flux measurement of trace species in an Arctic boundary layer. *Journal of Geophysical Research*, **97**, 16601–16626.

Robertson G.P. (1989) Nitrification and denitrification in humid tropical ecosystems: potential controls on nitrogen retention. In: Procter J. (ed) *Mineral Nutrients in Tropical Forest and Savanna Ecosystems*. Blackwell Scientific Publications, Oxford.

Sanhueza E., Hao W.M., Scharffe D., Donoso L. & Crutzen P.J. (1990) N_2O and NO emissions from soils of the northern part of the Guayana Shield, Venezuela. *Journal of Geophysical Research*, **95**(D13), 22, 48–22, 488.

Schlesinger W.H. (1991) *Biogeochemistry: An Analysis of Global Change*. Academic Press, New York.

Schmidt J., Seiler W. & Conrad R. (1988) Emission of nitrous oxide from temperate forest soils into the atmosphere. *Journal of Atmospheric Chemistry*, **6**, 95–115.

Sebacher D.I., Harriss R.C., Bartlett K.B., Sebacher S.M. & Grice S.S. (1986) Atmospheric methane sources: Alaskan tundra bogs, an alpine fen, and a subarctic boreal marsh. *Tellus*, **38B**, 1–10.

Sharkey T., Holland E. & Mooney H.A. (1991) *Trace Gas Emissions by Plants*. Academic Press, San Diego, California.

Steele L.P., Dlugokensky E.J., Lang P.M., Tans P.P., Martin R.C. & Masarie K.A. (1992) Slowing down of the global accumulation of atmospheric methane during the 1980s. *Nature*, **358**, 313–316.

Thompson A.M. (1992) The oxidizing capacity of the Earth's atmosphere: Probable past and future changes. *Science*, **256**, 1157–1165.

Tolba M.K. & El-Kholy O.A. (1992) *The World Environment 1972–1992*. Chapman & Hall, London.

Turner B.L. III, Clark W.C., Kates R.W., Richards J.F., Mathews J.T. & Meyer W.B. (1990) *The Earth Transformed by Human Action*. Cambridge University Press, Cambridge.

Vitousek P.M. & Matson P.A. (1993) Agriculture, the global nitrogen cycle, and trace gas flux. In: Oremland R.S. (ed) *The Biogeochemistry of Global Change: Radiatively Active Trace Gases*. Chapman & Hall, New York.

Vitousek P.M. & Sanford R.L. Jr (1986) Nutrient cycling in moist tropical forest. *Annual Review of Ecology and Systematics*, **17**, 137–167.

Vitousek P.M., Matson P.A., Volkmann C., Maass J.M. & Garcia-Mendez G. (1989) Nitrous oxide flux from seasonally-dry tropical forests: a survey. *Global Biogeochemical Cycles*, **3**, 375–382.

Watson R.T., Rodhe H., Oeschger H. & Siegenthaler U. (1990) Greenhouse gases and aerosols. In: Houghton J.T., Jenkins G.J. & Ephrams J.J. (eds) *Climate Change: The IPCC Scientific Assessment*, pp. 1–40. Cambridge University Press, Cambridge.

Whalen S.C., Reeburgh W. & Sandbeck K. (1990) Rapid methane oxidation in a landfill cover soil. *Applied Environmental Microbiology*, **56**, 3405–3411.

Whiting G.J., Chanton J., Bartlett D.S. & Happell J. (1991) Relationships between CH_4 emissions, biomass, and CO_2 exchange in a subtropical grassland. *Journal of Geophysical Research*, **96**, 13067–13071.

Williams E.J. & Fehsenfeld F.C. (1991) Measurement of soil nitrogen oxide emissions in three North American ecosystems. *Journal of Geophysical Research*, **96**, 1033–1042.

Williams E.J., Hutchinson G.L. & Fehsenfeld F.C. (1992) NO_x and N_2O emissions from soil. *Global Biogeochemical Cycles*, **6**(4), 351–388.

Zimmerman P.R., Greenberg J.P. & Westberg C.E. (1990) Measurements of atmospheric hydrocarbons and biogenic emission fluxes in the Amazon boundary layer. *Journal of Geophysical Research*, **93**, 1407–1416.

CHAPTER 2

Enclosure-based measurement of trace gas exchange: applications and sources of error

G.P. LIVINGSTON & G.L. HUTCHINSON

2.1 Introduction

A variety of techniques have been developed to measure surface–atmosphere gas exchange, including micrometeorological (see Chapter 5), enclosure and diffusion theory approaches (e.g. Fechner & Hemond, 1992). Each has advantages and disadvantages, and no single approach is applicable to all studies. Enclosure techniques are relatively low in cost, simple to operate, and especially useful for addressing research objectives served by discrete observations in space and time. In combination with appropriate sample allocations, they are adaptable to a wide variety of studies on local to global spatial scales and are particularly well suited to *in situ* and laboratory-based studies addressing physical, chemical and biological controls on surface–atmosphere trace gas exchange.

Although micrometeorological and diffusion theory techniques offer significant advantages for quantifying net gas exchange rates in some applications, there are many site conditions and logistical considerations for which these techniques are not appropriate. Micrometeorological approaches are not applicable, for example, in situations in which stationarity in the wind field cannot be assumed, conditions for which the exchange rate estimates are compromised by uncertainties in the required air density corrections (Webb *et al.*, 1980), or when suitable sensors or sampling mechanisms are either not available or too costly. Moreover, many research objectives require quantitative measures of trace gas exchange over defined areas or at spatial scales below the resolution possible with micrometeorological techniques. Applications of diffusion theory are also subject to considerable and unknown uncertainties when the trace gas sources or sinks are non-uniformly distributed in the soil or located too near the surface for gradients to be measured, or when non-diffusive transport is involved. In each of the above situations, enclosures are often applicable.

Our purpose here is to provide insight into the function and use of enclosure-based measurement systems applicable in soil–atmosphere trace gas exchange studies. We first review surface–atmosphere gas exchange mechanisms and then discuss the design and operation of various enclosure systems, including the calculations required to estimate gas exchange rates from observed chamber concentration data. In addition, we briefly discuss sampling issues related to extrapolating the exchange rate estimates beyond the temporal and spatial domains over which they were measured. Throughout this discussion, we identify common sources of error in the application of enclosures that can lead to inaccurate estimates of surface–atmosphere gas exchange rates. Although the focus of this chapter is on soil–atmosphere gas exchange, the concepts, concerns and guidelines presented are generally applicable to any enclosure system.

2.2 Exchange processes and controls

Although trace gases are readily exchanged between the Earth's surface and atmosphere, rates of exchange are highly variable over space and time. This variability is attributable to differences in the trace gas sources and sinks, as well as in the transport mechanisms involved. A general understanding of gas transport processes in near-surface soil, water and atmospheric environments is essential to evaluate the various techniques available for measuring surface gas exchange rates and to interpret the measurements they yield.

Gas transport within the atmosphere, across the surface–atmosphere boundary, and within subsurface environments differs with respect to both the processes and rates of transport. Within the mixed planetary boundary layer, horizontal transport of gases is dominated by pressure and density-driven advection, whereas vertical transport is dominated by eddy turbulence (Stull, 1988; see Chapter 5). As the surface is approached, the role of eddy turbulence diminishes rapidly and the relative importance of molecular transport increases. Gas transport within the atmospheric interfacial layer immediately adjacent the surface, across the surface–atmosphere boundary, and within the subsurface environment may be modelled as the sum of the individual contributions of advective processes, or mass flow (f_m), and molecular diffusion (f_d) (Kanemasu *et al.*, 1974; Nazaroff, 1992):

$$f = f_m + f_d \tag{2.1}$$

where f, the flux density or net rate of gas exchange per unit of cross-sectional surface area (e.g. $mg\,m^{-2}\,h^{-1}$) is defined positive upwards. The relative contributions of advection and diffusion vary greatly between ecosystems and with environmental conditions.

Turbulent transfer coefficients in the mixed layer above or within most plant canopies are on the order of 10^4–$10^2\,cm\,s^{-1}$. Within the interfacial layer, transfer coefficients are typically 10^{-1}–$10^{-2}\,cm\,s^{-1}$, with the lower limit defined by the coefficient of molecular diffusion in air (Kimball,

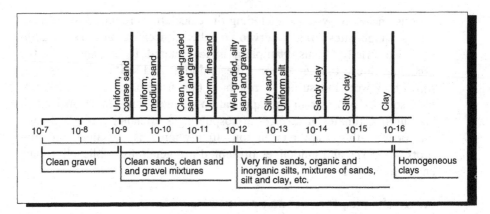

Fig. 2.1 Permeability (m^2) of representative soils. (From Nazaroff, 1992.)

1983). Rates of gas transport in soils and litter layers are greatly reduced relative to that in air, because the physical structure of these media reduces the cross-sectional area available for gas flow and increases the tortuosity, i.e. the effective path length a molecule must travel in its escape to the atmosphere. As a consequence, transfer coefficients within subsurface environments may be 10^2–10^5 times lower than in the atmospheric interfacial layer, i.e. 10^{-3}–10^{-7} cm s^{-1}.

2.2.1 Molecular diffusion
Molecular diffusion is the the net flow of a fluid under isothermal and isobaric conditions by random molecular motion along a gradient of decreasing concentration. Diffusive transport is described by Fick's law (Cambell, 1985; Ghildyal & Tripathi, 1987):

$$f_d = -D_0 dC/dz \qquad (2.2)$$

where D_0 is the binary molecular diffusion coefficient in air, C is the concentration of the species of interest and z is distance. Diffusivity depends on the properties of both the diffusing species and the medium through which it is diffusing, varying approximately with the square of absolute temperature and inversely with total air pressure (Bird *et al.*, 1960). Gaseous permeability, i.e. the ease with which gases move through soil, varies over several orders of magnitude (Fig. 2.1) in relation to the size, shape, orientation and water content of the soil pore spaces. These factors, in turn, define the fractional cross-sectional area and tortuosity of the air-filled pore space (Ghildyal & Tripathi, 1987; Nazaroff, 1992). Molecular diffusion in soils, therefore, is typically modelled by empirically defining an effective diffusivity (D_e) that accounts for the reduced gaseous permeability of soil relative to that of open air (Troch *et al.*, 1982; Kimball, 1983; Sallam *et al.*, 1984).

2.2.2 Advective transport

Advective transport, or mass flow, of gas across the soil surface occurs in response to a difference in total pressure between soil air and the overlying atmosphere. Darcy's law describes mass flow in soils (Kanemasu *et al.*, 1974; Nazaroff, 1992):

$$f_m = \frac{kC\rho}{\eta} \mathrm{d}P/\mathrm{d}z \tag{2.3}$$

where k is the intrinsic air permeability of the soil; C, ρ and η are the volumetric concentration, density and viscosity, respectively, of the species of interest; and $\mathrm{d}P/\mathrm{d}z$ is the pressure gradient. Pressure gradients across the soil–atmosphere interface can result from the volumetric expansion or contraction of air in response to temperature or barometric pressure changes (Clements & Wilkening, 1974; Schery *et al.*, 1984), from a change in the fraction of air-filled pore space due to infiltration or loss of soil water (Ghildyal & Tripathi, 1987), or through wind flow interactions with the surface topography (Wesley *et al.*, 1989). Small high-frequency pressure fluctuations also may occur in response to surface wind turbulence (Kimball & Lemon, 1972; Kimball, 1983). Characteristic times over which pressure gradients are propagated range from less than a minute to days in association with wind stresses, daily surface temperature cycles, or the passing of major fronts or storm events. In soils with low and high air permeability, respectively, vertical pressure gradients may develop over distances of centimetres to metres.

Regardless of their source, pressure gradients are rapidly attenuated with soil depth and decreasing air-filled porosity (Kimball, 1983). When permeability is low, as is characteristic of fine-textured soils such as silts and clays, molecular diffusion dominates gas transport. In more permeable, well-drained, porous soils or litter layers, however, the relative importance of advective transport increases, potentially enhancing exchange rates severalfold over that due to molecular diffusion alone.

2.2.3 Influence of soil water

Soil water dramatically affects gas transport in soils. When dry, gaseous diffusion rates are nearly linearly related to air-filled porosity over a wide range of textural conditions (Ghildyal & Tripathi, 1987). Wetting a soil reduces its air-filled porosity and increases the tortuosity, thus rapidly precluding advective transport and greatly reducing molecular diffusion. Diffusivity of gases in water is about 10^4 times smaller than in air, so pore spaces filled with water or isolated by water films represent effective

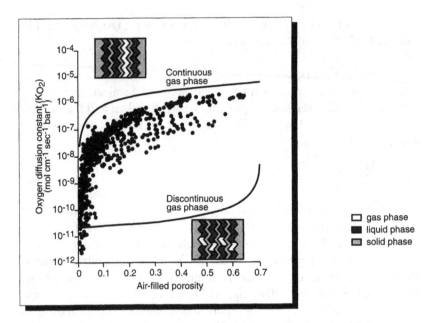

Fig. 2.2 Diffusivity of O_2 as a function of soil air-filled porosity. Gaseous diffusion at any level of porosity is determined by the continuity of the enclosed air spaces which, in turn, is determined by the texture and water content of the soil media. (From Focht, 1992.)

barriers to gas transport, even though the total air-filled porosity of the soil may remain high. In fine-textured soils, wetting reduces the continuity between pore spaces at a far greater rate than the fraction of air-filled pore space is reduced, so gas transport is significantly diminished long before air-filled porosity is eliminated (Fig. 2.2, discontinuous gas phase) (Currie, 1960, 1961; Ghildyal & Tripathi, 1987). Conversely, in coarse-textured soils continuity between pore spaces is maintained over a large range of soil water content (Fig. 2.2, continuous gas phase) and gas transport is less affected by reductions in air-filled porosity.

Differences in the water solubility of gas species also can result in significantly different concentration gradients at air–water interfaces, and thus to differences in their rates of molecular diffusion (Himmelblau, 1964). Precipitation and percolating water affect gas transport by physically displacing the air in soil pore spaces, by transporting dissolved atmospheric gases into the soil, and by vertically or laterally transporting soil gases in solution. These processes may lead to either an enhancement or suppression of trace gas exchange following rainfall (Glinski & Stepniewski, 1985), or to spatial or temporal displacement between source or sink areas and surface–atmosphere exchange areas (Dowdell *et al.*, 1979; Bowden & Bormann, 1986; Ronen *et al.*, 1988).

2.2.4 Vegetative transport

Vegetation significantly influences gas exchange across the surface–atmosphere interface through its influence on trace gas production, consumption and transport processes. Transport of gases to and from the canopy is the result of turbulent mixing, whereas migration across the leaf boundary layer occurs by molecular diffusion or mass flow governed by stomatal and cuticular conductance (Nobel, 1983; Jarvis & McNaughton, 1986). Plants may affect net trace gas production or consumption through alteration of the physical and chemical environment of the rhizosphere, by hosting microbial communities, by uptake or release of resources of value to soil microbial communities (Matson *et al.*, 1990), or via foliar exchange and metabolism of gases above the surface, e.g. non-methane hydrocarbons (NMHC), O_3, NH_3 (Fig. 2.3) (Denmead *et al.*, 1976; Lamb *et al.*, 1987). Plants may also function as a direct pathway for the flow of trace gases between the substrate and the atmosphere, particularly in wetland environments (Chanton & Dacey, 1991; Grosse *et al.*, 1991; Morrissey *et al.*, 1993; see also Chapter 4). For example, plant-mediated CH_4 transport bypasses oxidizing microbial communities at the anaerobic–aerobic interface in the soil substrate and greatly enhances CH_4 exchange

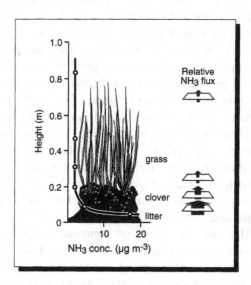

Fig. 2.3 A closed ammonia cycle in a pasture canopy. (From Wesley *et al.*, 1989; after Denmead *et al.*, 1976.) Ammonia released from the decomposition of surface litter is subsequently reabsorbed by foliage above.

compared to transport by diffusion. Over 90% of the observed CH_4 flux in herbaceous wetlands is via plant-mediated transport (see Chapter 4).

2.3 Enclosure design, construction and deployment

Enclosures function by restricting the volume of air available for exchange across the covered surface, so that any net emission or uptake of the enclosed gases can be measured as a concentration change. The measurement will yield a valid estimate of the trace gas exchange rate, or flux, only if the enclosure does not significantly perturb the production, consumption and transport processes responsible for its regulation. Recent reviews of enclosure construction and use for measuring soil or plant trace gas exchange include Wesley *et al.* (1989), Reicosky *et al.* (1990), IAEA (1992), Denmead & Raupach (1993) and Hutchinson & Livingston (1993).

2.3.1 Primary enclosure types

Enclosure design and operation can be fundamentally characterized on the bases of: (a) the operating conditions under which diffusive transport must proceed throughout the measurement period; and (b) whether or not advective transport is specifically addressed (Fig. 2.4).

In steady-state systems, the trace gas concentration gradient controlling molecular diffusion across the soil–atmosphere interface is assumed constant after an initial period of adjustment following deployment. This steady-state condition is maintained by passive regulation of the trace gas concentration in the enclosed air volume, typically by employing an open-path circulation system to sweep the enclosed volume using a constant flow of external air with known concentration of the species of interest. Although the steady-state concentration gradient that is established will likely differ from pre-deployment conditions, the perturbation can be minimized by optimizing the flow rate and make-up of the sweep air such that the difference in trace gas concentration inside and outside the enclosure is minimized. In contrast, the trace gas concentration gradient beneath non-steady-state systems is ever diminishing in response to continual concentration changes within the chamber air. Enclosure dimensions and deployment times, therefore, should be carefully selected in each application so that this negative feedback on the rate of molecular diffusion is minimized.

The terms 'dynamic' or 'open' are sometimes used synonymously to describe steady-state systems, and 'static' or 'closed' are often applied to non-steady-state systems. When used in conjunction with description of other chamber attributes, however, these terms often lead to confusion. For example, is a 'closed' chamber vented or not? Is a non-steady-state chamber with closed-air circulation through a non-destructive sensor

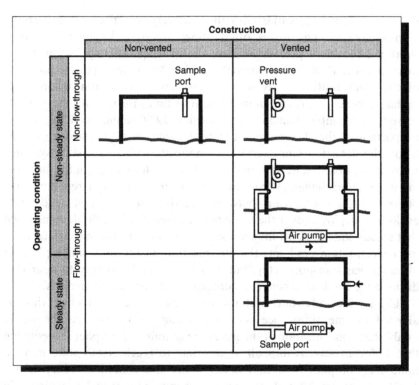

Fig. 2.4 Primary classification of enclosure configurations commonly employed to measure trace gas flux. The vertical axis represents steady-state and non-steady-state operating conditions under which molecular diffusion must proceed throughout the measurement period. The horizontal axis represents whether the enclosure is non-vented or vented and, therefore, whether advective transport beneath the enclosure is suppressed or permitted. Terms are defined in the text. Not all possible enclosure configurations are represented.

more appropriately classified as 'static' or 'dynamic'? To avoid such questions, we adhere to the more fundamental terms 'steady-state' and 'non-steady-state' to characterize enclosure types throughout this discussion.

Attention to advective transport is reflected in whether or not the enclosure is vented to the atmosphere. Vented enclosures are designed to effectively communicate atmospheric pressure changes and fluctuations to the enclosed air volume and thus minimally perturb mass flow across the soil–atmosphere interface. In contrast, advective transport in non-vented systems is suppressed. Steady-state systems are inherently vented because of their open-path circulation.

The advantages, limitations, assumptions and computation of exchange rates differ greatly between these primary enclosure designs. The design of choice, therefore, varies with the research objectives of each study. Non-steady-state enclosures often allow fluxes to be quantified over shorter deployment periods and are useful in quantifying low exchange rates, particularly if the minimum detectable flux is limited by the precision of the concentration analysis. In measuring surface–atmosphere N_2O exchange beneath a corn canopy at various times throughout the growing season, for example, Hutchinson & Mosier (1979) found that a substantial proportion of the fluxes measured by non-steady-state enclosures were below the detection limits reported obtainable with steady-state systems. In contrast, steady-state systems may be the preferred design for monitoring trace gas exchange at fixed locations over extended or repeated time periods. In a comparison between steady-state and non-steady-state enclosures deployed over the same period, Denmead (1979) demonstrated that steady-state systems induced smaller changes in the subsurface trace gas concentration gradient, thereby resulting in not only smaller bias in observed gas transport rates, but also more rapid recovery to near pre-disturbance conditions between consecutive measurement periods.

Users of steady-state systems must ensure that sufficient time is allowed for the subsurface trace gas gradient to adjust to the newly established concentration of the gas in the chamber atmosphere. Predicting this, unfortunately, is difficult as the time to reach steady-state may be quite long or poorly defined under many site conditions (Jury *et al.*, 1982). If not established, the true exchange rate may be significantly underestimated, so steady-state systems often incorporate an in-line sensor to monitor the concentration of the species of interest in real time. When there is no longer a measurable concentration change within the enclosure, steady-state is assumed. Sweep air containing a zero concentration of the species of interest should not be used if the resulting chamber concentration is less than ambient at steady-state (Castro & Galloway, 1991). For example, Fried *et al.* (1993) report that for OCS (with ambient concentration between 500 and 586 pptv), emissions from soils and vegetation were observed when OCS-free sweep air was used, whereas uptake was measured when ambient OCS sweep air was used.

Pressure vents are recommended for most enclosures not only to transmit atmospheric pressure changes to the enclosed air volume, but also to compensate for air sample withdrawal and possible reduction in chamber volume during deployment. Guidelines for calculating appropriate vent tube diameter and length as a function of chamber volume and wind speed are summarized in Fig. 2.5. These guidelines were defined to minimize resistance to air flow in response to atmospheric pressure fluctuations and to minimize the quantity of air exchanged by advective flow between the enclosed air volume and the atmosphere (Hutchinson & Mosier, 1981). Vented systems are especially recom-

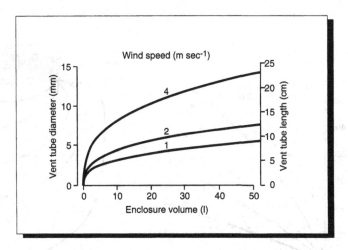

Fig. 2.5 Optimum vent tube diameter and length for selected wind speeds and enclosure volumes as described by Hutchinson & Mosier (1981).

mended whenever the underlying soils are highly permeable or the trace gas sources or sinks are located near the soil–atmosphere interface. Exchange due to molecular diffusion through the vent is assumed negligible over typical deployment times. The true exchange rate may be significantly underestimated if non-vented systems are used where mass flow is significant. In contrast, there is no disadvantage to employing a vented enclosure when mass flow is negligible. Location of the vent tube should be widely separated from the chamber's sampling port to avoid unintended interactions.

2.3.2 Other design considerations
In addition to the choice of general enclosure type, decisions are required with regard to deployment, chamber geometry, fabrication materials, temperature control and how concentration changes within the enclosed air volume are to be monitored. These considerations relate to minimizing potential disturbance effects not addressed by the primary enclosure design and optimizing the measurement process for the user. No single chamber design, geometry, deployment period or sampling and analytical approach is applicable in all situations. The measurement system *in toto* should be optimized relative to the site characteristics and magnitude of the exchange rates to be measured in each study.

Chamber geometry
Matthias *et al.* (1978) examined the effect of non-steady-state chamber geometry on soil gas exchange rates through use of a two-dimensional

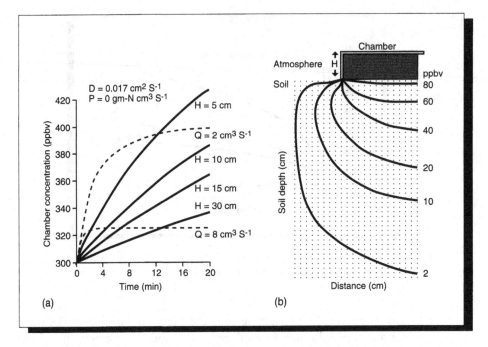

Fig. 2.6 Effect of enclosure dimensions on N_2O concentrations within and beneath the enclosed volume. (From Matthias *et al.*, 1978.) (a) Calculated concentrations within steady-state (---) and non-steady-state (——) enclosures for selected volume:basal area (H) configurations and sweep flow rates (Q). (b) Isolines of soil N_2O concentration change (ppbv) 20 min after deployment of a non-steady-state enclosure with $H = 10$ cm. The grid-point spacing (dot matrix) on which the calculations were made is 0.5 cm.

molecular diffusion model. They demonstrated that enclosures with small volume:basal area ratios ($V:A$) exhibit more rapid concentration increases (Fig. 2.6a), and thus more rapid feedback to the concentration gradient driving molecular diffusion across the surface (Fig. 2.6b), than enclosures with large ratios. Enclosures with large $V:A$ ratios also exhibit a more constant rate of concentration change within the enclosed air, but require longer sampling intervals to obtain a detectable concentration difference.

In general, a non-steady-state enclosure's $V:A$ ratio, or height if the cross-sectional area is constant, should be small enough that a change in the enclosed trace gas concentration can be measured over as short a time as logistically possible, yet large enough to minimize disturbance of the enclosed surface. For example, flux measurements on sites with large exchange rates are best served by chambers with large $V:A$ ratios and short deployment periods, whereas smaller chambers and longer deployments are often more applicable to low-flux sites. Reported $V:A$ ratios differ widely between studies, but are typically greater than 15 cm in field studies; overall measurement periods are generally in the range of 20–40 min. Whenever possible, measurement periods should be chosen such that the rate of concentration change can be assumed constant and, therefore, modelled using linear regression. This approach assumes that feedback to the subsurface concentration gradient is negligible over the measurement period, but also provides a means for objectively evaluating that assumption for each exchange rate estimate. Steady-state enclosure dimensions, flow rates and deployment periods likewise must be matched to the site characteristics and exchange rates to be measured, because of the influence of the $V:A$ ratio on the time required to achieve a steady-state concentration following deployment.

Rigid enclosures are commonly cylindrical or rectangular in cross-section, although any chamber geometry is acceptable that does not inherently restrict mixing within the enclosed air volume. The basal area of chambers designed for field use are typically 175 cm^2 to 1 m^2, although areas on the order of 500–900 cm^2 are most common. The selection of appropriate basal area is largely defined by the scale of the questions addressed by the research objectives and the sampling allocation of each study. In some situations the enclosure basal area should be small enough that environmental controls can be assumed uniformly distributed over the enclosed surface. For example, studies along steep environmental gradients may require relatively small basal areas to sample the range in variation of the controlling variable of interest. In other situations, it may be advantageous to integrate over larger basal areas to mask the spatial variability in exchange rates often encountered across distances as small as a centimetre.

Fabrication

Although available funds and deployment requirements influence the selection of both materials and construction, the fundamental requirement for chamber fabrication is that all materials (including sealants) be inert, i.e. non-permeable, non-reactive and not a source or sink for the gas species of interest. Aluminium, stainless steel and various plastics have been successfully employed in enclosures for non-reactive gas species like CH_4 and N_2O. For reactive species such as NO_x, NH_3, or sulphur gases, glass, Teflon or another inert material is required for all surfaces contacting the gases. The inert nature of an enclosure always should be confirmed under controlled laboratory conditions.

Enclosures may incorporate rigid or flexible materials, single- or multiple-component construction, and flow-through or non-flow-through air circulation. They are often fashioned from existing structures such as plastic irrigation pipe, carboys, etc. Enclosures constructed from Teflon film supported by a rigid support frame are well suited for situations requiring large enclosures or for measurement of reactive species (Steudler & Peterson, 1985; Khalil *et al.*, 1990; Klinger *et al.*, 1994; see also Chapter 4). Flexible enclosures not supported by a rigid frame, such as Teflon bags, have also been explored (Knapp & Yavitt, 1992). Their use, however, requires extreme care because of the increased potential for: (a) disturbance of the subsurface concentration gradient if the $V:A$ ratio is too small relative to the exchange rate; and (b) unwanted pressure perturbations that may influence natural exchange processes.

Single vs. multiple-component enclosures

Chamber construction must also reflect deployment considerations. Single-component chambers are deployed in a single step, i.e. they completely enclose a volume of air once sealed to the soil surface. They are perhaps most economical to fabricate, but also most difficult to deploy without error, as any physical disturbance of the measurement site during chamber placement may bias the observed exchange rate. Alternatively, multiple-component enclosures are deployed in two steps and are much less prone to such error. The first step, which can be carried out well in advance of actual exchange rate observations, is to seal an open base to the surface (Fig. 2.7). Completing the enclosure of a volume of air can thereafter be accomplished without further disturbance. The reduced risk of site disturbance offered by two-step deployment is particularly im-

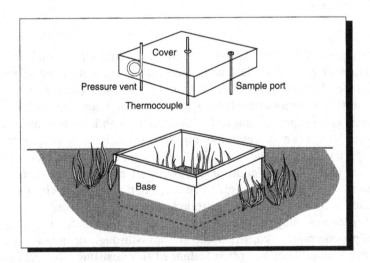

Fig. 2.7 Example of a multiple-component enclosure.

portant in studies requiring repeated observations at fixed locations. Other advantages of multiple-component enclosures are:

1 their bases may be designed for stacking (or removal) to accommodate changing exchange rates, vegetation height, or water/snow depths between sites or over time; and

2 they are adaptable to automated operation (Brumme & Beese, 1992; IAEA, 1992).

The enclosure described by Klinger *et al.* (1994) combines some of the advantages of both single- and multiple-component designs. Its flexible Teflon walls are permanently sealed to the chamber base, although enclosure is completed at the beginning of each measurement period only when the walls are drawn up over the internal frame and fastened with an elastic band.

The successful use of multiple-component enclosures is dependent upon achieving an effective seal between the cover and base. Some investigators use a compressible closed-cell foam gasket in conjunction with downward pressure from the weight of the cover, clamps, etc., while others use an air-tight overlapping (Luizão *et al.*, 1989) or abutting joint (Khalil *et al.*, 1990; Hutchinson & Brams, 1992) between the base and cover. A commonly employed seal involves a water-filled channel circumscribing the top of the enclosure base into which the walls of the overlying base or cover are inserted (Fig. 2.7). The extra effort and cost required in the fabrication of this design is often compensated by its ease and flexibility in deployment. The user, however, must assure that microbial activity in the water-filled channel does not represent a source or sink for the measured trace gas. A suitable substitute for water may be used, if measurements are planned at subfreezing temperatures (e.g. Whalen & Reeburgh, 1992).

Sampling ports

The exchange rate of some gases, such as CO_2, can be estimated by use of an appropriate chemical absorbent located within the enclosure (Rochette *et al.*, 1992), although the flux of most gas species must be quantified by sampling the enclosed air for analysis. In non-flow-through systems, a sampling port for removal of discrete gas samples is appropriate. Syringe sampling may be accommodated by a self-sealing septum or a valve with Luer-lok fittings, either of which can be mounted to a bulkhead union passing through the enclosure wall. Error in the measured exchange rate may occur if the rate of air sample withdrawal is not compensated by pressure venting, or if the sample volume is large relative to the total enclosure volume. Non-homogeneous mixing of the enclosed air is generally not an issue unless a considerable quantity of vegetation is enclosed, or the enclosure is tall relative to its basal area. Such cases can

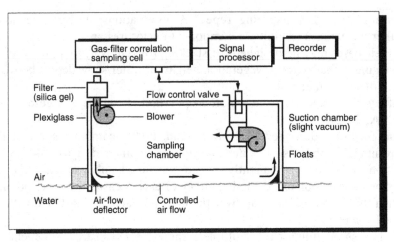

Fig. 2.8 Schematic of enclosure designed to examine the effects of wind stress at selected velocities on surface–atmosphere trace gas flux. (From Sebacher & Harriss, 1982.)

be addressed through use of a perforated sampling tube spanning the height of the chamber (Hutchinson *et al.*, 1992), internal fans or flow-through circulation systems. In systems incorporating internal fans, air flow may also be directed across the enclosed surface to simulate the effect on exchange rates of wind stress at selected velocities (Fig. 2.8) (Sebacher & Harriss, 1982; Moore & Roulet, 1991).

In steady-state systems, open-path flow-through circulation is an essential component of the flux estimation approach, but in non-steady-state systems, flow-through circulation is optional, serving only to facilitate mixing or sampling. In the latter case, circulation is induced in a closed loop (see Fig. 2.4) that may include provision for removal of discrete samples or a non-destructive sensor for monitoring the concentration of the species of interest (e.g. Sebacher & Harriss, 1982; Rochette *et al.*, 1992). In either open- or closed-path circulation systems, air entry and exit ports must be sufficiently large that the pressure gradient required to induce air flow is minimized. A pressure difference across these ports as small as 1 Pa has been shown to induce significant mass flow between soil air and the overlying enclosure, thereby causing a severalfold change in the apparent gas exchange rate (Kanemasu *et al.*, 1974). In all flow-through systems, the pump employed to induce air flow must be located and operated such that it does not alter the trace gas concentration of sample air; metal bellows pumps are often suitable in closed circulation applications.

Temperature controls

Mean ambient temperature, as well as fluctuations about the mean, should be preserved within the enclosure. Solar heating of the enclosure surface can rapidly lead to a substantial air temperature difference between the near-surface atmosphere and the enclosed volume. The magnitude of this difference depends strongly on the material employed in construction (Fig. 2.9), length of the deployment period and insolation levels. Although it affects the measurement of trace gas concentrations on a per-volume basis, a small temperature perturbation will not likely substantially perturb the rate of trace gas exchange across the soil–atmosphere boundary. Heat flux into the soil is rapidly attenuated with depth, so soil temperature change generally lags considerably behind that of air temperature. As safeguards, deployment times should be minimized, the enclosed air temperature monitored, and the enclosures should be shaded or fabricated from opaque (Matthias *et al.*, 1980), reflective (Hutchinson & Mosier, 1981) or insulating materials (Hutchinson & Brams, 1992). If the gas exchange process to be measured is related to physiological functioning of the enclosed plants (Knapp & Yavitt, 1992; Morrissey *et al.*, 1993), transparent construction materials may be used in conjunction with an external temperature controller (Whiting *et al.*, 1991; see also Chapter 4).

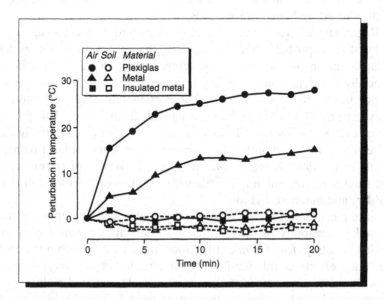

Fig. 2.9 Effect of enclosure construction materials and deployment time on enclosed air and soil temperatures. (From Hutchinson & Livingston, 1993; after Matthias *et al.*, 1980.)

2.3.3 Deployment

The goal in deploying an enclosure is to isolate a volume of air adjacent to the surface without perturbing the natural processes governing trace gas exchange across that surface. The process of establishing a seal between the surface and the enclosure, usually accomplished by inserting the walls of the enclosure (or base) into the soil, provides one of the greatest potentials for disturbance. The required insertion depth may be only a few millimetres in inundated or waterlogged soils, but may be as much as 10–20 cm in deep, well-drained moss or litter layers. If the lower edge of the enclosure does not easily penetrate the surface, it should not be forced; deployment will be greatly eased and soil compaction minimized if, instead, the enclosure is inserted into an incision cut into the soil with a sharp knife. Once the enclosure is in place, soil around its walls should be gently tamped to approximate the unperturbed soil bulk density.

Alternative methods for forming a disturbance-free seal between the surface and the enclosure have also been explored. Weighted non-permeable skirts extending from the base of the enclosure have been employed in both agricultural (Matthias *et al.*, 1980) and natural (Steudler & Peterson, 1985) ecosystems, but they may not form an acceptable seal unless the surface is free of vegetation, and surface winds are negligible. More recently, Castro & Galloway (1991) successfully deployed a baffled, double-walled enclosure that rests on the surface and is not affected by light surface winds.

Regardless of what type of seal is employed, some leakage of the enclosed air is probable across either the chamber walls or the enclosed surface. The measured flux is, therefore, likely to be underestimated, especially for deployments in highly porous substrate or during strong surface winds. Klinger *et al.* (1994) report that the underestimate may be as large as 30%, even under seemingly ideal conditions. This potential source of bias has received little attention, although it is relatively easily addressed by monitoring the concentration of an injected inert tracer gas (Khalil *et al.*, 1990; Klinger *et al.*, in press). Examples of tracers used in past studies include sulphur hexafluoride (SF_6), bromo-trifluoro-methane (CF_3Br) and butane (C_4H_{10}).

Care must be exercised during deployment not to influence artificially the soil permeability or induce pressure gradients through surface compaction resulting, for example, from foot traffic near or within the enclosure. Such effects could significantly alter transport pathways or induce mass flow into or away from the enclosure. Designated walkways and remote withdrawal of samples through at least 2–3 m of tubing attached to the sampling port are suggested for all sites where the weight or repeated presence of the investigator may affect observed exchange rates.

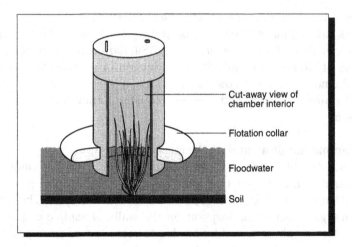

Cut-away view of chamber interior

Flotation collar

Floodwater

Soil

Fig. 2.10 Example of an enclosure with flotation collar.

Water-saturated soils are particularly susceptible to compaction, and even the weight of the enclosure itself can significantly bias observed exchange rates. In these situations, the enclosure should be suspended or its weight distributed over a large area, for example by use of flotation collars (Fig. 2.10).

Care should also be taken to minimize disturbance to the vegetation within or near enclosures because of its potential influence on the exchange processes under study. Physical damage resulting, for example, from cutting of roots during installation of the enclosure or severing stems and leaves of plants too large to fit within the enclosure may directly influence plant-mediated gas transport, affect resource availability to microbial communities involved in trace gas production or consumption, or stimulate a physiological response that could alter any of several controls on normal gas exchange processes. For example, Matson *et al.* (1990) demonstrated that within 1.5 h of severing the fine roots in the litter layer upon deployment of enclosure bases in a Brazilian rain forest, N_2O flux increased 2.8-fold relative to observations made 5 min after chamber deployment. Consideration must also be given to minimizing perturbation of the enclosure microclimate, e.g. CO_2 concentrations, light intensity, humidity, etc., because it influences plant functioning (Whiting *et al.*, 1991; Knapp & Yavitt, 1992; Morrissey *et al.*, 1993; see also Chapter 4). It is recommended that the effect of unavoidable vegetation disturbances on trace gas exchange rates be experimentally evaluated or otherwise addressed in each system under study.

Experimental manipulations may also result in spurious flux observations if not employed with care. For example, elimination of natural precipitation or limitation of experimental irrigation to the enclosed surfaces can significantly modify soil permeability and induce subsurface lateral gas flow (Guenzi *et al.*, 1994). Similarly, amendments for the purpose of enhancing or inhibiting production or consumption processes must also not be limited to the enclosed area, as this may induce horizontal concentration gradients and subsequent lateral transport by molecular diffusion.

2.4 Sample handling and analysis

The rigorous safeguards and controlled laboratory testing employed in chamber design and operation must be repeated in defining protocol for the collection, transport, storage and analysis of air samples. The potential for trace gas reaction or sorption on the walls of sample containers or tubing, for sample loss or dilution due to leakage, for trace gas dissolution in water condensed from the sample, and for artefact sources and sinks of the trace gas attributable to chemical, biological or photochemical reactions in the sample air should all be considered. Analyses of sample air should be completed within a few hours following collection whenever possible to minimize adverse storage effects. Loss or dilution of samples stored in syringes is often 2% or more per day. Glass containers may permit longer storage times, but not without added potential for contamination by or leakage through the stoppers, as well as added difficulty associated with the extraction of samples at near ambient pressures. Brooks *et al.* (1993) evaluated five common containers for sample storage prior to N_2O analysis. Leakage through and degassing of N_2O differed greatly between septa and container type, with red rubber stopper serum vials ranking worst and screw-cap tubes with Hungate septa ranking the best over the 14-day test period. As a general practice, gas standards with known concentration should be handled, stored and analysed regularly in the same manner as actual samples. Concurrent measures of other trace gas concentrations and exchange rates are also invaluable for tracking sample handling efficiencies.

Errors associated with the analytical determination of trace gas concentration depend on both the properties of the gas and the technique chosen for its analysis (e.g. Edwards & Sollins, 1973). The potential for error from this source is substantial, but like sample handling error, is not unique to chamber-based techniques and can be minimized by meticulously following accepted analytical protocol (see Chapter 6).

2.5 Flux estimation

To estimate trace gas flux density, i.e. the rate of gas exchange per unit surface area, from observed chamber air concentrations, the following protocol is required: (a) assuming a model of the relation between concentration and time in the enclosure air; (b) fitting the model to the observed concentration data; (c) predicting the exchange rate; and (d) testing the credibility of the prediction.

These guidelines assure rigour in reported exchange rates and promote the design of measurement protocols with improved efficiency and sensitivity.

Trace gas concentration of the enclosed air as a function of time, enclosure basal area and volume of the enclosed air are the minimum parameters required to compute flux density. However, because the concentrations are usually measured as a mixing ratio, i.e. in units of volume per unit volume of air (e.g. ppmv), they must be converted to a mass or molecular basis using the ideal gas law, which in turn, requires knowledge of the temperature and pressure of the enclosed air at the time of sampling:

$$C_i = \frac{q_i M_i P}{RT} \tag{2.4}$$

where C_i, q_i and M_i represent the mass/volume concentration, volume/volume concentration and molecular weight of species i, respectively; P is barometric pressure; T is temperature (°K); and R is the universal gas constant. For example, the atmospheric concentrations of CH_4 (1.7 ppmv) and N_2O (310 ppbv) are 1.13 and 0.57 mg m^{-3}, respectively, at NTP (20°C, 1 atm).

In steady-state systems, trace gas flux density (f) may be modelled as:

$$f = \frac{s}{A}(C_0 - C_i) \tag{2.5}$$

where s represents the sweep flow rate, A the chamber basal area, and C_i and C_0 the concentrations of the trace gas of interest in incoming and outgoing sweep air. As previously noted, it is important to assure that the system is at steady state and that perturbation of the natural exchange rate has been minimized. Precision of the estimated exchange rate is derived from replicate concentration observations over time.

In most cases where non-steady-state systems are employed, the applicable exchange model should predict flux density at the moment of deployment ($t = 0$), because this represents the only time at which the true exchange rate is unaffected by the presence of the enclosure. If post-emission reactions occur within the enclosure, however, an estimate at some other time in the measurement period may be more appropriate. For example, NO emitted into a non-steady-state system immediately following deployment is rapidly oxidized by ambient O_3 in the enclosed air volume to NO_2, which is strongly sorbed by soil, water and plant surfaces. The rate of change in NO concentration within the chamber head space, therefore, does not reflect the true surface–atmosphere NO exchange rate until the O_3 captured during chamber deployment has been destroyed or sorbed (Anderson & Levine, 1987).

Both linear and non-linear models have been proposed to describe the relation between trace gas concentration and time in non-steady-state systems. If the enclosure dimensions, deployment period and measurement protocol are suitably matched to the rate of trace gas exchange and site characteristics, a linear model may be adopted, which assumes a constant exchange rate over the period of observation:

$$f = \frac{V}{A}\frac{dC}{dt} \qquad (2.6)$$

where V represents enclosure volume, A its basal area and dC/dt the observed rate of concentration change. Measurement variability in the concentration observations may be accounted for using linear regression, but only if three or more observations are available.

In many common applications, particularly when working with low exchange rates, highly permeable soils or when the trace gas sources or sinks are very near the surface, analytical limitations may preclude using a measurement period sufficiently short that a linear model is appropriate. In these situations, i.e. when the data do not support the assumption that the net rate of exchange is constant over the measurement period, a non-linear model of concentration change over time must be employed. Matthias et al. (1978) and Hutchinson & Mosier (1981) proposed separate non-linear models, each based on the theory of molecular diffusion in soils. The iterative approach required by the Matthias et al. (1978) model makes it difficult to apply. The solution described by Hutchinson & Mosier (1981) applies to the special case defined when observations over two successive time periods of equal length are available:

$$f = \frac{V(C_1 - C_0)^2}{A(t_1 - t_0)(2C_1 - C_2 - C_0)}\ln\left[\frac{C_1 - C_0}{C_2 - C_1}\right]$$

$$\text{for } t_2 = 2t_1 \quad \text{and} \quad \frac{C_1 - C_0}{C_2 - C_1} > 1 \qquad (2.7)$$

where variables are defined as in Eqn. 2.6. Subscripts refer to the measurement event, where event 0 represents ambient conditions. Due to the limited number of observations over time, the Hutchinson & Mosier (1981) solution is highly sensitive to measurement imprecision in the concentration data, although this can be partially overcome by using replicate observations at each measurement time. Neither non-linear

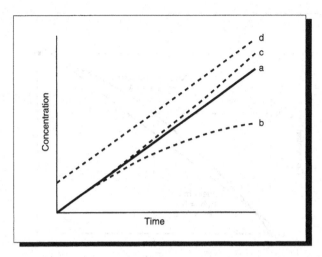

Fig. 2.11 Concentration change over time in a non-steady-state enclosure in response to typical disturbance effects: (a) linear response when the enclosure dimensions and measurement period are appropriately matched to the rate of exchange; (b) continuously decreasing exchange rate resulting either when the enclosure dimensions or measurement period are improperly matched to the actual exchange rate, or due to failure of the enclosure seal; (c) increasing exchange rate due to increasing soil temperature over the measurement period; and (d) elevated initial concentration resulting from disturbance during deployment.

approach includes provision for objectively testing the predictive capability of the model.

Once a model characterizing concentration change within the enclosure has been adopted, it is essential to verify for each flux measurement that the model applies to the observational data. Qualitative examination of a scatter plot of actual vs. model-predicted concentrations over time is often the most effective means of judging model fit when the number of observations is limited. Minimally, this examination should be used to both identify and screen observations objectively that do not meet model expectations. These may occur, for example, as a result of site disturbance, an improperly sealed enclosure, sample handling, storage or analytical difficulties, enclosure dimensions or deployment times that are not suited to the exchange rate being measured, as well as situations in which the model adopted simply does not apply (Fig. 2.11). If a linear model is assumed, one can compare, in addition, the observed and predicted concentrations at time $t = 0$ when the observations at that time are excluded from the analysis. Any substantial difference between the actual and predicted values suggests either that the site was disturbed during deployment, or that the exchange rate was not constant over the measurement period.

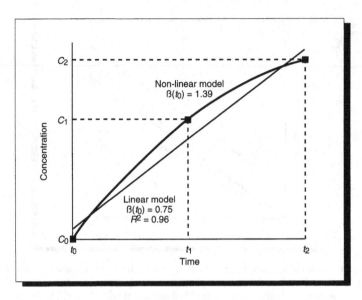

Fig. 2.12 Example of the error resulting from application of an inappropriate model of
concentration change over time when estimating the rate of trace gas exchange in non-
steady-state systems. The error in any given situation may be greater or less, depending
upon the observed concentration data. Exchange rates (β) were predicted at $t = 0$, i.e. at
the moment of deployment, using linear regression and the Hutchinson & Mosier (1981)
non-linear models applied to simulated diffusive gas exchange data. The data were defined
such that $(C_2 - C_1) = 0.5(C_1 - C_0)$ and $t_2 = 2t_1$, in arbitrary units. (From Hutchinson &
Livingston, 1993.)

Figure 2.12 illustrates the potential error in estimating exchange rates
that may result from adopting an inappropriate model of concentration
change over time. This simulation assumes that the trace gas concentration varies as predicted by the Hutchinson & Mosier (1981) model.
Support for this assumption lies in the successful application of the model
in several N_2O emission studies (Mosier & Hutchinson, 1981; Mosier *et
al.*, 1982; Folorunso & Rolston, 1984; Hutchinson & Brams, 1992). Note
that, despite the appearance of 'reasonable' linearity ($R^2 = 0.96$), the
linear model underestimated the true exchange rate by 46%. The magnitude of error in actual application, of course, will vary with each situation. Although the adoption of an inappropriate exchange model may
not alter the conclusions of experiments designed to test possible controls
on exchange rates, its contribution to error in scaling from local to
regional–seasonal or global–annual domains is readily apparent. The
development of a non-linear regression approach that is based on gas
transport theory, accounts for measurement variability, and is adaptable
to significance testing would substantially improve estimates of exchange

rates from highly permeable soils, but at this time requires further research.

Evaluation of the predictive capability of the linear model also requires testing the significance of the slope parameter (β) in the linear regression analysis against the hypothesis, H_0: $\beta = 0$, i.e. testing that time is an acceptable predictor of concentration in the enclosed air volume. The hypothesis is tested against a t-test statistic with $n-2$ degrees of freedom and an α significance level:

$$t = \frac{b - \beta}{\text{SE}} \tag{2.8}$$

where b is the estimated slope, and SE is the standard error of the slope estimate, defined as the ratio of $S_{y \cdot x}$ is the standard deviation of the residuals ($S_{y \cdot x}$) to the root mean square of the sampling time intervals $\sqrt{S_{xx}}$:

$$S_{xx} = \sum_{i=1}^{n} (x_i - \bar{x})^2 \tag{2.9}$$

where x_i is the time at which observation i was made and $\bar{x}$ is the mean of the time periods over which all concentrations were measured. The t-test is described in detail in introductory statistics texts (e.g. Dowdy & Weardon, 1983) and is computed by default in linear regression software packages. Estimates of significant trace gas exchange should be reported only when the null hypothesis is rejected; otherwise, the rate of exchange should be reported as below the detection limit of the measurement process.

The sample allocation employed to monitor concentration change within an enclosure must balance the need for sensitivity with the need to verify applicability of a particular exchange model. Table 2.1 reflects the sensitivities of selected sampling schemes in a linear regression application. Note that increasing the root mean square of the sampling time intervals ($\sqrt{S_{xx}}$) increases the power of the significance test, i.e. permits detection of smaller exchange rates at the significance level chosen. All sources of variability in sample handling and analysis ($S_{y \cdot x}$) should be minimized. Increasing the range of the measurement period (e.g. from 20 to 30 min), increasing the number of independent concentration observations within the measurement period, or employing replicate observations at selected measurement times greatly enhances sensitivity, particularly if the replicates are allocated at each end of the measurement period. However, sensitivity must be balanced against the need to verify that a linear model is applicable, so a sampling scheme that distributes at least some observations throughout the measurement period should be

Table 2.1 Sensitivity of selected sample allocations when employing linear regression to estimate the rate of trace gas exchange in a non-steady-state enclosure. The larger the root mean square of the sampling time intervals ($\sqrt{S_{xx}}$), the greater the sensitivity of the sample allocation to detect a given rate of exchange. $\sqrt{S_{xx}}$ is defined in the text; n is the total number of observations. The significance of the resultant slope estimate is testable using the t-test statistic listed in the last column, where t is the student's central t for $n - 2$ degrees of freedom at an $\alpha = 0.05$ significance level

Number of observations at time (min)									
0	5	10	15	20	25	30	n	$\sqrt{S_{xx}}$	$t_{\alpha,\,n-2}$
3	—	—	—	—	—	3	6	36.7	2.776
2	—	1	—	1	—	2	6	30.8	2.776
2	—	—	2	—	—	2	6	30.0	2.776
2	—	—	—	—	—	2	4	30.0	4.303
1	1	1	1	1	1	1	7	26.5	2.571
1	1	—	1	—	1	1	5	25.5	3.182
1	—	1	—	1	—	1	4	22.4	4.303
1	—	—	1	—	—	1	3	21.2	12.706
1	—	—	—	—	—	1	2	21.2	*
2	—	1	1	2	—	—	6	20.5	2.776
2	—	1	—	2	—	—	5	20.0	3.182
1	1	1	1	1	—	—	5	15.8	3.182
1	—	1	—	1	—	—	3	14.1	12.706

* Zero degrees of freedom; linear regression is not applicable to only two observations over time.

selected. Note also that allocations based upon only two observations are not recommended. Linear regression is not applicable in such cases and, as a result, measurement variability cannot be accounted for, the significance of the resultant exchange rates cannot be tested, nor can the applicability of a linear model be verified. Studies that combine an ambient air analysis with only a single enclosure concentration measurement, as is sometimes reported, also have no means to identify disturbance effects during chamber deployment.

Sample allocations in support of the measurement of concentration change should be optimized for each study, or even each site when possible. For example, allocations for sites with expected low flux demand more replicates or longer measurement periods than sites with expected high exchange rates. Site-by-site flexibility in the allocation of concentration observations, as well as in enclosure dimensions, not only maximizes the precision of the resultant flux estimates, but also reduces the time and cost of the overall study by eliminating superfluous observations.

Because measurement protocols and analytical precisions vary, a 'minimum significant flux' (MSF) should be reported as a measure against which observations can be compared between studies. Hutchinson & Livingston (1993) proposed that the t-test statistic used to evaluate the significance of the linear regression slope parameter could be used for this purpose:

$$MSF = t \cdot SE \tag{2.10}$$

given $n-2$ degrees of freedom and an $\alpha/2$ significance level, and SE is the standard error of the estimated slope (see Eqn. 2.8). This statistic accounts for the imprecision associated with the analytical determination of trace gas concentration, that associated with sample handling, and that associated with the number and timing of the concentration observations used to estimate the exchange rate. Summaries of this statistic may then be prepared for any selected subset of the data using normal descriptive techniques.

The overall precision of individual flux density estimates, however, is dependent upon the pooled precision of all parameters involved in their calculation. In practice, the accuracy and precision of variables other than concentration are seldom measured and can be assured only by prudent measurement protocol. For example, the accuracy of enclosure volume measures can be significantly compromised, particularly if soil surfaces are irregular, a significant volume of plant material is enclosed by the chamber, or chambers are not of simple geometry. The potential error in the resultant exchange rate estimates is inversely proportional to the chamber volume:basal area ratio. Thus, a 1 cm error in the measurement of height for enclosures 20 or 50 cm tall, corresponds to a 5 or 2% error in the flux density estimate. Where estimation of enclosure volume is critical, a known volume and concentration of an inert tracer gas may be injected into the enclosure upon deployment (Khalil *et al.*, 1990; Klinger *et al.*, 1994). The ideal gas law and subsequent observations of the tracer concentration may then be used to estimate enclosure volume. Concentration measurements employed in exchange rate computations are relatively insensitive to errors in the measurement of temperature and pressure. Only a 1% error in concentration results from a 3°C error in the estimate of air temperature or a 1 kPa (7.6 mmHg) error in the estimate of total pressure.

2.6 Sampling design and data analyses

Preceding sections focused on quantifying trace gas exchange at the spatial and temporal scales of an individual enclosure. However, by far the largest challenge in the use of enclosures pertains to how best to represent trace gas exchange over spatial or temporal domains beyond that measured. Inappropriate sample allocations and data analyses, or extension of data interpretations beyond the time and space domains for which they were designed, can lead to gross misrepresentation of the true exchange rates for the area and period of interest. Moreover, our current understanding of surface–atmosphere trace gas exchange is particularly sensitive to such errors because of the limited number of observations and ecosystems represented in existing inventories of measured exchange rates. Guidelines for addressing sampling design and data analysis issues are well framed in the theory of statistics; we offer here only a cursory overview of specific issues common to trace gas exchange studies.

2.6.1 Process studies

Process and inventory research objectives can rarely, if ever, be addressed by a single sampling design. The primary effects and interactions of trace gas exchange controlling factors are best examined in laboratory and *in situ* experiments where factors other than those of interest are either actively controlled or held constant by sampling design. Such studies are essential to understand controlling processes (see Chapter 10), develop mechanistic trace gas exchange models (see Chapter 11), and define and improve sample allocations in inventory studies. Enclosures are often uniquely applicable to these situations because of the typically small domains over which observations are required. Sampling designs applicable to process level studies are well defined in classical Gaussian and distribution-free approaches (Cochran & Cox, 1957; Conover, 1971; Hicks, 1973; Day & Quinn, 1989). Replicate observations within each combination of time, location and other controlled variable are required as differences between factors can only be demonstrated by comparison to the within-factor variability (Green, 1979; Hurlbert, 1984). If a study involves repeated observations at fixed locations, a repeated measures analysis must be employed, as opposed to an analysis of variance; the observations over time are not statistically independent.

Field studies are essential both for confirming the results of controlled experiments and for establishing the range of variability under natural settings. Such studies are particularly challenging because of the potential for multiple interactions between exchange rate controlling factors. Sample allocations and data analyses that isolate response to the factor of interest from response to potential confounding factors are often difficult to define if the factors are not directly or readily observable, or if the factors co-vary. In these situations, some factors may be qualitatively evaluated by monitoring readily observable surrogate factors to which those of interest are closely associated, such as landscape position, veg-etation community type and soil water content. Only when the effects of confounding factors are removed by appropriate sample allocation and data analysis, can the response to independent factors of interest be observed; otherwise, the response may be lost in the overall variability of the observations. For example, to evaluate the effect of soil temperature on CH_4 exchange rates in northern ecosystems, the confounding effect of competing microbial production and consumption processes must first be removed. In field studies, this is most readily approached by stratifying the flux data on the basis of whether the sites measured were inundated at the time of sampling, i.e. on whether CH_4 oxidation was a significant factor affecting the observed emissions (e.g. Crill *et al.*, 1988; Morrissey & Livingston, 1992).

Random sample allocations are well suited to both experimental and inventory approaches and should be employed unless alternative allocations can be explicitly justified. *Ad hoc* selection of 'representative' or 'typical' sites must be avoided as this can lead to significant misrepresentation of the observations and subsequent conclusions. Spatial and geostatistical approaches (e.g. Isaaks & Srivastava, 1989) offer distinct advantages over random allocations to meet interpolation and extrapolation objectives, but only if their intense data requirements can be met. Pilot studies designed to assess the spatial or temporal autocorrelation between observations are invaluable for predicting the scales over which controlling factors influence trace gas exchange, or conversely, the scales over which measured exchange rates may be considered spatially independent and, therefore, as true replicates in subsequent analyses (e.g. Robertson, 1987, 1993).

2.6.2 Inventory studies

The use of enclosures to support large area inventories of trace gas sources and sinks is most challenging and potentially subject to great uncertainty. Success at the global scale is dependent upon how well the Earth is represented by the subareas selected for study, and how well each subarea is represented by the sample allocation chosen. Although large area inventories may be best addressed by sampling schemes that combine enclosure and micrometeorological approaches over a variety of scales of observation (e.g. Wesley *et al.*, 1989; Robertson, 1993; Schimel *et al.*, 1993), limitations in funding or in instrumentation often dictate that enclosures remain the primary approach applicable for many trace gas species and geographic areas of study. Global scale inventories require extrapolation of enclosure-based observations by factors as great as 10^{12}–10^{14}, thus making sampling and analysis choices at both the subarea and global scales particularly critical to the credibility of the estimates.

If the area to be sampled has a coarse-scale environmental pattern, a stratified sampling approach offers many advantages (Cochran & Cox, 1957). A simple example entails partitioning the study area into relatively homogeneous subareas (strata) known or hypothesized to have similar exchange rates of the species of interest (e.g. Matson & Vitousek, 1987,

1990; Livingston *et al.*, 1988; O'Neill, 1988; Bartlett *et al.*, 1989, 1992; Matson *et al.*, 1989; Stewart *et al.*, 1989). The resulting strata need not be contiguous. The number of samples assigned to each stratum is typically proportional to its areal extent or to its contribution to the regional variance (e.g. Morrissey & Livingston, 1992). Within each stratum, samples are allocated randomly to spatially independent sites. The mean exchange rate ($\bar{F}$) for the area of study is defined as the sum of the area-weighted mean exchange rates of each stratum:

$$\bar{F} = \sum_{i=1}^{m} \rho_i \bar{f}_i \tag{2.11}$$

where ρ_i is the fractional area and $\bar{f}_i$ the mean exchange rate of stratum i, and m is the total number of strata representing the study area. The variance associated with the overall mean exchange rate ($V(\bar{F})$) is similarly computed as the sum of the area-weighted individual strata variances:

$$V(\bar{F}) = \sum_{i=1}^{m} \rho_i^2 V(\bar{f}_i) \tag{2.12}$$

and

$$V(\bar{f}_i) = \frac{s_i^2}{n_i} \tag{2.13}$$

where $V(\bar{f}_i)$ is the within-stratum population variance given n_i samples; other variables are defined as in Eqn. 2.11. These equations assume that the area of each stratum is measured without error.

Although the accuracy of a large area exchange rate estimate based on a stratified random approach is not necessarily any greater than that obtainable using a simple random sampling approach, its precision is potentially far greater given the same level of effort. Even if between-strata differences in exchange rates prove non-significant, the resultant overall precision of the stratified random allocation is no less than that of a simple random approach. In addition, this approach can be repeated at different scales of hierarchical stratification or combined with other sample allocations towards the goals of acceptable precision and as few observations as necessary.

For measurement sites defined as spatially independent locations within a given stratum, the question arises in inventory studies as to how many individual flux estimates are required to characterize the within-site vs. the between-site variability. Within-site replicates are essential only if the research objectives require knowledge of the within-site variability or between-site differences. Within-site multiple observations are rarely spatially independent at the overall inventory scale and, as such, are of little value in between-strata comparisons or overall inventory estimates. To address the latter, it is generally far more efficient to allocate a single observation to each site and include as many spatially independent sites within each stratum as can be afforded. This approach, although purposefully confounding within-site and between-site variability, maximizes the precision of the resulting large area flux estimate for a fixed number of observations (see Eqn. 2.10). In reality, more complex sample allocations based on hierarchical stratifications or other multistage approaches may be required to characterize large area emissions affordably at the precision required. The above prioritization of effort, however, remains valid. Misuse of within-site observations as independent observations in extrapolation efforts remains a serious error in many trace gas exchange and other ecological studies (Hurlbert, 1984).

Temporal variability in trace gas exchange rates must also be explicitly considered in inventory studies. A separate sample allocation designed solely to characterize this variability at the appropriate scale is generally required. As with spatial sampling, temporal strata may be defined such that within-strata variability is less than that between strata. Minimally, diel (24 h) variability should be assessed at fixed locations and the results used to guide subsequent spatial sampling efforts. For example, if hourly-to-daily temporal variability is small relative to the expected spatial variability at the scale of interest, it may often prove effective to confound hourly-to-daily variability with between-site variability. As a result, exchange rates for spatial replicates need not be measured simultaneously, nor on the same day, nor even at the same time of day. The potential to increase the number of spatially replicate observations in support of a large area flux estimate, therefore, is greatly expanded. Seasonal variability must also be addressed if annual flux estimates are required, because differences between seasons are often as large as differences between ecosystems (e.g. Matson *et al.*, 1989, 1992; Whalen & Reeburgh, 1992). Periodic events that may significantly affect trace gas exchange rates within the planned sampling period, such as precipitation, should be considered in advance so that a separate sample allocation to address these events can be employed if necessary.

Automated enclosures that combine inexpensive data acquisition and control systems with multi-component chambers, offer substantial advantages for characterizing temporal variability at fixed locations on daily to seasonal scales. Brumme & Beese (1992), IAEA (1992) and Loftfield *et al.* (1992) describe successful approaches.

2.6.3 Data summaries
The heterogeneity inherent in field-measured trace gas exchange rates makes their summarization difficult. Even the underlying population dis-

tribution of the flux estimates must often be assumed, because the number of estimates is typically limited. If the data appear to be normally distributed or only negligibly skewed, it is appropriate to estimate population parameters using the mean and standard deviation of the sample data. However, the frequency distribution of many soil properties, including their trace gas exchange rates, is often positively skewed, i.e. although most observed values are small, there are a few observations that are extremely large. As a result, the mean of the sample distribution is greater than the median and the coefficient of variation (CV = mean/ standard deviation) is often greater than 100%. If the number of observations is small, there may be no clear best estimates of the population central tendency and variability and conclusions drawn from such data should be accepted only with appropriate caution.

Classical descriptive statistics based on the arithmetic mean and variance are unbiased regardless of the underlying distribution and are thus recommended when the number of observations is sufficient ($n >$ 20–30), even if the underlying sample distribution is unknown. The variance (imprecision) associated with the mean may be high, however, if the sample distribution is skewed. Stratification, for example on the basis of environmental factors, will substantially reduce the skewness of the within-stratum exchange rates in many situations. Ideally, the underlying population distribution can be established, and the mean estimated accordingly with lower variance.

The log-normal distribution is often a reasonable approximation for many soil processes (Parkin & Robinson, 1992), although it is not appropriate for all skewed data. For 100 or fewer observations with a population coefficient of variation of at least 100%, Parkin et al. (1988) demonstrated that the uniformly minimum variance unbiased estimators are superior to either the method of moments or maximum likelihood techniques for characterizing log-normally distributed data. The method estimates the population mean (m) and variance (s^2) as follows:

$$m = \exp(\bar{y})\Psi(\hat{\sigma}^2/2) \tag{2.14}$$

$$s^2 = \exp(2\bar{y})\left\{\Psi(2\hat{\sigma}^2) - \Psi\left[\frac{(n-2)}{(n-1)}\hat{\sigma}^2\right]\right\} \tag{2.15}$$

where $\bar{y}$ and σ^2 are the estimated population mean and variance, respectively, of the log-transformed data:

$$\bar{y} = \frac{1}{n}\sum_{i=1}^{n}\ln(x_i) \tag{2.16}$$

$$\hat{\sigma}^2 = \frac{1}{n-1}\sum_{i=1}^{n}[\ln(x_i) - \bar{y}]^2 \tag{2.17}$$

given x_i = the untransformed i^{th} observation, n = the number of observations, and Ψ_z = the following power series operating on z:

$$\Psi_z = 1 + \frac{z(n-1)}{n} + \frac{z^2(n-1)^3}{n^2(n+1)2!} + \frac{z^3(n-1)^5}{n^3(n+1)(n+3)3!}$$

$$+ \frac{z^4(n-1)^7}{n^4(n+1)(n+3)(n+5)4!} + \frac{z^5(n-1)^9}{n^5(n+1)(n+3)(n+5)(n+7)5!}$$

$$+ \ldots \tag{2.18}$$

Parkin *et al.* (1988) suggested evaluating this series until the final term accounts for <1% of the sum of the preceding terms, which usually requires 6–10 terms. Confidence intervals for the mean are asymmetric; their calculation is defined by Parkin *et al.* (1990).

The median or the geometric mean of the sample distribution is sometimes reported rather than the mean in summaries of trace gas exchange data on the premise that these statistics are less influenced by the extreme values often observed in field studies. The sample geometric mean, defined as the antilogarithm of the mean of the log-transformed data [GM = $\exp(\bar{y})$], is an estimator of the median, not the mean, of the log-normal distribution and is positively biased by an amount inversely proportional to sample size (Parkin & Robinson, 1993). Although both the mean and the median are measurements of central tendency, Gilbert (1987) and Parkin & Robinson (1992) illustrate that because the median systematically underestimates the total mass of the sample distribution, it is an inappropriate estimator when it is the inventory of the measured variable that is of interest. Thus in trace gas exchange studies where an integrated estimate of the total mass exchanged is required, the mean is the preferred estimator.

Areal and temporal summaries of flux observations should explicitly note the assumed underlying population distribution and define the transformation and summary statistics employed. It is imperative that all observations that pass objective quality-control criteria be included in the reported summaries, even those exchange rate estimates that do not differ significantly from zero and those for which the observed direction of exchange cannot be explained mechanistically. Outliers should be ignored only if it can be demonstrated that they resulted from an obvious error in data collection; they are otherwise important in characterizing the population sampled. This reporting of observations *in toto* is necessary to prevent bias in the summary statistics. To facilitate comparisons between studies and extrapolations over larger domains, summary statistics should be reported for both non-transformed and transformed data.

2.7 Concluding remarks

Trace gas exchange between soil–plant systems and the atmosphere is a complex phenomenon whose measurement by chamber systems is subject to many potential sources of error. Bias due to physical and biological disturbances associated with the measurement process can be mostly overcome by using appropriate enclosure design, relatively short deployment times, and reasonable care to minimize site disturbances. Similarly, bias introduced during trace gas concentration measurement and flux estimation can be minimized by following a carefully conceived measurement protocol and statistically sound procedures for choosing an appropriate model to estimate flux from the observed concentrations. Exploratory surveys and experience are indispensable in both establishing the expected range of exchange rates for any particular setting, and identifying situations particularly vulnerable to disturbance. Although potential sources of error in individual chamber-based estimates of trace gas exchange are numerous, the use of sampling designs inappropriate for quantifying the exchange processes of interest and subsequent analyses that fail to consider the original sample allocation represent the most serious critiques of chamber-based studies.

Enclosures offer great potential to contribute to both process level and global–annual scale understanding of trace gas exchange. They are not applicable in all situations because of inherent limitations in the technique (Trevitt *et al.*, 1988; Wesley *et al.*, 1989; Baldocchi & Meyers, 1991) or sampling logistics. They do, however, represent a powerful technique applicable to many experimental and survey objectives and, if the sample allocation is suitable, a cost-effective means of characterizing trace gas exchange processes on many spatial and temporal scales.

Acknowledgements

We thank P. Matson and R. Harriss for forethought and guidance in the preparation of this chapter, as well as L. Morrissey, L. Klinger, and the anonymous reviewers whose critical reviews and discussions were so valuable in the evolution of this chapter. We also acknowledge Merin McDonnell for the timely preparation of graphic materials. This work was supported by the NASA Interdisciplinary Research in Earth Sciences Program and the Agricultural Research Service, US Department of Agriculture.

References

Anderson I.C. & Levine J.S. (1987) Simultaneous field measurements of biogenic emissions of nitric oxide and nitrous oxide. *Journal of Geophysical Research*, **92**, 965–976.

Baldocchi D.D. & Meyers T.P. (1991) Trace gas exchange above the floor of a deciduous forest 1. Evaporation and CO_2 efflux. *Journal of Geophysical Research*, **96**, 7271–7285.

Bartlett D.S., Bartlett K.B., Hartmann J.M. *et al.* (1989) Methane emissions from the

Florida Everglades: Patterns of variability in a regional wetland ecosystem. *Global Biogeochemical Cycles*, **3**, 363–374.

Bartlett K.B., Crill P.M., Saas R.L., Harriss R.C. & Dise N.B. (1992) Methane emissions from tundra environments in the Yukon–Kuskokwim Delta, Alaska. *Journal of Geophysical Research*, **97**, 16645–16660.

Bird R.B., Stewart W.E. & Lightfoot E.N. (1960) *Transport Phenomena*. John Wiley & Sons, New York.

Bowden W.B. & Bormann F.H. (1986) Transport and loss of nitrous oxide in soil water after forest clear-cutting. *Science*, **233**, 867–869.

Brooks P.D., Atkins G.J., Herman D.J., Prosser S.J. & Barrie A. (1993) Rapid, isotopic analysis of selected soil gases at atmospheric concentrations. In: Harper L.A., Mosier A.R., Duxbury J.M. & Rolston D.E. (eds) *Agricultural Ecosystem Effects on Trace Gases and Global Climate Change*, pp. 193–202. American Society of Agronomy Special Publication No. 55. Madison, Wisconsin.

Brumme R. & Beese F. (1992) Effects of liming and nitrogen fertilization on emissions of CO_2 and N_2O from a temperate forest. *Journal of Geophysical Research*, **97**, 12851–12858.

Campbell G.S. (1985) *Soil Physics with Basic, Transport Models for Soil–Plant Systems*. Elsevier Science, New York.

Castro M.S. & Galloway J.N. (1991) A comparison of sulfur-free and ambient air enclosure techniques for measuring the exchange of reduced sulfur gases between soils and the atmosphere. *Journal Geophysical Research*, **96**, 15427–15437.

Chanton J.P. & Dacey J.W.H. (1991) Effects of vegetation on methane flux, reservoirs, and carbon isotopic composition. In: Sharkey T., Holland E., & Mooney H. (eds) *Trace Gas Emissions from Plants*, pp. 65–92. Academic Press, San Diego, California.

Clements W.E. & Wilkening M.H. (1974) Atmospheric pressure effects on ^{222}Rn transport across the Earth–air interface. *Journal of Geophysical Research*, **79**, 5025–5029.

Cochran W.G. & Cox G.M. (1957) *Experimental Designs*. John Wiley & Sons, New York.

Conover W.J. (1971) *Practical Nonparametric Statistics*, 2nd edn. John Wiley & Sons, New York.

Crill P.M., Bartlett K.B., Harriss R.C. *et al.* (1988) Methane flux from Minnesota peatlands. *Global Biogeochemical Cycles*, **2**, 371–384.

Currie J.A. (1960) Gaseous diffusion in porous media: Part 2. Dry granular materials. *British Journal of Applied Physics*, **11**, 318–324.

Currie J.A. (1961) Gaseous diffusion in porous media: Part 3. Wet granular materials. *British Journal of Applied Physics*, **12**, 275–281.

Day R.W. & Quinn G.P. (1989) Comparison of treatments after an analysis of variance in ecology. *Ecological Monograph*, **59**, 433–463.

Denmead O.T. (1979) Chamber systems for measuring nitrous oxide emission from soils in the field. *Soil Science Society of America Journal*, **43**, 89–95.

Denmead O.T. & Raupach M.R. (1993) Methods for measuring atmospheric gas transport in agricultural and forest systems. In: Harper L.A., Mosier A.R., Duxbury J.M. & Rolston D.E. (eds) *Agricultural Ecosystem Effects on Trace Gases and Global Climate Change*, pp. 19–43. American Society of Agronomy Special Publication No. 55. Madison, Wisconsin.

Denmead O.T., Freney J.R. & Simpson J.R. (1976) A closed ammonia cycle within a plant canopy. *Soil Biology and Biochemistry*, **8**, 161–164.

Dowdell R.J., Burford J.R. & Crees R. (1979) Losses of nitrous oxide dissolved in drainage water from agricultural land. *Nature*, **278**, 342–343.

Dowdy S. & Wearden S. (1983) *Statistics for Research*. John Wiley & Sons, New York.

Edwards N.T. & Sollins P. (1973) Continuous measurement of carbon dioxide evolution from partitioned forest floor components. *Ecology*, **54**, 406–412.

Fechner E.J. & Hemond H.F. (1992) Methane transport and oxidation in the unsaturated zone of a *Sphagnum* peatland. *Global Biogeochemical Cycles*, **6**, 33–44.

Focht D.D. (1992) Diffusional constraints on microbial processes in soil. *Soil Science*, **154**, 300–307.

Folorunso O.A. & Rolston D.E. (1984) Spatial variability of field-measured denitrification gas fluxes. *Soil Science Society of America Journal*, **48**, 1214–1219.

Fried A., Klinger L.F. & Erickson D.J. III (1993) Atmospheric carbonyl sulfide exchange in bog microcosms. *Geophysical Research Letters*, **20**, 129–132.

Ghildyal B.P. & Tripathi R.P. (1987) *Soil Physics*. John Wiley & Sons, New York.

Gilbert R.O. (1987) *Statistical Methods for Environmental Pollution Monitoring*. Van Nostrand Reinhold, New York.

Glinski, J. & Stepniewski W. (1985) *Soil Aeration and its Role for Plants*. CRC Press, Boca Raton, Florida.

Green R.H. (1979) *Sampling Design and Statistical Methods for Environmental Biologists*. John Wiley & Sons, New York.

Grosse W., Buchel H. & Tiebel H. (1991) Pressurized ventilation in wetland plants. *Aquatic Botany*, **39**, 89–98.

Guenzi W.D., Hutchinson G.L. & Beard W.E. (1994) Nitric and nitrous oxide emissions and soil nitrate distribution in a center-pivot-irrigated cornfield. *Journal of Environmental Quality*, **23**, 483–487.

Hicks C.R. (1973) *Fundamental Concepts in the Design of Experiments*. Holt, Rinehart & Winston, New York.

Himmelblau D.M. (1964) Diffusion of dissolved gases in liquids. *Chemical Reviews*, **64**, 527–550.

Hurlbert S.H. (1984) Pseudoreplication and the design of ecological field experiments. *Ecological Monographs*, **54**, 187–211.

Hutchinson G.L. & Brams E.A. (1992) NO versus N_2O emissions from an NH_4^+-amended Bermuda grass pasture. *Journal of Geophysical Research*, **97**, 9889–9896.

Hutchinson G.L. & Livingston G.P. (1993) Use of chamber systems to measure trace gas fluxes. In: Harper L.A., Mosier A.R., Duxbury J.M. & Rolston D.E. (eds) *Agricultural Ecosystem Effects on Trace Gases and Global Climate Change*, pp. 63–78. American Society of Agronomy Special Publication No. 55. Madison, Wisconsin.

Hutchinson G.L. & Mosier A.R. (1979) Nitrous oxide emissions from an irrigated cornfield. *Science*, **205**, 1125–1127.

Hutchinson G.L. & Mosier A.R. (1981) Improved soil cover method for field measurement of nitrous oxide fluxes. *Soil Science Society of America Journal*, **45**, 311–316.

Hutchinson G.L., Beard W.E., Vigil M.F. & Halvorson A.D. (1992) NO and N_2O emissions from perennial grass and winter wheat in the semiarid Great Plains. In: *Agronomy Abstracts*, p. 260. American Society of Agronomy, Madison, Wisconsin.

IAEA (International Atomic Energy Agency) (1992) *Manual on Measurement of Methane and Nitrous Oxide Emissions from Agriculture*. INIS Clearinghouse, IAEA-TECDOC-674, Vienna.

Isaaks E.H. & Srivastava R.M. (1989) *An Introduction to Applied Geostatistics*, Oxford University Press, New York.

Jarvis P.G. & McNaughton K.G. (1986) Stomatal control of transpiration: scaling up from leaf to region. *Advances in Ecological Research*, **15**, 1–49.

Jury W.A., Letey J. & Collins T. (1982) Analysis of chamber methods used for measuring nitrous oxide production in the field. *Soil Science Society of America Journal*, **46**, 250–256.

Kanemasu E.T., Powers W.L. & Sij J.W. (1974) Field chamber measurements of CO_2 flux from soil surface. *Soil Science*, **118**, 233–237.

Khalil M.A.K., Rasmussen R.A., Wang M.-X. & Ren L. (1990) Emissions of trace gases from Chinese rice fields and biogas genetators: CH_4, N_2O, CO, CO_2, chlorocarbons, and hydrocarbons. *Chemosphere*, **20**, 207–226.

Kimball B.A. (1983) Canopy gas exchange: gas exchange with soil. In: Taylor H.M., Jordon W.R. & Sinclair T.R. (eds) *Limitations to Efficient Water Use in Crop Production*.

American Society of Agronomy, Madison, Wisconsin.

Kimball B.A. & Lemon E.R. (1972) Theory of soil air movement due to pressure fluctuations. *Agricultural Meteorology*, **9**, 163–181.

Klinger L.F., Zimmerman P.R., Greenberg J.P., Heidt L.E. & Guenther A.B. (1994) Carbon trace gas fluxes along a successional gradient in the the Hudson Bay lowland. *Journal of Geophysical Research*, **99**, 1469–1494.

Knapp A.K. & Yavitt J.B. (1992) Evaluation of a closed-chamber method for estimating methane emissions from aquatic plants. *Tellus*, **44B**, 63–71.

Lamb B., Guenther A., Gay D. & Westbert H. (1987) A national inventory of biogenic hydrocarbon emissions. *Atmospheric Environment*, **21**, 1695–1705.

Livingston G.P., Morrissey L.A., Card D.H. & Kasting J.F. (1988) Estimating regional methane flux in high latitude ecosystems. *EOS Transactions of the American Geophysical Union*, **69**, 1084.

Loftfield N.S., Brumme R. & Beese F. (1992) Automated monitoring of nitrous oxide and carbon dioxide flux from forest soils. *Soil Science Society of America Journal*, **56**, 1147–1150.

Luizão F., Matson P., Livingston G., Luizão R. & Vitousek P. (1989) Nitrous oxide flux following tropical land clearing. *Global Biogeochemical Cycles*, **3**, 281–285.

Matson P.A. & Vitousek P.M. (1987) Cross-system comparisons of soil nitrogen transformations and nitrous oxide flux in tropical forest ecosystems. *Global Biogeochemical Cycles*, **1**, 163–170.

Matson P.A. & Vitousek P.M. (1990) Ecosystem approach to a global nitrous oxide budget. *BioScience*, **40**, 667–672.

Matson P.A., Vitousek P.M. & Schimel D.S. (1989) Regional extrapolation of trace gas flux based on soils and ecosystems. In: Andreae M.O. & Schimel D.S. (eds) *Exchange of Trace Gases Between Terrestrial Ecosystems and the Atmosphere*, pp. 97–108. John Wiley & Sons, New York.

Matson P.A., Vitousek P.M., Livingston G.P. & Swanberg N.A. (1990) Sources of variation in nitrous oxide flux from Amazonian ecosystems. *Journal of Geophysical Research*, **95**, 16789–16798.

Matson P.A., Gower S.T., Volkman C., Billow C. & Grier C. (1992) Soil nitrogen cycling and nitrous oxide flux in a Rocky Mountain Douglas-fir forest: effects of fertilization, irrigation and carbon addition. *Biogeochemistry*, **18**, 101–117.

Matthias A.D., Blackmer A.M. & Bremner J.M. (1980) A simple chamber for field measurement of emissions of nitrous oxide from soils. *Journal of Environmental Quality*, **9**, 251–256.

Matthias A.D., Yarger D.N. & Weinbeck R.S. (1978) A numerical evaluation of chamber methods for determining gas fluxes. *Geophysical Research Letters*, **5**, 765–768.

Moore T.R. & Roulet N.T. (1991) A comparison of dynamic and static chambers for methane emission measurements from subarctic fens. *Atmosphere-Ocean*, **29**, 102–109.

Morrissey L.A. & Livingston G.P. (1992) Methane emissions from Alaska Arctic tundra: an assessment of local spatial variability. *Journal Geophysical Research*, **97**, 16661–16670.

Morrissey L.A., Zobel D.B. & Livingston G.P. (1993) Significance of stomatal control on methane release from *Carex*-dominated wetlands. *Chemosphere*, **26**, 339–356.

Mosier A.R. & Hutchinson G.L. (1981) Nitrous oxide emissions from cropped fields. *Journal of Environmental Quality*, **10**, 169–173.

Mosier A.R., Hutchinson G.L., Sabey B.R. & Baxter J. (1982) Nitrous oxide emissions from barley plots treated with ammonium nitrate or sewage sludge. *Journal of Environmental Quality*, **11**, 78–81.

Nazaroff W.W. (1992) Radon transport from soil to air. *Reviews of Geophysics*, **30**, 137–160.

Nobel P.S. (1983) *Biophysical Plant Physiology and Ecology*. W.H. Freeman, San Francisco, California.

O'Neill R.V. (1988) Hierarchy theory and global change. In: Rosswall T., Woodmansee

R.G. & Risser P.G. (eds) *Scales and Global Change*, pp. 29–45. John Wiley & Sons, New York.

Parkin T.B. & Robinson J.A. (1992) Analysis of lognormal data. *Advances in Soil Science*, **20**, 193–235.

Parkin T.B. & Robinson J.A. (1993) Statistical evaluation of median estimators for lognormally distributed variables. *Soil Science Society of America Journal*, **57**, 317–323.

Parkin T.B., Chester S.T. & Robinson J.A. (1990) Calculating confidence intervals for the mean of a lognormally distributed variable. *Soil Science Society of America Journal*, **54**, 321–326.

Parkin T.B., Meisinger J.J., Chester S.T., Starr J.L. & Robinson J.A. (1988) Evaluation of statistical estimation methods for lognormally distributed variables. *Soil Science Society of America Journal*, **52**, 323–329.

Reicosky D.C., Wagner S.W. & Devine O.J. (1990) Methods of calculating carbon dioxide exchange rates for maize and soybean using a portable field chamber. *Photosynthetica*, **24**, 22–38.

Robertson G.P. (1993) Fluxes of nitrous oxide and other nitrogen trace gases from intensively managed landscapes: A global perspective. In: Harper L.A., Mosier A.R., Duxbury J.M. & Rolston D.E. (eds) *Agricultural Ecosystem Effects on Trace Gases and Global Climate Change*, pp. 95–108. American Society of Agronomy Special Publication No. 55. Madison, Wisconsin.

Robertson P. (1987) Geostatistics in ecology: interpolation with known variance. *Ecology*, **68**, 744–748.

Rochette P., Gregorich E.G. & Desjardins R.L. (1992) Comparison of static and dynamic closed chambers for measurement of soil respiration under field conditions. *Canadian Journal of Soil Science*, **72**, 605–609.

Ronen D., Margaritz M. & Almon E. (1988) Contaminated aquifers are a forgotten component of the global N_2O budget. *Nature (London)*, **335**, 57–59.

Sallam A., Jury W.A. & Letey J. (1984) Measurement of gas diffusion coefficient under relatively low air-filled porosity. *Soil Science Society of America Journal*, **48**, 3–6.

Schery S.D., Gaeddert D.H. & Wilkening M.H. (1984) Factors affecting exhalation of radon from a gravelly sandy loam. *Journal of Geophysical Research*, **89**, 7299–7309.

Schimel D.S., Davis F.W. & Kittel T.G.F. (1993) Spatial information for Extrapolation of Canopy Processes: Examples from FIFE. In: Ehleringer J.R. & Field C.B. (eds) *Scaling Physiological Processes, Leaf to Globe*, pp. 21–38. Academic Press, New York.

Sebacher D.I. & Harriss R.C. (1982) A system for measuring methane fluxes from inland and coastal wetland environments. *Journal of Environmental Quality*, **11**, 34–37.

Steudler P.A. & Peterson B.J. (1985) Annual cycle of gaseous sulfur emissions from a New England *Spartina alterniflora* marsh. *Atmospheric Environment*, **19**, 1411–1416.

Stewart J.W.B., Aselmann I., Bouwman A.F. *et al.* (1989) Extrapolation of flux measurements to regional and global scales. In: Andreae M.O. & Schimel D.S. (eds) *Exchange of Trace Gases Between Terrestrial Ecosystems and the Atmosphere*, pp. 155–174. John Wiley & Sons, Berlin.

Stull R.B. (1988) *An Introduction to Boundary Layer Meteorology*. Kluwer Academic Publishers, Boston, Massachusetts.

Trevitt A.C.F., Freney J.R., Denmead O.T., Zhao-Liang Z., Gui-Xiu C. & Simpson J.R. (1988) Water–air transfer resistance for ammonia from flooded rice. *Journal of Atmospheric Chemistry*, **6**, 133–147.

Troch F.R., Jabro J.D. & Kirkham D. (1982) Gaseous diffusion equations for porous materials. *Geoderma*, **27**, 239–253.

Webb E.K., Pearman G.I. & Leuning R. (1980) Correction of flux measurements for density effects due to heat and water vapour transfer. *Quarterly Journal of the Royal Meteorological Society*, **106**, 85–100.

Wesley M.L., Lenschow D.H. & Denmead O.T. (1989) Flux Measurement Techniques. In:

Lenschow D.H. & Hicks B.B. (eds) *Global Tropospheric Chemistry: Chemical Fluxes in the Global Atmosphere*, pp. 31–46. National Center for Atmospheric Research, Boulder, Colorado.

Whalen S.C. & Reeburgh W.S. (1992) Interannual variations in tundra methane emission: a 4-year time series at fixed sites. *Global Biogeochemical Cycles*, **6**, 139–159.

Whiting G.J., Chanton J., Bartlett D. & Happell J. (1991) Methane flux, net primary productivity and biomass relationships in a sub-tropical grassland community. *Journal of Geophysical Research*, **96**, 13067–13071.

CHAPTER 3

Trace gas exchange across the air–water interface in freshwater and coastal marine environments

S. MACINTYRE, R. WANNINKHOF & J.P. CHANTON

3.1 Introduction

Gas exchange in freshwater and coastal marine ecosystems proceeds by three primary pathways:
1 diffusive and turbulent transfer across the air–water interface;
2 bubble ebullition; and
3 through emergent aquatic plants.
In this chapter we will discuss transfer across the air–water interface, the mechanism by which gas is exchanged directly between the water column and the atmosphere. Bubble ebullition and the transport of gases between the atmosphere and sediments or water-logged soils via rooted plants is the subject of Chapter 4. We will discuss the factors controlling gas transfer across the air–water interface and methodologies for measurement of fluxes.

Gas exchange occurs across the air–water interface in all aqueous environments from the headwaters of rivers to the pelagic ocean. The magnitude and direction of the gas flux is dependent on the concentration difference between the atmosphere and the water and upon transport processes at the interface, with turbulence being important in most aquatic environments. Gas concentrations between air and water can be in disequilibrium due to changes in temperature and salinity, or because gases are produced or consumed by biological or chemical processes. Trace gases released from the water to the atmosphere can be produced within the water column (e.g. dimethyl sulphide) or within the sediments (e.g. CH_4). The water column can also serve as a source or sink for gases such as CO_2 or O_2.

The transport processes at the air–water boundary are parameterized by a gas transfer velocity also known as a gas transfer coefficient or piston velocity and, for streams and rivers, as a reaeration coefficient. Considerable effort has gone into determining empirical relations between gas transfer velocity and wind speed, which has a major effect on gas transfer (see Wanninkhof, 1992 for a summary). Scatter in the relations, discrepancies at high and low wind speeds, and the knowledge that turbulence is caused by more than wind stress alone have maintained interest in obtaining other parameterizations (Hunt, 1984; Kitaigorodskii &

Donelan, 1984; Csanady, 1990; Banerjee, 1991). Analysis of turbulence measurements in the upper metre indicates that turbulence can be estimated from wind speeds, heat fluxes and wave characteristics (Kitaigorodskii *et al.*, 1983; Imberger, 1985; Agrawal *et al.*, 1992). We will discuss progress currently being made in relating these parameters to the gas transfer velocity as well as types of research that would enable these relations to be better quantified.

3.2 Gas exchange at the air–water interface

Exchange across the air–water boundary region is the rate-limiting step for gas transfer. The boundary region consists of a boundary layer several millimetres thick in the atmosphere above the air–water interface, and an aqueous boundary layer less than 0.5 mm thick in the water below the air–water interface. The uppermost part of the aqueous boundary layer is in equilibrium with air and has a gas concentration of αC_a where C_a is the concentration of the gas in air and α is the Ostwald solubility coefficient; the lower portion of the boundary layer has a concentration of C_w (Fig. 3.1). For soluble gases such as NH_3 and SO_2, the rate-limiting step is transfer across the boundary layer in the air. For slightly soluble gases, such as DMS, CH_4, CO_2 and O_2, the aqueous boundary layer retards transport. This chapter will emphasize processes relevant to slightly soluble gases. Transport of slightly soluble gases can be viewed as occur-

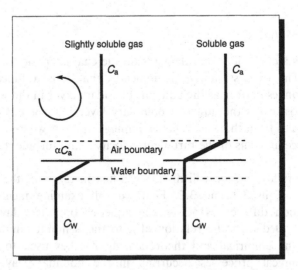

Fig. 3.1 Schematic of gas concentrations across the air–water interface. The left-hand side depicts the concentration gradient of a slightly soluble gas. The gradient, which for illustrative purposes is drawn as linear, is concentrated in the aqueous or water boundary. Models discussed in this chapter address the transport across this gradient. For soluble gases the resistance to gas transfer is in the air boundary layer (right-hand side).

ring in two stages. In the first, turbulent eddies transport fluid to and within the aqueous boundary layer in the water. In the second, dissolved gases transported by these eddies diffuse through a thin laminar layer at the top of this boundary layer, the diffusive sublayer, which is on the order of 10–100 µm thick. A good summary of the processes governing gas transfer and the distinction between exchange of soluble gases and insoluble gases can be found in Liss (1983).

The diffusive flux, F, of slightly soluble gases across the boundary can be calculated from the expression

$$F = k(C_w - \alpha C_a) \tag{3.1}$$

where k is the gas transfer velocity. The term diffusive flux is used here to mean flux occurring by a combination of molecular diffusive and turbulent diffusive processes. The concentration gradient across the aqueous boundary layer, $C_w - \alpha C_a$, is influenced by the sources and sinks of gas in the water and air and by variations of solubility with temperature and salinity. k is a function of the physical processes at the boundary, in particular the turbulence, the kinematic viscosity of the water (v) and the molecular diffusion coefficient (D) of the gas. The Schmidt number ($Sc = v/D$) is the ratio of the last two terms and is gas specific. The dependence of the gas transfer velocity on the thermodynamics as expressed through the Schmidt number and on the hydrodynamics in the uppermost part of the water column can be seen by writing k in terms of the Schmidt number,

$$k = Sc^{-n} f(u, l) \tag{3.2}$$

where n is a coefficient that depends on the characteristics of the water's surface. The expression $f(u, l)$ indicates that k is a function of the hydrodynamics in or near the aqueous boundary layer in the water. Given that in most cases the aqueous boundary layer is turbulent or that turbulent eddies from the waters below impinge upon it, we here express the dependence in terms of the turbulent velocity u and turbulent length scale l.

In the following, we will discuss gas flux in terms of the parameters defined in Eqns 3.1 and 3.2. First we will examine factors affecting concentration differences across the aqueous boundary layer (Section 3.3). We will discuss the relation of k to the Schmidt number (Section 3.4) and the empirical and theoretical approaches used to relate k to hydrodynamical processes occurring in the boundary layer in water (Section 3.5). Later in the chapter, we will discuss procedures to measure gas flux and the gas transfer velocity (Section 3.6).

3.3 Dependence of gas flux on concentration differences

3.3.1 Dependence of solubility on temperature and salinity
The temperature (T) dependence of gas solubility (Wilhelm *et al.*, 1977) has the following form:

$$\ln \alpha = A1 + A2/T + A3 \ln T \tag{3.3}$$

where *A1*, *A2* and *A3* are all coefficients which depend upon the gas of interest. The temperature dependence of the solubility of CO_2, CH_4, DMS and N_2O in freshwater is shown in Fig. 3.2 (Weiss, 1974; Wiesenburg & Guinasso, 1979; Weiss & Price, 1980; Dacey *et al.*, 1984). Over the range of ambient temperatures solubility changes by a factor of two. The temperature of the upper millimetre of the water column can be up to 1°C cooler than the mixed layer below it. Striking differences in calculated CO_2 flux, as well as flux of other gases, from the ocean to the atmosphere can be obtained by including this difference (Robertson & Watson, 1992).

The influence of salinity on solubility, the Setschenow effect, can be expressed in the form: $\ln \alpha = B S$, where S is the salinity and B is a constant. The solubility of gas in seawater is approximately 20% less than in freshwater.

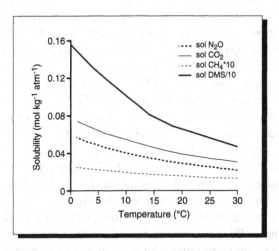

Fig. 3.2 The change of solubility of N_2O, CO_2, DMS and CH_4 with temperature (Weiss, 1974; Wiesenburg & Guinasso, 1979; Weiss & Price, 1980; Dacey *et al.*, 1984). The solubility is expressed in units of $mol\,kg^{-1}\,atm^{-1}$. The solubility of methane is multiplied by a factor of 10 and that of DMS divided by 10 to fit on the same scale. Methane is an order of magnitude less soluble than CO_2 or N_2O. The solubility of DMS is much greater than that of CO_2. Due to the low solubility of methane, release from sediments is often dominated by bubble ebullition; CO_2 and DMS dissolve in the pore waters and escape by aqueous diffusion.

3.3.2 Special case: CO_2 exchange

Exchange of CO_2 is different from other slightly soluble gases as its exchange can be enhanced by chemical reactions:

$$CO_2 + H_2O = H_2CO_3$$

and:

$$CO_2 + OH^- = HCO_3^-$$

and it is buffered in the water column. Theoretical and laboratory studies have verified the chemical enhancement process. The gas transfer rate is proportional to the gradient across the aqueous boundary layer. When CO_2 molecules react to form carbonic acid and bicarbonate in the boundary, the gradient steepens and exchange is enhanced. The reaction with water has a low rate constant and does not increase the rate of exchange significantly, however, the reaction with hydroxide can enhance the exchange for alkaline solutions. The enhancement is inversely proportional to the surface turbulence. For quiet waters enhancement can be large while for rough waters the enhancement has little effect (Bolin, 1960; Hoover & Berkshire, 1969; Quinn & Otto, 1971; Smith, 1985). Chemical enhancement will be particularly important for alkaline lakes. Oxburgh et al. (1991) estimate an enhancement of 2.5 for Mono Lake, California, an alkaline lake with a pH of 9.8. Enhancement is also important for freshwater with little buffer capacity. During phytoplankton blooms, the dissolved CO_2 (CO_2(aq)) is depleted due to photosynthesis causing the pH of surface water to increase, a process which enhances the CO_2 exchange (Emerson, 1975a; Herczeg, 1987).

The inorganic carbon species, CO_3^{2-}, HCO_3^- and CO_2, form a buffer system which causes the concentration of CO_2(aq) in water to change more slowly in response to gas exchange than do concentrations of other gases. The slow change is due to the reaction $CO_2 + CO_3^{2-} + H_2O = 2HCO_3^-$. Changes in CO_2(aq) in the ocean are about 10 times slower than for a corresponding gas without a buffering mechanism. For inland waters the buffering factor will depend on the total carbon content and the alkalinity of the water. Detailed descriptions of aquatic carbon chemistry can be found in Stumm & Morgan (1981) and Butler (1982). Because of the buffering, some methodological changes must be kept in mind when performing the procedures described in Section 3.6.2.

3.3.3 Physical processes affecting gas concentrations in the mixed layer

Concentrations of gas in the mixed layer depend not only on flux across the air–water interface but also upon transport to the mixed layer from the thermocline, deep waters or hypolimnion, and from the sediments (see discussion of Crill *et al.*, 1988 in Section 3.5.6). Wind mixing and convective motions due to heat loss at the surface can cause entrainment of waters with different concentrations to the mixed layer, and processes such as upwelling also transport dissolved gases. If upwelling or any other process causing localized transport occurs, gas flux may be spatially variable. Imberger & Patterson (1990) discuss processes causing spatial variability in lakes. If such processes are occurring, sampling must be conducted at a number of locations. Several tools exist to indicate the likelihood of spatial heterogeneity. Remote sensing will indicate different temperatures or chlorophyll concentrations at the surface of larger water bodies; these differences indicate that different hydrodynamical processes may be occurring at the different sites. Non-dimensional numbers may also be used to predict the likelihood of spatial heterogeneity. For instance, upwelling and depth of upwelling can be predicted from the Wedderburn number and the Lake number which can be assessed from wind speed, density stratification and lake morphometry (Imberger & Patterson, 1990).

The flux of gases into the mixed layer from deeper waters can be assessed in two ways. The flux due to turbulent mixing can be calculated as $J = K_z \, dC/dx$ where K_z, the coefficient of eddy diffusivity, is obtained either by the heat transfer method (Jassby & Powell, 1975; Jellison & Melack, 1993) or by the methods using microstructure profilers discussed on pp. 86–89. Alternatively, if the depth of the mixed layer is observed to increase over time and concentrations of gas increase with depth, the increased concentrations in the mixed layer due to entrainment can be calculated (e.g. Jellison & Melack, 1993; Crill *et al.*, 1988). The importance of turbulent mixing of epilimnetic waters with deeper water was illustrated in a lake in the Amazon floodplain. The calculated fluxes of methane from the thermocline to the epilimnion by turbulent diffusion over a period of several days was very close to the range of the measured diffusive flux across the air–water interface (Crill *et al.*, 1988).

3.4 Dependence of gas transfer velocity on Schmidt number

The gas transfer velocity k is proportional to Sc^{-n}. Recent wind tunnel work indicates k is proportional to $Sc^{-2/3}$ at low wind speeds, and to $Sc^{-1/2}$ at higher wind speeds (Jähne *et al.*, 1987b). Deacon (1977), who had the initial insight for this approach, found that flow over a smooth surface gives $n = 2/3$. Csanady (1990) showed analytically that the $Sc^{-2/3}$ dependence is correct when the flow is adjacent to a solid boundary and the $Sc^{-1/2}$ dependence is correct for a fluid boundary. For estimates of the gas transfer velocity in the field, field and laboratory studies have shown that in most cases a 1/2 power dependence is appropriate (Holmen & Liss, 1984; Ledwell, 1984; Upstill-Goddard *et al.*, 1990). These data

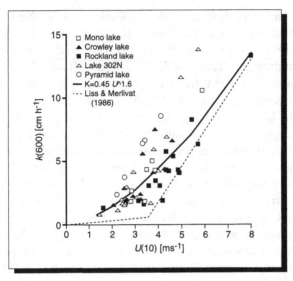

Fig. 3.3 Summary of several experiments on lakes to determine the relation between wind speed and gas exchange using the deliberate tracer SF_6. Mono Lake has a surface area of $200\,km^2$; Crowley Lake is $20\,km^2$; Rockland Lake is $1\,km^2$; Lake 302 N is $0.13\,km^2$; Pyramid Lake is $500\,km^2$. Wind speeds are normalized to 10 m height and the gas transfer velocity is normalized to $Sc = 600$ (i.e. $k(600)$ refers to normalizing k to CO_2 at 20°C using Eqn. 3.4). The thick solid line is a least squares power law fit through all the experimental data (Eqn. 3.7). The dashed line is the relation of Liss & Merlivat (1986). The average transfer velocity for the duration of each experiment (1–3 months) shows gas transfer velocities increasing with lake size.

indicate the lack of a physical basis for the stagnant film model of gas transfer for which $n = 1$ (see Section 3.5).

The dependence of gas transfer velocity on the Schmidt number offers a way to convert gas transfer velocities from one gas to another. For instance, if the gas transfer velocity is known for one gas, it can be computed for the other using the Schmidt numbers for the respective gases:

$$k_1/k_2 = (Sc_1/Sc_2)^{-n} \tag{3.4}$$

where the subscripts 1 and 2 refer to gases 1 and 2. Gas exchange velocities over lakes and over the ocean are frequently normalized to $Sc = 600$ which is the Schmidt number of CO_2 at 20°C in freshwater and is written as $k(600)$. In the normalization, both the right and left hand sides of Eqn. 3.2 are multiplied by Sc^n which gives $kSc^n = f(u, l)$. For $n = 1/2$ and $Sc = 600$, '$k(600)$' implies $k_x * (Sc_x/600)^{1/2}$ where x refers to the gas in question, (e.g. the y axis in Fig. 3.3). In stream work, reaeration coefficients, defined as the gas transfer velocity divided by average depth, are normalized to oxygen exchange at 20°C.

The kinematic viscosity of water is defined as the dynamic viscosity divided by density. For freshwater the dynamic viscosity and density can be found in reference tables such as the *CRC Handbook of Chemistry and Physics* (Weast & Astle, 1980). The viscosity of seawater at selected salinities and temperatures is listed in Millero (1974), Riley & Skirrow (1975) and in Table 3 of Hartman & Hammond (1985).

Diffusion coefficients for many gases have been determined in laboratory studies. Most studies use the diffusion of CO_2 or CH_4 as a reference, which facilitates intercomparison between investigations. Experimental work by Jähne *et al.* (1987a) covers several gases of environmental interest. In addition to the experimental work, empirical equations have been developed to determine diffusion coefficients (Wilke & Chang, 1955; Hayduk & Laudie, 1974). The equation of Wilke & Chang (1955), adapted by using the revised association factor of water suggested by Hayduk & Laudie (1974), yields values within 5% of experimental results for gases other than hydrogen and helium. Diffusivities are expressed by Wilke & Chang (1955) as:

$$D = (7.4 \ 10^{-8} \ (qM_b)^{0.5}T)/(\nu_B V_a^{0.6})$$ (3.5)

where q is the association factor of water ($=2.26$), ν_B is the dynamic viscosity, M_b is the molar weight of water and V_a is the molar volume at normal boiling point of the gas of interest. Replacing the constants with the appropriate values for water and substituting V_a with $0.285 \ V_c^{1.048}$ where V_c is the critical volume (Reid *et al.*, 1987) we obtain:

$$D = (10^{-6}T)/(\nu_B V_c^{0.629})$$ (3.6)

where temperature T is measured in degrees Kelvin. Critical volumes of gases and viscosities of water can be found in reference books such as the *CRC Handbook of Chemistry and Physics* and the *Matheson Gas Data Book* (Braker & Mossman, 1975; Weast & Astle, 1980).

The Schmidt numbers of N_2O, CH_4 and CO_2 in freshwater and saltwater ($S = 35‰$) at different temperatures are given in Table 3.1. The difference in Schmidt number for these gases at a particular salinity is small but the temperature effect is pronounced; consequently temperature must be measured at the surface of the water body for correct determination of ν and D. Schmidt numbers for DMS are not listed due to the uncertainty in the value. Diffusion coefficients of DMS have recently been determined experimentally but have not appeared in the literature (Saltzman *et al.*, 1993). Their experimental values are in accord with theoretical estimates from Eqn. 3.6.

Table 3.1 Schmidt numbers for CH_4, CO_2, and N_2O in freshwater and seawater

Temp (°C)	CH_4, freshwater	CH_4, seawater	CO_2, freshwater	CO_2, seawater	N_2O, freshwater	N_2O, seawater
0	1907	2049	1921	2084	2071	2246
1	1789	1925	1799	1954	1927	2093
2	1680	1810	1686	1834	1795	1952
3	1579	1704	1582	1723	1675	1824
4	1486	1605	1486	1620	1564	1705
5	1399	1513	1396	1524	1463	1597
6	1318	1427	1313	1436	1369	1497
7	1242	1347	1236	1353	1283	1405
8	1172	1273	1164	1276	1204	1320
9	1107	1204	1097	1204	1131	1242
10	1046	1139	1035	1138	1064	1169
11	989	1078	977	1075	1001	1102
12	936	1022	923	1017	943	1039
13	886	969	873	963	890	981
14	840	919	825	912	840	928
15	796	872	781	864	793	877
16	755	828	740	819	750	831
17	717	787	701	777	710	787
18	681	748	665	738	673	746
19	647	712	631	701	638	708
20	616	678	599	666	605	672
21	586	645	569	633	574	639
22	558	615	541	602	546	608
23	531	586	514	573	519	578
24	506	559	490	546	494	550
25	483	533	466	520	470	524
26	460	509	444	495	448	500
27	439	486	423	472	427	477
28	420	464	403	451	407	455
29	401	444	385	430	389	435
30	383	424	367	410	372	415

3.5 Dependence of the gas transfer velocity on hydrodynamics

Considerable effort has been expended in relating k to the hydrodynamics at the aqueous boundary layer, in particular the turbulence, and to environmental parameters such as wind speed which affect the hydrodynamics. Several types of models have been developed to describe the processes occurring. The earliest models for calculating k were developed assuming the boundary acts as a stagnant film over which gas exchange occurs by molecular diffusion; $k = D/z$ where z, the thickness of the film (Lewis & Whitman, 1924), is a function of wind speed (Broecker & Peng, 1974; Emerson, 1975(b); Broecker et al., 1978; Peng & Broecker, 1980). This model implies that k is proportional to Sc^{-1}, but, as discussed above, field and laboratory studies have shown that in most cases $Sc^{-1/2}$ or $Sc^{-2/3}$ is appropriate.

At present, k is determined from empirical relationships which depend upon wind speed alone (see Section 3.5.1) or models based on the characteristics of the turbulence in the boundary (see Section 3.5.5). The differences in approach partially reflect the background of the experimenters, with oceanographers and ecologists favouring empirical relations and fluid dynamacists and chemical engineers favouring process-orientated studies which lead to models appropriate to particular types of flows. In the following sections, we will bring together these disparate literatures, find analogues in nature for the processes occurring in the laboratory, and suggest future directions for research based on the insights from both approaches.

We will discuss the success of empirical relationships based on wind speed in Section 3.5.1. As only measurements of wind speed are required, studies based on these relations are relatively easy to conduct. Possible improvements to this approach include development of paramaterizations based on wind stress which will take into account the effects of atmospheric stability and, to a certain extent, the development of the wave field (Sections 3.5.3 & 3.5.4). We will also discuss the various models for the gas transfer velocity. Summaries are presented in Theofanous (1984), Sonin et al. (1986) and Brumley & Jirka (1988). Much of the model development in the last 25 years has been based upon the surface renewal model (Table 3.2). This model and its variations require either measurements of the characteristics of the turbulence in the surface waters, a task for which the instrumentation and analytical tools have only recently been developed for studies in nature, or of the shear stress. When this approach is applied in field situations, a diverse set of measurements may be required, but other processes than wind induced mixing will be included in the calculation of k.

3.5.1 Relationship between gas transfer velocity and wind speed
Empirical relationships for k have largely been based upon wind speeds as they are an easily measured parameter and, to a certain extent, incorporate wave breaking and shear stress. Wind tunnel studies have shown a strong correlation between the gas transfer velocity and wind speed (Liss, 1983; Broecker et al., 1978; Merlivat & Memery, 1983; Broecker & Siems, 1984; Jahne et al., 1984, 1985; Ledwell, 1984; Wanninkhof & Bliven, 1991). For most studies, the general shape of the curve between gas transfer and wind speed is similar. Compilation of detailed studies by Broecker et al. (1978) and Broecker & Siems (1984) shows a trilinear relationship between wind speed and gas transfer velocity with two discontinuities. They found that gas transfer velocities have a very weak

dependence on wind speed when the water surface is smooth, that the dependence becomes stronger when capillary waves form, and that when significant wave breaking occurs (at about $13\,\mathrm{m\,s^{-1}}$ in the wind tunnel) the dependence increases even more.

Based on a compilation of field studies, a number of equations relating k and wind speed have been proposed (Hartman & Hammond, 1985; Smethie *et al.*, 1985; Liss & Merlivat, 1986; Wanninkhof, 1992). The most widely used for ocean work is that of Liss & Merlivat (1986) which uses a trilinear relation with the slope of each line adjusted to field results. Recent work suggests that this relation underestimates gas transfer on lakes at low wind speeds and overestimates it at higher winds (Upstill-Goddard *et al.*, 1990; Wanninkhof *et al.*, 1991).

Field studies using deliberate tracers have found relations similar to those in wind tunnels at wind speeds up to $8\,\mathrm{m\,s^{-1}}$, and the few data points at high wind speeds suggest a steepening of the curve (Wanninkhof *et al.*, 1985, 1987, 1991; Upstill-Goddard *et al.*, 1990). The field data show more scatter than the results from the wind tunnels, in part because of experimental uncertainties, in part because variations in wind speed and wind direction over the measurement interval will cause variations in the surface turbulence, and in part because of other causes of turbulence. A least squares power law fit through the results of five lake experiments with deliberate tracers (Fig. 3.3) yields the following relation:

$$k(600) = 0.45u_{10}^{1.6}(Sc/600)^{-0.5} \qquad (r^2 = 0.66) \qquad (3.7)$$

where u_{10} is the wind speed at $10\,\mathrm{m}$ height above the water surface and r^2 is the correlation coefficient. Sc is divided by 600 to normalize k to CO_2 at $20°C$ in freshwater (see Eqn. 3.4). The lakes in which the studies were conducted range in size from 0.13 to $500\,\mathrm{km^2}$, had water temperatures of $4–23°C$, and had salinities up to 73‰. The gas transfer velocity determined over intervals ranging from days to weeks is dependent on wind speed for a wide range of lake sizes.

Gas exchange from water bodies in the absence of wind (e.g. especially stagnant wooded swamps or swamp flood water between emergent plants) may be controlled primarily by convective motions caused by heat loss which occurs, for instance, when the water surface is warmer than the air. Gas flux from still water bodies may be greater at night (cf. Crill *et al.*, 1988) but will still occur in the day due to heat loss in the aqueous boundary layer (see Section 3.5.5). Alternatively, if currents are produced in these water bodies by processes such as differential heating (Imberger & Patterson, 1990), turbulence may be generated by current shear due to variations in current velocity with depth. These processes are more fully described and rigorous methods are given for quantifying gas exchange under conditions of low or zero wind speed in Sections 3.5.5, 3.5.6, 3.6.5

and 3.6.6. Empirical studies in small ponds and swamps which have defined gas exchange velocities from determinations of CH_4 emissions with chambers at 0 wind speed and measurements of CH_4 concentrations in the water have obtained $k(600)$ values of $1.7\,cm\,h^{-1}$ (Sebacher et al., 1983) and $0.8 \pm 0.5\,cm\,h^{-1}$ (Happell, 1992; Happell et al., 1993). Careful laboratory measurements of CO_2 flux indicate that k is independent of wind speed below $3\,m\,s^{-1}$ (Ocampo-Torres et al., 1994). The data of Ocampo-Torres et al. and the non-zero values of k at zero wind speed suggest that the curve on Fig. 3.3 terminating at a u_{10} value of $1.5\,m\,s^{-1}$ be continued parallel to the x axis, without further decrease.

3.5.2 Effects of surface slicks

Surfaces of inland waters and coastal waters are frequently covered by natural or contaminant surface slicks which reduce the gas transfer velocity (Broecker et al., 1978; Goldman et al., 1988). Hydrophobic organic compounds are ubiquitous in the aqueous boundary layer; amounts less than necessary to form a molecular film on the surface will reduce the gas transfer velocity (Goldman et al., 1988). Laboratory studies have shown that nearly all natural waters have enough surface organic material to reduce the gas transfer velocity (Asher & Pankow, 1989). If, when surface slicks occur, the surface has characteristics of a solid surface, the Schmidt number dependence may change (Hunt, 1984). For strong winds, surface active materials will be mixed with deeper waters, as a result the effects of slicks on gas transfer will only be noticed at low to moderate wind speeds (Donelan, 1990). Surface slicks have been found to reduce the wind stress at the air–water interface by 16% (Wei & Wu, 1992) and reduce the development of the wave field such that the mean square slope (see Section 3.5.4) is reduced by 69% compared to a clean surface at wind speeds of $2.6\,m\,s^{-1}$ and by 11% at wind speeds of $6\,m\,s^{-1}$ (Tang & Wu, 1992).

3.5.3 Relationship between gas transfer velocity and wind stress

Relationships between the gas transfer velocity and wind stress τ should be more accurate than ones based on wind speed because the actual force of the wind per unit area on the water's surface is quantified by the wind stress. For instance, when air temperatures just above the water's surface are warmer than those 10 m above the water's surface, the wind stress will be higher for a given wind speed.

Wind stress can be obtained from a series of wind speed measurements taken at a number of heights between the water's surface and ca. 10 m above the surface. When the wind speeds are plotted against the logarithm of height above the surface, the slope of the line is called the friction velocity u_{*a} where the subscript 'a' implies the measurement is in air.

Wind stress τ is related to the friction velocity through the following expression:

$$\tau = \rho_a u_{*a}^2 \tag{3.8}$$

where ρ_a is the density of the air. Because the wind stress on either side of the air–water interface is equal, the friction velocity in water u_{*w} can also be defined:

$$\tau = \rho_w u_{*w}^2. \tag{3.9}$$

Relations between gas transfer velocity and wind stress are generally expressed in terms of u_{*a} or u_{*w}.

Wind stress is most commonly computed using a drag coefficient C_d and measurements of wind speed U at a height x above the water's surface:

$$\tau = \rho_a C_{dx} U_x^2. \tag{3.10}$$

By convention, C_{dx} and U_x are given at 10 m height. Drag coefficients have been obtained from micrometeorological studies where wind speeds and temperature profiles are measured at a number of heights; data on the wave field may be collected as well. From such studies, we have learned that C_d depends on wind speed and, as mentioned above, atmospheric stability, and the wave field (Large & Pond, 1981; Liu & Schwab, 1987). For instance, when compared to its value at neutral stability, the drag coefficient is larger when the air tens of centimetres above the water surface is colder than the air just above the water surface (unstable stratification) and the drag coefficient is less when the water surface, and the air adjacent to the surface is colder than the air further from the surface (stable stratification) (Deardorff, 1968; Hicks, 1972; Donelan et al., 1974). For intermediate wind speeds a constant value in the range of 1.1×10^{-3} to 1.5×10^{-3} is often adapted for C_d. Erickson (1994) found k to vary by up to a factor of 3 for a wide range of atmospheric stabilities.

Given that $\tau = \rho_a u_{*a}^2 = \rho_w u_{*w}^2$, conversion between u_{*a} or u_{*w} can be done somewhat readily for a given wind speed. The accuracy of the conversion will depend on whether temperatures have been measured in air and water immediately above and below the interface and in air at the height of the wind measurements and whether humidity has been obtained at the height of the wind measurements. Measurement of u_{*w} is useful for obtaining k in the turbulence models to be discussed in Section 3.5.5. When turbulence is generated by the wind stress alone and the turbulence is isotropic, u_{*w} is proportional to u, the turbulent velocity scale (see Section 3.5.5).

To our knowledge, accurate relations between gas transfer velocity and wind stress have only been obtained in laboratory settings. Figure 3.4

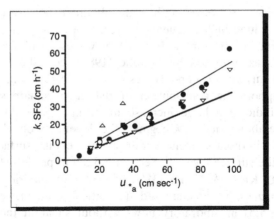

Fig. 3.4 Gas transfer coefficient from the 100 m Delft Tank (Wanninkhof & Bliven, 1991) plotted against u_{*a}, the air friction velocity. The data is for SF_6 with $Sc = 1300$ (in this case k, SF_6 is equivalent to $k(1300)$). The wind-only cases are: (●), data on gas flux at 80 m in the tank and data are whole tank averages. Wind plus mechanically generated waves cases are: (△), 0.7 Hz large amplitude waves; (■), 0.7 Hz small amplitude waves; and (○), 0.4 Hz small amplitude waves. The solid lines are based on Eqn. 3.19 after conversion to u_{*a}. The outliers at c. 20 and 34 cm s^{-1} occurred in runs with larger amplitude waves. (Adapted from S. Banerjee, personal communication, 1991.)

shows the results of such a study in a 100 m wind-wave tank. The relation between k and u_{*a} is clearly linear for a given wave field. As wave amplitudes change, there is more variation in k for a given wind speed.

3.5.4 Relationship between gas transfer velocity and surface waves

The development of the wave field may be the reason for the changing dependence between k and wind speed noted by Broecker *et al.* (1978), Broecker & Siems (1984) and by Liss & Merlivat (1986). Merlivat & Memery (1983) attributed the increase in k at wind speeds ranging from 10 to 14 m s^{-1} to enhanced gas transport via bubbles caused by breaking waves. The increase in k in Fig. 3.4 for the larger amplitude wave case is also attributed to breaking waves. However, waves can also contribute to production of turbulence by generation of vorticity. This mechanism of mass transfer will be discussed in Section 3.5.5.

Jahne *et al.* (1987), who use an expression of the form $k = a_1 u_{*w} Sc^{-1/2}$ to express the dependence of k upon wind stress, demonstrated with data from wind-wave tanks that this expression failed above a u_{*w} value of 0.3 cm s^{-1} ($U \sim 2$ m s^{-1}). They found that the mean square slope of the total wave field needed to be included in the expression for k but did not feel this could be developed based on laboratory measurements. They used a laser gauge over the wind-wave tank to measure the slope of the waves (Lange *et al.*, 1982). Slope is often measured with wire gauges from which amplitude a and wavelength λ are obtained, but are problematic because they exclude high frequency waves. Slope, also known as wave steepness, is a/λ. Their work indicates that waves of all sizes, not just capillary waves as proposed by Coantic (1986), contribute to the turbulence at the surface. Ocampo-Torres *et al.*'s (1994) laboratory experiments also support the contribution of the mean square slope to k. Their results indicate a large increase in mean square slope as wind speeds increase above $3\,\mathrm{m\,s^{-1}}$; k begins to be dependent on wind speed at this value. We will discuss recent measurements relating surface waves to turbulence in the surface layers in Section 3.5.5 and pp. 89–90.

In contrast, Khoo & Sonin (1992) found that the slope a in the expression $k = a_1 u_{*\mathrm{w}} Sc^{-1/2}$ increased at a critical turbulent velocity scale (see Section 3.5.5) in laboratory flows without shear at the air–water interface, a situation analogous to one without wind and ensuing waves. The point at which k is augmented depends upon the Reynolds number. They suggest that another mechanism than wave breaking may be important in enhancing mass transfer and that it is likely due to the mechanism of turbulent transport in the liquid.

3.5.5 Relationship between gas transfer velocity and turbulence parameters

The gas transfer velocity does not depend only on wind stress and wave breaking. This is particularly true for streams and rivers (Jirka & Brutsaert, 1984), and can be true for wetlands, lakes and the ocean under some conditions. For instance, at night in an Amazon floodplain lake, wind stress contributed less to mixing than did convective motions in the water column caused by heat loss from the lake to the atmosphere (S. MacIntyre, unpublished data). The surface renewal model has been used to express the results of many of the laboratory studies exploring the relation between the gas transfer velocity and turbulence near the air–water interface. We will review this model, discuss the various processes causing mixing near the boundary region in naturally occurring water bodies, and discuss the applicability of results from laboratory and analytical studies to naturally occurring situations.

The surface renewal model is based on the assumption that the gas transfer velocity depends upon the time t for turbulent eddies to renew the water at the interface with water whose concentration is that of the bulk fluid (i.e. $k = (D/t)^{1/2}$) (Higbie, 1935; Danckwerts, 1951). This model predicts that k is proportional to $Sc^{-1/2}$. The earliest variations on this model attempted to estimate t based on the proportion of the interface which is turbulent (Dobbins, 1956; O'Connor & Dobbins, 1958; Harriot, 1962) or the time scale for different size classes of eddies to bring water to the interface. Although the most recent versions of the surface renewal model appear complex, they are elaborations on these earlier concepts. The underlying parameters for the model are the turbulent velocity scale u and the turbulent length scale l.

The turbulent velocity scale $u = <u'^2>^{1/2}$ where u' are the fluctuations in velocity from the mean and the angle bars denote averaging. This velocity scale is also known as the root mean square (rms) velocity and is the standard deviation of the velocity fluctuations (Tennekes & Lumley, 1972).

Turbulent flows have several important length scales. One is the size of the largest eddies; these perform most of the mixing. Generally, they scale with the thickness of the boundary layer of the flow; where the boundary layer is the portion of the water column affected by the presence of a boundary, be it the atmosphere or sediments. The aqueous boundary layer is one of the subdivisions of the boundary layer caused by shear at the air–water interface. In a lake, the largest eddies may be the size of the uppermost mixing layer (~ 1–10 m); in a stream, assuming no stratification, they may scale with the depth of the water column. In the thermocline, they will scale with the size of overturning regions (~ 1–10 m). Technically, l is based on velocity measurements at two points, one of which is fixed and the other of which is moved progressively farther from the first. When velocity measurements are made at two nearby points, the velocity fluctuations will be highly correlated; as the two points are separated, the correlation of the velocity fluctuations will decrease. The turbulent length scale is based on the distance over which velocity fluctuations are correlated with each other (Tennekes & Lumley, 1972; Eqn. 3.21).

Large eddies will keep the concentration of gases uniform in the uppermost portion of the water column. Their energy is supplied to the smallest eddies at a rate proportional to u^3/l. Assuming that the rate at which turbulent kinetic energy is dissipated (ε) is equivalent to the supply rate,

$$\varepsilon = u^3/l \tag{3.11}$$

(Taylor, 1935). Energy dissipation rates provide an indication of the intensity of the turbulence, with order of magnitude changes being meaningful. Energy dissipation rates of 10^{-4} to $10^{-6}\,\mathrm{m^2\,s^{-3}}$ are found in energetic mixed layers, with the higher values found near the surface of the water column and at the initiation of overturning events. Values of $10^{-7}\,\mathrm{m^2\,s^{-3}}$ are found in less active portions of the mixed layer and in more active parts of the thermocline. Values of 10^{-8} and $10^{-9}\,\mathrm{m^2\,s^{-3}}$ are often found in the thermocline with the latter value indicating low levels of turbulence.

The size of the smallest eddies, the Kolmogoroff microscale, can be estimated from $(v^3/\varepsilon)^{1/4}$ where v is kinematic viscosity; these scales are 0.3 and 3 mm for energy dissipation rates of 10^{-4} and $10^{-8} m^2 s^{-3}$ respectively. In a turbulent flow, eddies cover a spectrum of sizes from the smallest to the largest. Measurement of the rate of energy dissipation and determination of u and l will be discussed on pp. 84–90. From these parameters the turbulent Reynolds number can be calculated

$$Re_t = u\,l/v. \tag{3.12}$$

Fortescue & Pearson (1967) formulated the gas transfer velocity based on the velocity and length scales of the large eddies,

$$k = c_1(D\,u/l)^{1/2} \tag{3.13}$$

where u/l, the inverse of the time scale of the large eddies, is the rate of strain, and a is a constant determined empirically. Lamont & Scott (1970) and Banerjee et al. (1968) formulated k in terms of the smallest eddies, obtaining

$$k = c_2 D^{1/2}(\varepsilon/v)^{1/4}. \tag{3.14}$$

Laboratory data exists to support both approaches (Komori et al., 1990; Banerjee, 1991; Komori, 1991). To separate the thermodynamic effects which are expressed by the Schmidt number from the hydrodynamic effects on gas tranfer, we have rewritten the models with $kSc^{1/2}$ on the left-hand side leaving the hydrodynamic component on the righthand side. The equations are summarized in Table 3.2. When rewritten, Eqns 3.13 and 3.14 become Eqns 3.15 and 3.16 in Table 3.2. To define k, simply divide the right side of each equation by $Sc^{1/2}$. In these formulations, the gas transfer velocity depends upon the turbulent velocity scale and Re_t. Theofanous et al.'s (1976) results suggested that the large eddy model be selected for flows which are less turbulent and the small eddy model be selected for turbulent flows with higher Reynolds numbers (Table 3.2, Eqn. 3.17). More recent work indicates that a wide range of eddy sizes impinges upon the aqueous boundary layer (Hunt & Graham, 1978; Brumley & Jirka, 1988) implying that the above distinctions are arbitrary. Banerjee (1990) computes k when the contribution from the spectrum of eddy sizes impinging on the aqueous boundary layer is included (Table 3.2, Eqn. 3.18). Again the expression shows the dependence of k upon u and Re_t; the results are consistent with Theofanous et al.'s two regime model. The latest versions of the surface renewal model will be discussed below concurrently with the mechanisms causing turbulence.

The various mechanisms that are known to cause renewal of water at the aqueous boundary layer in naturally occurring water bodies are illustrated in Fig. 3.5 and the models developed to account for these

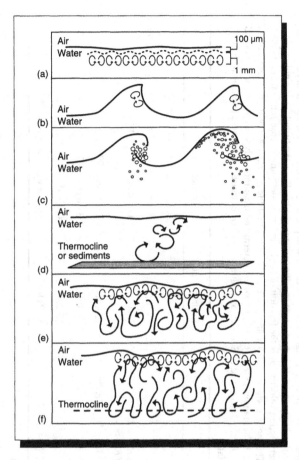

Fig. 3.5 Generation of renewal events at the aqueous boundary layer. (a) Interfacial shear and penetrative convection due to heat loss create regions with enhanced surface renewal. (---) Indicates the boundary of the diffusive sublayer. Shown here are paired eddies that create regions of divergence (flow is up, then outward) and convergence (downward flow). The upward flow divergence creates pressure forces that thin the diffusive sublayer. Other mechanisms of renewal are possible. (b) Rollers in breaking wavelets. (c) Breaking waves with production of bubbles and turbulence. (d) Turbulence generated at an interface propogates to the aqueous boundary layer. (e) Turbulence extends beyond the aqueous boundary layer due to wind mixing, large scale penetrative convection, and shear at the thermocline or benthic boundary. (f) Turbulence extends into the thermocline.

processes are included in Table 3.2. Wind stress causes shear at the air–water interface and subsequent mixing. Convective motions are caused by an increase in density at the air–water interface due to cooling or evaporation of water. Shear at the air–water interface and convection cause regions of enhanced renewal in the aqueous boundary layer. Figure 3.5a illustrates the flow within the aqueous boundary layer. When shear

Table 3.2 Surface renewal model presented as $kSc^{1/2}$ on the left side of each equation to show the flow dependence of the model. To calculate k, simply divide the right side of each equation by $Sc^{1/2}$; u is turbulent velocity scale, C_i is a constant specific for each model, Re_t is the turbulent Reynolds number, Rf is the flux Richardson number, u_{*w} is the water friction velocity, p_1 is the fraction of the boundary undergoing intense renewal and p_2 is based on the probability distribution of renewal events. The models which explicitly apply to flows with shear at the interface are written in terms of u_{*w}. Coefficients C_1 and C_2 are presented in Isenogle (1985) and Theofanous (1984); $C_3 \sim 1$; P_1 will be determined from remotely sensed data or from wind speeds (Monahan & O'Muircheartaigh, 1980); C_4, C_5 and P_2 are determined in Soloviev & Schluessel (1994). Because few of these coefficients have been applied in field situations, further testing is appropriate

Model	$kS_c^{1/2} =$	Equation no.	Reference
Large eddy	$C_1 u Re_t^{-1/2}$	(3.15)	Fortescue & Pearson, 1967
Small eddy	$C_2 u Re_t^{-1/4}$	(3.16)	Banerjee et al., 1968; Lamont & Scott, 1970
Two regime	Large eddy applies for $Re_t < 500$ Small eddy applies for $Re_t > 500$	(3.17)	Theofanous et al., 1976
Surface divergence	$C_3 u[0.3(2.83 Re_t^{3/4} - 2.14 Re_t^{2/3})]^{1/4} Re_t^{-1/2}$	(3.18)	Banerjee, 1990
Interfacial shear	$0.108\, u_{*w}$ to $0.158\, u_{*w}$	(3.19)	Banerjee, 1991
Eddy resolving analytical	$C_4 p_1 u_{*w}$	(3.20)	Csanady, 1990
Surface processes	$C_5 p_2 u_{*w}(1 + Rf/Rf_{cr})^{1/4}(1 + Ke/Ke_{cr})^{-1/2}$	(3.21)	Soloviev & Schluessel, 1994

Eqns 3.18 and 3.19 are in SI units $(\mathrm{m\,s^{-1}})$ other equations are in $\mathrm{cm\,h^{-1}}$.

occurs at the air–water interface, the gas transfer velocity is proportional to u_{*w} (Table 3.2, Eqn. 3.19) (Banerjee, 1991). The spacing of the renewal events equals u_{*w}/v (Lam & Banerjee, 1992); for wind speeds of 3 m s^{-1} they are ~3 cm apart and for wind speeds of 10 m s^{-1} they are ~1 cm apart. The contribution of heat loss to k is also included in Eqns 3.15–3.18 as the convective movements contribute to u and l. Soloviev & Schluessel (1994) have parameterized the effects of heat flux on k by using the flux Richardson number Rf (Table 3.2, Eqn. 3.21). Use of this equation and the dimensionless numbers used in it (i.e. Rf and Ke, the Keuligan number) are discussed in Oldham & Imberger (1992). Heat fluxes in the aqueous boundary layer depend upon the sum of latent heat exchange, sensible heat exchange, net long wave radiation and short wave radiation. The first three processes occur at the surface, whereas significant solar radiation is absorbed in the uppermost layer (e.g. $100 \, \mu\text{m}$) only during periods of intense heating (Soloviev & Schluessel, 1994). Because the sum of the first three terms is often negative, instabilities caused by heat loss are believed to be frequent with a periodicity dependent upon the heat loss.

As wind speeds increase, waves are generated at the interface, and turbulence may be caused by a number of mechanisms. Vorticity induced by capillary waves may contribute substantially to the wave induced component of the shear stress via turbulence production near the water's surface (Sanjoy Banerjee, personal communication). The energy dissipation rate can be calculated from the vorticity, $\varepsilon = -v\omega^2$ where vorticity $\omega = 2(a\kappa)^2\sigma$ where a is wave amplitude, κ is wave number, and σ is the radian frequency of the waves. Vorticity is the local rotation in a fluid due to velocity shear. In research on falling liquid films, production of vorticity is the dominant mechanism of mass transfer. The contribution of vorticity to the tangential stress on the water is $\tau = 2\mu(a\kappa)^2\sigma$ and μ is dynamic viscosity. As κ: $2\pi/r$ the term $a\kappa/2\pi$ is wave steepness, which is high for capillary waves, consequently, as $a\kappa$ is squared, capillary waves make a considerable contribution to the vorticity and to the tangential stress. Up to wind speeds of 4 m s^{-1}, the tangential stress due to vorticity production by capillary waves is as large as the wind stress. Consequently, whenever waves are generated, especially capillary waves, the contribution of the wave stress to gas flux may be considerable. However, as capillary waves are patchily distributed, this mechanism for gas flux needs further exploration, particularly at low wind speeds.

At moderate wind speeds and sufficient fetch as in the open ocean or large lakes, rollers are formed in breaking wavelets (Fig. 3.5b) that act similarly to the surface divergences in Fig. 3.5a (Csanady, 1990). The gas flux is enhanced by the vertical flow in the divergences and the diffusional boundary layer is thinned directly above them by pressure forces. k is proportional to u_{*w} and the fraction of the surface influenced by the rollers (Table 3.2, Eqn. 3.20). If this fraction is also dependent upon u_{*w} as in shear flows without waves (e.g. Lam & Banerjee, 1992), k will be proportional to u_{*w}^2. Monahan & O'Muircheartaigh (1980) found that the percentage of the water's surface covered with whitecaps is proportional to $U^{3.5}$, leading to the unlikely result that k will be proportional to u_{*w}^4.

At higher wind speeds, long wave breaking and bubble formation become more important (Fig. 3.5c). Soloviev & Schluessel (1994) found that k is proportional to u_{*w} in this case (Table 3.2, Eqn. 3.21). Woolf & Thorpe (1991) suggest modifying the equation for the direct flux of gas (Eqn. 3.1) to account for that introduced by bubbles:

$$F = -k[\alpha C_a(1 + \Delta) - C_w] \tag{3.22}$$

where Δ is the equilibrium fractional supersaturation of the gas and depends upon wind speed.

Surface waters can be renewed by turbulence generated away from the water boundary. For instance, turbulence can be caused by shear at the benthic boundary layer or at the thermocline in lakes and at the sediment–water interface in streams and rivers (Fig. 3.5d). For example, the energy dissipation rate at the surface in streams and rivers due to the upward propagation of the turbulence produced at the interface is

$$\varepsilon \sim= 0.4\,u_{*b}^3/H \tag{3.23}$$

where the shear velocity u_{*b} caused by shear at the sediment water interface is written:

$$u_{*b} = (gHS)^{1/2} \tag{3.24}$$

where S is the slope of the bed, H is the stream depth and g is gravity (Jirka & Brutsaert, 1984). In this formulation, u_{*b} represents the turbulent velocity scale and H the scale of the large eddies.

The surface divergence model (Eqn. 3.18, Table 3.2) applies to these situations as it was derived (Banerjee, 1991) based on theory (Hunt & Graham, 1978) and experiments where turbulence originated away from the air–water interface (Brumley & Jirka, 1988). One of the major results of Hunt & Graham's and Brumley & Jirka's work is that eddies of all sizes contribute to the surface renewal and effect the gas transfer velocity. For this reason, Eqn. 3.18 is more general than Eqns 3.15–3.17 although it gives comparable results at each Re_t (Banerjee, 1991). As the surface divergence model has been tested in flows with Re_t only up to 1600 (Banerjee, 1991), and as Re_t in lakes typically ranges from 10 to 2000 but can reach 10 000, the model will need testing in more energetic field situations.

It will be useful to explore if Eqn. 3.18 applies when turbulence is produced at the air–water interface. When values of Re_t that have frequently been found in the surface waters of lakes during wind mixing are substituted into Eqn. 3.18, a linear dependence between u and k is obtained for each Re_t Given that C_3 ~1, for $Re_t = 1200$, $kSc^{1/2} = 0.0914\,u$; for $Re_t = 100$, the coefficient is 0.19. The similarity of the coefficients, 0.0914 and 0.19, to those in Eqn. 3.19 and the near equivalence of u to u_{*w} within boundary layers when turbulence is isotropic and when the turbulence is induced by wind shear shows that Eqns 3.18 and 3.19 have features in common. However, Re_t will change with increases in forcing at the surface. Eqn. 3.18 is more general than Eqn. 3.19 because u can include the effects of penetrative convection and wave breaking as well as wind speed.

When wind speeds are high enough or heat losses large enough, gases introduced through the interface will be mixed throughout the upper part of the water column so that the gas concentrations are uniform within it (Fig. 3.5e). The limiting step for gas flux will be transport through the aqueous boundary layer. However, if the water column below the aqueous boundary is not turbulent, introduced gases will accumulate within and just below the aqueous boundary layer. The rate-limiting step will be transport between the aqueous boundary layer and the waters immediately adjacent to it and the waters below. In both cases, a model that incorporates turbulence throughout the mixing layer will be accurate (Table 3.2, Eqns 3.15–3.18). Oldham & Imberger's (1992) fine scale oxygen measurements provide evidence for accumulation of gases below the aqueous boundary layer, although in their case the gases were produced there. At low wind speeds oxygen concentrations due to biological production increased above saturation in the upper 30 cm with a pronounced concentration gradient in this region; when wind speeds increased, vertical mixing distributed the oxygen throughout the upper 2 m. The saturation in the upper part of the water column was reduced, partially because of gas exchange across the aqueous boundary layer, but primarily because the oxygen was mixed deeper into the lake. The non-uniform oxygen concentrations that they observed in the upper 30 cm when wind speeds are low attest to the importance of sampling for C_w near the air–water interface.

Figure 3.5f illustrates mixing below the thermocline. Deeper mixing does not affect the gas transfer velocity, but may change the concentration of gases in the mixed layer as will be discussed further in Section 3.5.6. The flux of gases into the mixed layer can be estimated using the coefficient of eddy diffusivity (Kz, Section 3.3.3 and p. 89).

3.5.6 Application of the surface renewal model to a tropical lake

The small eddy version of the surface renewal model has been tested on a lake on the Amazon floodplain (L. Calado) where wind speeds were low (Crill *et al.*, 1988). Turbulence parameters were calculated from the surface energy budget and density stratification in the lake (see pp. 85–86). Because of surface heat losses and low wind speeds, convective cooling contributed strongly to the mixing, particularly at night. In fact, during a 2-month period prior to the gas flux study, of the total flux of turbulent kinetic energy into the lake at a given instant of time, 85–100% was due to buoyancy flux (i.e. heat loss) at night, 50–100% was due to buoyancy flux at sunrise (06.00 hours) and sunset (18.00 hours), and 0–80% was due to buoyancy flux during daylight from 07.00 to 17.30 hours (S. MacIntyre, unpublished data). Based on the available data, heat was being lost from the upper mixed layer *ca.* 40% of the time between 07.00 and 17.30 hours, indicating a substantial contribution of convective motions to the turbulent velocity scale during the day when these motions are usually assumed to be suppressed. The turbulence parameters during the gas flux study were available only at dusk and dawn when wind speeds were less than $2\,\mathrm{m\,s}^{-1}$ except at one period when they reached $4\,\mathrm{m\,s}^{-1}$. The coefficients for Eqn. 3.14, $k = c_2 D^{1/2}(\varepsilon/v)^{1/4}$, have previously only been obtained for laboratory studies and range from 0.12 to 0.56 (Isenogle, 1985). When calculations of gas flux using these coefficients to obtain k were compared to gas fluxes from the same time periods measured with a continuous-sampling chamber, the best comparison was obtained using the larger coefficient. The correspondence between the calculated and measured fluxes was good, with the calculated fluxes ranging from 2.8 to $20\,\mathrm{mg\,CH_4\,m}^{-2}\,\mathrm{day}^{-1}$ with an average flux of $7\,\mathrm{mg\,CH_4\,m}^{-2}\,\mathrm{day}^{-1}$ ($n = 12$) and the measured fluxes ranging from 0 to $34\,\mathrm{mg\,CH_4\,m}^{-2}\,\mathrm{day}^{-1}$ and averaging $9.3\,\mathrm{mg\,CH_4\,m}^{-2}\,\mathrm{day}^{-1}$.

This field study was the first application of the surface renewal model in a natural environment. Further field studies designed to determine the coefficients in Eqns 3.13 and 3.18 are essential. Convective motions are likely to be important for mixing whenever heat loss occurs, and may be dominant in any sheltered, warm water lake.

In the L. Calado study, estimates of gas flux based on estimating k from wind speeds were lower than those obtained using the chamber, often by a factor of 2–5. Neglect of the contribution due to free convection may have contributed to the discrepancy. Alternatively, the agitation of the surface of the water by the fan within the chamber may have been too large.

Wind mixing made its largest contribution to the budget of turbulent kinetic energy of L. Calado during the day when density stratification was greatest. Because of solar heating, the depth of the thermocline typically decreased from *ca.* 6 m at dawn to 0.2–1.5 m at midday when wind speeds were highest, and then began deepening again in late afternoon due to both wind mixing and convective motions due to heat loss. The depth of the thermocline increased throughout the night due to convection from heat loss. Methane concentrations increased with depth in the water column. In this type of situation, where the highest concentrations of dissolved gases are deeper in the water column and entrainment of these deep waters typically occurs only at night, diffusive gas fluxes are likely to be greatest at night. We observed this pattern for methane, where over the course of a 24-h study, diffusive fluxes were greatest at sunrise, lowest at midday, and progressively increased from sunset to sunrise (Crill *et al.*, 1988). Knowledge of processes occurring in the water column is critical for designing sampling programs and for interpreting the results of gas flux experiments.

3.5.7 Relationship between the surface renewal model and empirical expressions for the gas transfer velocity

The variation of the surface renewal model proposed for interfacial shear (Table 3.2, Eqn. 3.19) is plotted against results from wave tank experiments (Wanninkhof & Bliven, 1991) in Fig. 3.4. The model results encompass those from the wind tunnel when there was a steady wind field. Gas transfer velocities were higher and outside the range predicted by the shear model for the larger amplitude waves which were mechanically induced ($\triangle$). This discrepancy for the larger amplitude waves highlights the problem of using a shear model alone when the wave field is evolving or for different wave fields.

When the interfacial shear model (Table 3.2, Eqn. 3.19) is plotted against the results of the deliberate tracer experiments on lakes, results are only in agreement at the higher wind speeds (Fig. 3.6). We related u_{*w} to wind speed using u_{*w} as $\sim 0.001\,U$ which is based on typical values of the drag coefficient and densities but which neglects atmospheric stability and variations in surface roughness. The large discrepancy may result because a number of factors (e.g. variations in fetch, development of the wave field, atmospheric stability) contribute to k which are not included when k is regressed against wind speed or wind stress. The variations in the dependence of k upon wind speed for different sizes of experimental tanks (Ocampo-Torres *et al.*, 1994) highlights the difficulties of extrapolating laboratory data to the field.

Given that turbulence in the boundary region may be caused by convective cooling, wind shear, breaking waves, and can be augmented by turbulence that originates at the thermocline or at the benthic boundary layer, estimates of k based on measurements of turbulence parameters (e.g. Crill *et al.*, 1988) or multipart models such as Soloviev & Schluessel's would appear most appropriate whereas those based on wind speed alone least appropriate. Parameterizing k using wind stress instead of wind

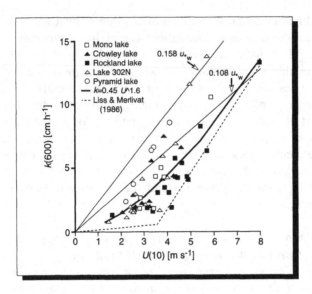

Fig. 3.6 Comparison of interfacial shear model (Eqn. 3.19) with results in Fig. 3.3.

speed will be an improvement; inclusion of wave parameters, particularly on larger bodies of water, should lead to significant improvements (see pp. 89–90 and Erickson (1993)).

3.6 Direct and indirect measurement of gas fluxes and gas transfer velocities

Use of floating helmets, enclosures in the water and injection of trace gas into the water in conjunction with measurements of wind speed and other environmental parameters are the prevalent means to determine the relation between gas transfer velocity and wind speed in the field. Indirect methods include obtaining the gas flux by difference in chemical budgets or determining of the gas transfer velocity using the models discussed above and using Eqns 3.1 and 3.2. Descriptions and methods for each procedure follow after a brief description of the merits of the most widely used procedures.

Advantages of estimating gas flux using a gas transfer velocity calculated from empirical relations (e.g. Eqn. 3.7) includes the relative ease of measuring wind speeds and the ability to conduct measurements in rough as well as calm weather. As it is generally beyond the scope of most field studies to measure all parameters influencing gas transfer velocities; flux estimates based on wind speeds will be subject to uncertainties on the order of the variance of the data points in Fig. 3.3. Improvement in the estimates is likely by including the turbulence characteristics, with the easiest enhancement being to calculate wind stress. Chamber methods remove the uncertainty due to estimates of k but cannot be used when wind speeds are high. They isolate the water surface from the wind shear and convective cooling that cause renewal events at the very surface which, as mentioned earlier, occur over horizontal distances of centimetres. Because turbulence in the water column can be generated away from the surface, and because large-scale mixing from the surface downward can be advected, chambers will measure diffusive fluxes generated by these events. Tracer experiments encompass all types of environmental forcing.

3.6.1 Direct measurements using the helmet or chamber method

The helmet or chamber method (see Chapter 2), which is also extensively used in soil flux studies, consists of placing an enclosure over the water surface and measuring the change in concentration in the head space. The helmet can be either moored or free floating with the lip just below the water surface so the water below the helmet can freely exchange with the main body of water. Different designs have been employed, including ones with removable tops (Frankignoulle, 1988), and designs in which wind is generated over the headspace with a fan (Sebacher et al., 1983). The head space can be ambient air, or it can be enriched or depleted with the gases to be measured. By minimizing the height of the box, greater concentration changes will be observed over time which improves the accuracy of the measurement.

With the helmet method gas fluxes can be determined over a period of minutes to hours. The flux of gas of interest is determined directly rather than by extrapolating a result from a different gas using the Schmidt number dependency. The disadvantage is that the surface turbulence under the helmet differs from the surrounding water. The gas transfer velocity is likely to be different too. The bobbing motion of the helmet will create pressure fluctuations inside the box which can influence the gas transfer velocity as well. No systematic study of the impact of the dimensions or deployment methods has been performed but helmets appear to give values that are higher than other measurements (Hartman & Hammond, 1984). More comparisons between helmet deployments and independent estimates are necessary to assess the accuracy of the method. The helmet method is an excellent way to determine the relative rates of exchange for different gases (Conrad & Seiler, 1988). This is of particular interest in determining enhancement of exchange of 'acidic' gases such as CO_2 in alkaline waters.

The gas transfer velocity is determined from the accumulation or depletion of gas in the head space under the helmet. Eqn. 3.1 can be rewritten:

$$k = F/(C_w - \alpha C_a) \tag{3.25}$$

and:

$$F = dM/dt A_H^{-1} \tag{3.26}$$

where A_H is the surface area of the helmet and dM is the change in mass of the gas in question. Since $dM = V_H dC_a$, and $h_H = V_H/A_H$ where V_H is the volume of head space and h_H is the average height of the helmet above the water surface, by substitution Eqn. 3.25 can be written as:

$$k dt = h_H dC_a (C_w - \alpha C_a)^{-1}. \tag{3.27}$$

C_w will not change with time if the water under the helmet is rapidly replaced. Eqn. 3.27 can be integrated to yield:

$$k = (h_H/\alpha)\ln[(C_w - \alpha C_a)_1/(C_w - \alpha C_a)_2](t_2 - t_1)^{-1} \tag{3.28}$$

where t_1 and t_2 refer to times 1 and 2.

During deployment of a helmet, a water sample and a series of head space samples is taken. After correcting for the volume of head space removed, the slope of the line $\ln(C_w - \alpha C_a)$ versus time multiplied by h_H yields the gas transfer velocity. This method will work for most gases with water concentrations in disequilibrium with the head space. Larger concentration changes facilitate the measurement of the gas transfer velocity. Changes in head space concentration will be a function of solubility with more soluble gases having larger changes in concentration per unit time.

3.6.2 Direct measurement of gas fluxes using enclosures

The enclosure method in the water is similar to the helmet method except that concentration changes in water are measured rather than in air. A parcel of water is enclosed using a submerged tub or flexible tube extending to the sediment interface. The flux can be determined from the change in concentration of gas in the isolated parcel of water measured when the gas concentrations in water and air are out of equilibrium. If air and water are in equilibrium, the water concentration can be perturbed by, for instance, deoxygenating the water, by addition of sodium sulfate (Yu & Hamrick, 1984), or adding a trace (Emerson, 1975). The flux is determined from the rate that equilibrium is re-established as calculated from the change in mass of the gas in the enclosure over time. Work by Torgersen et al. (1982) suggests that the in-water enclosure methods yield higher than average fluxes. This is probably because of pumping action of the flexible sides on the water column. Determination of fluxes by this method is similar to the helmet method except that the air concentration remains constant while water values change. When the flux of CO_2 is being determined, the change in total carbon content in the water must be measured in addition to the gradient of gaseous CO_2 over the interface (Section 3.3.2).

3.6.3 The deliberate tracer method: applications to streams, rivers and lakes

In the deliberate tracer method a gaseous tracer is added to the water body and the concentration change with time is monitored. Ideal tracers are non-toxic, non-reactive and measurable at low concentrations such that a minimal amount needs to be added to the water body.

Tracer studies to determine gas transfer have been performed over the past several decades in rivers (Tsivoglou, 1967; Rathbun, 1977, 1979; Rathbun & Grant, 1978; Holley & Yotsukura, 1984; Wilcock, 1988; Wanninkhof et al., 1990). Unless the total mass of tracer can be measured accurately at several locations along the reach of the stream or river, a non-volatile tracer is added along with the volatile tracer to separate the change in concentration due to dispersion and gas transfer. Dispersion of a tracer cloud is determined from the decrease in concentration of the non-volatile tracer; the concentration of the volatile tracer decreases due to gas exchange and dispersion. Non-volatile tracers include tritium and rhodamine dyes. Volatile tracers include krypton, ethylene, propane and methyl chloride (Rathbun & Grant, 1978; Kwasnik & Feng, 1979; Wilcock, 1984). Krypton and tritium can be measured simultaneously at low levels by β counting but their radioactivity is a safety concern. Ethane, propane and, in particular, methyl chloride are measurable at low levels but are not stable in the water column over long periods of time.

In the first whole-lake experiment, radium and tritium were injected into a lake and the daughters ^{222}Radon and ^{3}Helium were used to measure the exchange (Torgersen et al., 1982). This method has limited application since there are concerns about releasing radioactive compounds into the environment. ^{3}Helium can be measured at very low concentrations but involves costly extraction procedures and mass spectroscopy.

Halogenated trace gases offer an attractive alternative. They can be measured easily at low levels by head space analyses using electron capture gas chromatography (Lovelock & Watson, 1978; Lovelock & Ferber, 1982). Gas transfer studies on lakes ranging in size from 0.13 to $500\,km^2$ have been performed using the trace gas SF_6 (Wanninkhof et al., 1985, 1987, 1991; Crusius & Wanninkhof, 1990; Upstill-Goddard et al., 1990). A small amount of the gas is injected by bubbling SF_6 into the lake through a dispersing stone near the base of the epilimnion or by injecting water saturated with SF_6. It is mixed throughout the lake by wind and wave action. Mixing occurs over a period of a week or less. Once the concentration of trace gas is homogeneous, samples are taken to determine the mass decrease over time. Gas transfer velocities are determined according to:

$$k = F/(C_w - \alpha C_a). \tag{3.29}$$

For experiments with SF_6, $\alpha C_a \ll C_w$ and $F = h(dC_w/dt)$ where h is the effective depth of water exchanging with the atmosphere. Substitution and integration yields:

$$k = h/(t_2 - t_1)\ln(C_{w,t1}/C_{w,t2}). \tag{3.30}$$

When the water is stratified during the summer little exchange occurs between the hypolimnion and the epilimnion. The depth of exchange is the effective depth of the mixed layer where the effective depth takes into account the shallow areas near shore using a bathymetric map.

Experiments on small lakes have several advantages for determining relationships between gas transfer and forcing functions which can change on short time scales. They are often shallow so greater concentration changes over time will occur and gas transfer can be measured for shorter time intervals. Concentration variations due to differences in wind stress and due to variation in depth are homogenized rapidly. If edge effects and fetch influence gas transfer significantly, empirical relations obtained on small lakes might be biased. Work on lakes of different sizes suggests that larger lakes show slightly greater rates of gas transfer (see Fig. 3.3).

With the deliberate tracer method the time interval can be chosen to optimize the objectives of the study. Short time intervals are advantageous for relating gas transfer to rapidly changing environmental forcing functions. Longer time periods lead to larger concentration changes and gas transfer can be measured with greater precision. Figure 3.7a shows the concentration decrease with time for the first lake experiment performed with SF_6 (Wanninkhof et al., 1985). The variation of trace gas with depth and location was minimal during the experiment (Fig. 3.7b). The relationship between gas exchange and wind speed for this experiment is shown in Fig. 3.7c.

3.6.4 Indirect methods based on chemical budgets: applications to lakes and streams

Opportunities to determine gas fluxes may arise as a part of studies of geochemical cycles of gases. For instance, in studies of photosynthesis and respiration cycles in aquatic systems, the oxygen flux across the air–water interface is one of the unknowns. C. Langdon of Lamont–

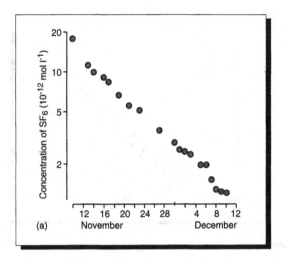

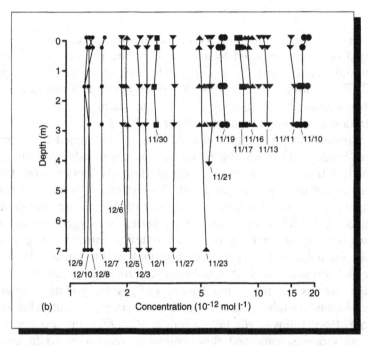

Fig. 3.7 Results of a deliberate tracer experiment on Rockland Lake (November 10–December 10, 1983). (a) Change in average concentration with time. (b) Concentration of SF_6 with depth in the lake for several days. (c) Relationship between gas transfer and wind speed. The wind speed is measured at 1 m height and the exchange coefficient is at *in situ* temperatures. (---) Indicates the results of a wind tunnel experiment of Broecker *et al.* (1978) normalized to Rockland Lake conditions. (——) is the least squares fit through the lake data, and the (– – –) is the relationship between gas exchange and wind speed for steady winds deduced from the Rockland Lake data (see Wanninkhof *et al.*, 1985) (*continued on p. 82*).

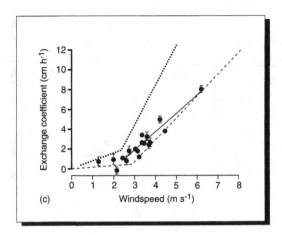

Fig. 3.7 Continued

Doherty Geological Observatory (personal communication) uses a sub-merged enclosure tethered to a buoy to isolate the photosynthesis and respiration cycle from the gas flux. The oxygen concentration change inside the enclosure, dC_{box} is the difference between respiration, R, and photosynthesis, P, $dC_{box}/dt = P - R$. The change in the oxygen concentration in the water column is $dC/dt = P - R - F$, where P, R and F are expressed in rate units. Once the flux is determined by comparing values inside and outside the enclosure, the gas transfer velocity can be estimated from the air–water concentration difference. Gas transfer velocities can only be measured with this method if air and water are in disequilibrium and other sources and sinks of the gas are accounted for. Wallace & Wirick (1992) use a time series of oxygen measurements at two depths from a buoy in coastal water to estimate oxygen fluxes. Rapid excursions of oxygen concentrations in the surface layer were observed which are attributed to bubble-enhanced gas transfer.

For streams and rivers, point sources of gaseous pollutants or ground water inputs containing excess gases can serve as inputs (Elsinger & Moore, 1983; Ellins *et al.*, 1990; Clark *et al.*, 1992). From concentration decreases downstream and flow estimates, reaeration coefficients are calculated. This method requires good estimates of discharge rate and changes in discharge over the section where gas transfer is measured.

3.6.5 Indirect estimates of gas flux: estimation of gas transfer velocity using wind speed and wind shear

Wind speeds are measured at a height above the water's surface chosen to be within the logarithmic boundary layer of the atmosphere (Peixoto &

Oort, 1992). In oceanic conditions and for large lakes, the height selected is by convention 10 m. For lakes with smaller fetch, the height may only be 1 or 2 m above the water's surface and data are corrected to 10 m using logarithmic scaling (cf. Liu & Schwab, 1987). Gas transfer velocities in lakes can be estimated using wind speed and the appropriate Schmidt number. At present, use of Eqn. 3.7 is an appropriate approach for calculating the gas exchange velocity in lakes at low to moderate wind speeds when the gas transfer velocity is strongly dependent upon wind speed. For oceanic work and for work at higher wind speeds, appropriate choices are the equations of Smethie *et al.* (1985), Liss & Merlivat (1986) and Wanninkhof (1992). Once k has been derived, it can then be utilized in Eqn. 3.1.

Improved flux estimates should be obtained if wind stress is taken into account as discussed in Section 3.5.3. There are a number of approaches; at a minimum wind speed would be measured at one height, air temperatures would be required at the same height and just above the water's surface, and relative humidity obtained at the same height as wind speed, using either a sling psychrometer at short intervals or a recording humidity sensor. Equations for calculating the drag coefficient as a function of atmospheric stability are included in a report by the TVA (1972). Other references are provided in Section 3.5.3 and pp. 85–86. Donelan (1990) provides a detailed review of procedures for measuring surface fluxes of momentum, heat and mass.

Gas transfer velocities will be decreased in situations in which the turbulence above the air–water interface is decreased. For instance, Liss *et al.* (1981) showed in a wind tunnel experiment that under net evaporative conditions (unstable atmospheric boundary layer) or neutral stratification, the gas transfer velocities were 30% higher than under condensing conditions at moderate wind speeds. When evaporation occurs or when the atmosphere is neutrally stratified, heat loss may occur within the aqueous boundary layer with accompanying convective motions that promote gas transfer. Such motions would not occur with condensation. In addition, the upper layers of the water column may become stably stratified with effects on the larger scale turbulence (see pp. 86–89).

When wind speeds are less than about $2.5\,\mathrm{m\,s^{-1}}$, k is only slightly dependent on wind speed (e.g. Figs 3.3 & 3.7c). The importance of shear and waves even at low wind speeds is suggested by the high frequency fluctuations in current speed found at 1 cm depth but not at 5 or 10 cm depth at wind speeds of $<2\,\mathrm{m\,s^{-1}}$ when ripples of between 1 mm and 1 cm height were present in a shallow lake (MacIntyre, 1984). The available data (Sebacher *et al.*, 1983; Happel, 1992; Happel *et al.*, in press) suggest that k is ~1 for wind speeds less than $1.5\,\mathrm{m\,s^{-1}}$. However, more precise values will be obtained when the other processes affecting surface renewal are considered. For instance, convective cooling is likely to be important and should be included in estimations of k as in Eqn. 3.21 or by computing turbulent velocity and length scales using the surface energy flux method described below and using the results in Eqn. 3.18. Estimating gas flux in the absence of wind, and from fairly stagnant waters, has been problematic but should be tractable with this approach, assuming heat losses in the upper $100\,\mu m$ are generating renewal events. For streams, the turbulence generated by shear stress at the benthic boundary should be incorporated (Eqns 3.23 & 3.24).

3.6.6 Indirect estimates of gas flux: estimation of gas transfer velocity using turbulence parameters: ε, u, l and Re_t

To estimate the gas exchange velocity and include many of the processes causing turbulence in the boundary region requires measurement of the turbulence parameters used in Eqns 3.15–3.18. Three different approaches can be used. The first approach, which we call the surface energy budget approach, estimates the energy dissipation rate ε from meteorological measurements and profiles of irradiance, temperature and conductivity (Imberger, 1985). This method can be used whenever a good meteorological station is located near or on the body of water under study. The meteorological data and irradiance measurements are used to compute the energy budget at the surface of the water body, depth of the mixing layer l near the surface is inferred from the density structure, and ε is proportional to the energy flux divided by l. The turbulent velocity scale u is obtained from the relation $\varepsilon = u^3/l$, and Re_t can be obtained from u and l (see Section 3.5.5). The second approach is to use turbulent microstructure profilers, a subject reviewed in Gregg (1987). These profilers allow direct estimation of ε and l. Although the use of these instruments has been restricted to a small scientific community, they are becoming commercially available. Fast response current meters moored in the upper metre provide a third approach (Kitaigorodskii *et al.*, 1983; Agrawal *et al.*, 1992), but are best done as a collaborative study because of the complexity of the data analysis and measurement techniques. To our knowledge, the surface energy budget approach has only been applied in one instance to test the surface renewal model (Crill *et al.*, 1988) and data from turbulent microstructure profilers or moored current meters have not yet been used for this purpose. These latter techniques would be most appropriate in collaboration with a physical limnologist or oceanographer. The surface energy budget approach and microstructure profiling are discussed in greater detail below.

*Estimates of ε and l based on surface energy fluxes and
density structure of the water column*

Imberger (1985) found that the average turbulent kinetic energy dissipation rate ε_{av} in the upper mixing layer can be calculated from the surface energy flux E_F and depth of the upper mixing layer h:

$$\varepsilon_{av} = 0.82\, E_F/h. \tag{3.31}$$

Shear and penetrative convection were the dominant mixing processes in the upper mixing layer when these relations were tested. Although mixing layers are best demarcated by changes in dynamic properties of the flow such as ε or by overturning scales (Thorpe, 1977; Imberger, 1985), measurements by Shay & Gregg (1986) indicated that temperature differences as small as 0.02°C also could demarcate such regions. Consequently, h can be obtained in freshwater lakes by temperature profiling or from thermistor chains when the resolution of the sensors is at least 0.01°C and individual sensors have been calibrated to within 0.01°C of each other. In saline waters, high resolution conductivity data would be needed as well. Density is calculated for freshwaters and slightly saline waters using the equations in Chen & Millero (1977) and for oceanic waters using the equation of state for seawater (UNESCO, 1983). Spiking, the pronounced overestimates of salinity and density due to the mismatch in time constants of the temperature and conductivity sensors, can be removed following procedures in Fozdar *et al.* (1985).

The surface energy flux per unit density is defined as

$$E_F = 1/2(q_*^{3}) = 1/2(w_*^{3} + c_n^{3} u_{*w}^{3}) \tag{3.32}$$

where q_*^{2} is the turbulent kinetic energy per unit density, w_* is the penetrative convection velocity scale, and c_n is a constant whose value is 1.33. The energy flux has units $m^3\,s^{-3}$.

$$w_* = (Bh)^{1/3} \tag{3.33}$$

where the buoyancy flux

$$B = g\alpha H/\rho C_p \tag{3.34}$$

where g is gravity, α is the coefficient of thermal expansion, C_p is heat capacity, ρ is density, and H is the surface heat flux

$$H = \rho C_p <w'\theta'(D)> + q(D) + q(e) - 2/h \int_e^D q(z)dz. \tag{3.35}$$

The first term on the right is the sum of latent heat flux, sensible heat flux, and long wave back radiation, $q(D)$ is the net short wave radiation at the surface, $q(z)$ is the radiation in the water column at a depth z, and $q(e)$ is the radiation remaining at the base of the mixed layer (Imberger,

1985). The various approaches (i.e. eddy correlation, gradient flux, bulk transfer, energy balance and combined methods) for obtaining the different terms of the heat flux are discussed in Peixoto & Oort (1992). Detailed equations are found in TVA (1972) and Strub (1983). The TVA report and Large & Pond (1982) include equations for calculating the drag coefficient as a function of atmospheric stability. Ryan & Harleman (1973), Weisman & Brutsaert (1973) and Strub (1983) provide methods to calculate evaporation when the air column over the water body is unstable as is likely in tropical and other warm water lakes at night. Sensible heat can be obtained from the Bowen ratio (Strub, 1983). The upward long wave radiation can be measured directly with pyrgeometers. Alternatively, when only surface temperatures are available, it is given by the Stefan–Boltzman law $I = \xi\sigma T_s^4$ where ξ is the emissivity of the water's surface (0.97), σ is the Stefan–Boltzman constant ($\sigma = 5.6701 \times 10^{-8}\,W\,m^{-2}\,K^{-4}$), and T_s is the absolute temperature of the water's surface. This term must be corrected for vapour pressure and cloud cover (Strub, 1983) to give net long wave radiation. Net short wave radiation is obtained from an Eppley pyrannometer corrected for upward reflectance (albedo) by measuring the ratio of downward to upward irradiance throughout the day by a downward- and an upward-looking Eppley pyranometer.

To obtain the heat absorbed at each depth, the attenuation of irradiance must be determined for different wavelength bands in the range 0.28–2.8 micrometers. Attenuation is described by $I_z = I_o e^{(-\lambda z)}$, where I_o and I_z are the irradiance incident at the surface and at a depth z and λ is the attenuation coefficient. The attenuation of light from 400 to 700 nm can be obtained using a PAR sensor (photosynthetically available radiation). The mean attenuation of ultraviolet light (300–400 nm) can be obtained spectrophotometrically from water filtered through a 0.45 m Millipore filter. Hale & Query (1973) determined the attenuation of infrared light in pure water for five intervals from 700 to 2400 nm. The energy content for these seven wavelength bands can be obtained from the curve of solar spectral distribution (cf. Gates, 1966). Results from these calculations are in Jellison & Melack (1993).

Application of turbulent microstructure profiling to assessment of gas transfer velocity

Turbulent microstructure profilers provide the *in situ* data required to estimate ε, u, l and Re_t as needed to test the surface renewal model. Profilers in use either collect data on the upcast or the downcast; data from the upcast is required for gas transfer measurements in order to obtain measurements near the surface at a distance from the research vessel (cf. Carter & Imberger, 1986). Turbulent microstructure profilers are of two basic types, those that measure shear in the water column (Osborne, 1974) and those that measure the small scale temperature gradients caused by turbulent mixing (Dillon & Caldwell, 1980). When shear is estimated, energy dissipation rates are estimated following Oakey (1982). When temperature gradient profiles are measured, ε is estimated following Dillon & Caldwell (1980) and Imberger & Boashash (1986). When concurrent profiles of temperature and conductivity are available with resolution on the centimetre scale, the size of overturning regions l can be ascertained based on instabilities in the calculated density profile (Thorpe, 1977; Dillon, 1982). Both ε and l have been obtained using temperature gradient microstructure profilers (Imberger & Boashash, 1986; Imberger & Ivey, 1991; MacIntyre, 1933). The rms value of the displacement lengths, the depths parcels of water would be transported to make the unstable density profile increase monotonically with depth, is the displacement scale and provides a direct estimate of the turbulent length scale. The turbulent velocity scale can be obtained from the relation $u = (\varepsilon l)^{1/3}$ (Section 3.5.5) which applies when buoyancy does not greatly affect the turbulence (Taylor, 1935).

Figure 3.8 illustrates the results from temperature gradient microstructure profiling (S. MacIntyre & J. Imberger, unpublished data). The data were obtained from a productive 2.5 m deep lake about 1 h after wind speeds had risen from 3 to 8 m s^{-1} and turbulent mixing had created an upper mixing layer and reduced the overall thermal stratification in the lake. The temperature profile (Fig. 3.8a) shows a *ca.* 10 cm thick layer of cooler water at the top of a 60 cm near isothermal layer. Below this layer is a thermocline where temperatures decrease but in which there are numerous instabilities most likely due to shear from internal waves. Figure 3.8b shows the temperature gradients. It is the frequency of changing gradients per unit depth that determines ε. Displacement lengths, calculated after converting the temperature profile to density, are indicated in Fig. 3.8c; maximum displacements are nearest the surface and are *ca.* 60 cm in vertical extent and delineate the upper mixing layer. When these displacements are filtered and the maximum displacement centred in the overturning region (Fig. 3.8d), three large overturning regions can be observed in the profile. The upper 0.6 m is one overturning region; it is comprised of eddies ranging from less than a millimetre up to 0.6 m in size which will renew the surface causing transport of gases and preventing accumulation or depletion of gases just below the boundary. The energy dissipation rates in Fig. 3.8e were obtained from the Wigner–Ville transform (Imberger & Boashash, 1986). Rates are high in the upper 60 cm, averaging 7.2×10^{-7} m^2 s^{-3}, decrease abruptly just below the surface mixing layer, and increase again indicating turbulence within the thermocline. If concentrations of dissolved gas increase with depth

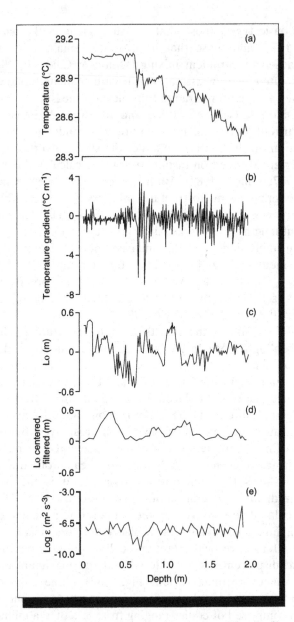

Fig. 3.8 Results from temperature gradient microstructure profiling. Profiles of temperature (a), temperature gradients (b), displacement lengths (c), centred and filtered displacement lengths (d), and log of energy dissipation rate (e) obtained as the log of the Wigner–Ville transform. (Data are from North Lake in Western Australia.)

within the thermocline, mixing would reduce the concentration gradients and would increase the concentrations near the top of the thermocline. When the upper mixing layer deepens, waters enriched with gases would be entrained into the surface waters. The reverse would be true if gases were more concentrated in the upper mixing layer.

Using l/u to estimate the renewal time at the water's surface assumes that every velocity fluctuation will transport solutes or gases; an assumption that may not be valid in stably stratified waters. As stratification increases, the frequency and magnitude of fluctuations transporting dissolved gases may decrease (Itsweire & Helland, 1989). The ratio of the overturning time scale l/u to the diffusional time scale, $l^2/8K_z$ (where K_z is the coefficient of eddy diffusivity), was constant where turbulence was active but decreased in more stratified waters where buoyancy forces retarded mixing (MacIntyre, 1993). This decrease in the ratio of the two time scales suggests that in stratified fluids the ratio l/u may not correctly estimate the time scale for renewal of the surface water with dissolved gasses; $l^2/8K_z$ may be a better choice, and should be substituted into Eqn. 3.13. The coefficient of eddy diffusivity K_z can be obtained as in Osborn & Cox (1972) or when ε and the density stratification are known (Osborn, 1980; Ivey & Imberger, 1991).

Micro-oxygen electrodes have recently been added to turbulent microstructure profilers. In oligotrophic waters, or in situations where productivity rates have been measured, these will provide a method to measure gas fluxes concurrently with turbulence parameters (Oldham, in press; Oldham & Imberger, 1992).

Bowers & Simpson (1987) provide the methodology for obtaining the rate of energy dissipation in coastal waters and estuaries where tidal mixing is important. MacKenzie & Leggett (1991) provide an example of their approach. As much of the turbulence induced by tides is generated near the benthic boundary layer, tidal mixing would not be expected to influence turbulence near the surface in stratified flows.

Estimates of ε from wind speeds under conditions of breaking waves
Using surface meteorology to estimate ε has been tested in lakes with wind speeds up to $6\,\mathrm{m\,s^{-1}}$ (Imberger, 1985). As wind speeds increase above this value, the contribution of breaking waves to the energy budget will increase. In the first field experiment in which energy dissipation rates were obtained in a field of breaking waves, Kitaigorodskii *et al.* (1983) obtained values of ε near the surface that were up to two orders of magnitude greater than if only the shear stress generated by the wind were considered (i.e. $\varepsilon > u_{*w}^3/\kappa l$ for $U > 8\,\mathrm{m\,s^{-1}}$ and κ, von Karman's constant, has a value of 0.4). Agrawal *et al.* (1992) found similarly high values. The high values of ε occurred intermittently and apparently as a

result of the intermittency of breaking waves. Future work on gas flux at higher wind speeds should include estimates of wave parameters and surface meteorology.

Kitaigorodskii *et al.*'s (1983) work provides empirical coefficients to determine horizontal and vertical turbulent velocities when wind speed, wave heights, and wave spectra are available; and recent work by M. Donelan and colleagues (personal communication) indicates that ε will scale with rms wave slope, rms wave height, the corresponding phase speed, depth and the friction velocity in water. Although the analysis of wave parameters to the detail they invoke is likely to be beyond the scope of most investigations, there is some hope of simpler relations. For instance, E. Terray (personal communication, Woods Hole Oceanographic Institution) observed that in a field of breaking waves, the rms vertical turbulent velocities w' depended upon wind shear and the depth of the wave influenced layer L which was *ca.* 10 times the rms wave amplitude. He and his colleagues found mean square velocities could be expressed as

$$<w'^2> = 1.25^2\, u_{*w}^2\, [1 + (L/z)^2] \qquad (3.36)$$

where z is the depth of interest but should not be arbitrarily small. Using u_* and rms wave heights from Kitaigorodskii *et al.* (1983), we computed

$$\varepsilon = [1/2(w'^2 + u'^2)]^{3/2}/L \qquad (3.37)$$

where w' was calculated from Eqn. 3.36. Note that Eqn. 3.37 is functionally equivalent to Eqn. 3.11. u' as $u' \sim 2.3\, u_{*w}$ and $w' \sim 1.25\, u_{*w}$, consequently, we calculated $<u'^2> = 2.3^2\, u_{*w}^2[1 + (L/z)^2]$. Values of ε were *ca.* five times higher than those assumed based on wind shear alone (i.e. $\varepsilon = u_{*w}^3/\kappa l$). However, ε values were two to 10 times less than when u' and w' were obtained using the empirical relations in Kitaigorodskii *et al.* (1983) These relations developed by E.A. Terray and colleagues for u' and w' require further testing but suggest that when surface waves are breaking improved estimates of ε and subsequently of k may be obtained from basic measurements of wind speed and wave height in neutral stratification.

Acknowledgements
We thank E.A. Terray and Sanjoy Banerjee for insightful discussions and Cheryl Kelley, Carolyn Oldham, John Melack and an anonymous reviewer for helpful comments on the manuscript.

References
Agrawal Y.C., Terray E.A., Donelan M.A. *et al.* (1992) Enhanced dissipation of kinetic energy beneath surface waves. *Nature*, **359**, 219–220.

Asher W. & Pankow J.F. (1989) Direct observation of concentration fluctuations close to a gas–liquid interface. *Chemical Engineering Science*, **44**, 1451–1455.

Banerjee S. (1990) Turbulence structure and transport mechanisms at interfaces. *Proceedings of the 9th International Heat Transfer Conference*, pp. 395–417 Jerusalem, Israel.

Banerjee S. (1991) Turbulence/interface interactions. In: Hewitt G.F., Mayinger F., Riznic J.R. (eds) *Phase–Interface Phenomena in Multiphase Flow*. Hemisphere Publishing Corporation, New York.

Banerjee S., Rhodes E. & Scott K.S. (1968) Mass transfer to falling wavy liquid films in turbulent flow. *Industrial Engineering and Chemistry Fundamentals*, **7**, 22–27.

Bolin B. (1960) On the exchange of carbon dioxide between the atmosphere and the sea. *Tellus*, **12**, 274–281.

Bowers D.G. & Simpson J.H. (1987) Mean position of tidal fronts in European-shelf seas. *Continental Shelf Research*, **7**, 35–44.

Braker D. & Mossman J. (1975) *Matheson Unabridged Gas Data Book*. Matheson Gas Products, East Rutherford, New Jersey.

Broecker H.C. & Siems W. (1984) The role of bubbles for gas transfer from water to air at higher windspeeds; Experiments in the wind-wave facility in Hamburg. In: Brutsaert W. & Jirka G.H. (eds) *Gas Transfer at Water Surfaces*, pp. 229–238. Reidel, Hingham, Massachusetts.

Broecker H.C., Peterman J. & Siems W. (1978) The influence of wind on CO_2 exchange in a wind-wave tunnel, including the effects of mono layers. *Journal of Marine Research*, **36**, 595–610.

Broecker W.S. & Peng T.H. (1974) Gas exchange rates between air and sea. *Tellus*, **26**, 21–35.

Brumley B.H. & Jirka G.H. (1988) Air–water transfer of slightly soluble gases: turbulence, interfacial processes and conceptual models. *Physico Chemical Hydrodynamics*, **10**, 295–319.

Butler J.N. (1982) *Carbon Dioxide Equilibria and their Applications*. Addison-Wesley, Reading, Massachusetts.

Carter G.D. & Imberger J. (1986) Vertically rising microstructure profiler. *Journal of Atmospheric and Oceanic Technology*, **3**, 462–471.

Chen C.T. & Millero F.J. (1977) The use and misuse of pure water PVT properties for lake waters. *Nature*, **266**, 707–708.

Clark J.F., Simpson H.J., Smethie W.M. & Toles C. (1992) Gas exchange in a contaminated estuary inferred from chlorofluorocarbons. *Geophysical Research Letters*, **19**, 1133–1136.

Coantic M. (1986) A model of gas transfer across air-water interfaces with capillary waves. *Journal of Geophysical Research*, **91**, 3925–3943.

Conrad R. & Seiler W. (1988) Influence of the surface microlayer on the flux of non-conservative trace gases (CO, H_2, CH_4, N_2O) across the ocean–atmosphere interface. *Journal of Atmospheric Chemistry*, **6**, 83–94.

Crill P.M., Bartlett K.B., Wilson J.O. *et al.* (1988) Tropospheric methane from an Amazonian floodplain lake. *Journal of Geophysical Research*, **93**, 1564–1570.

Crusius J. & Wanninkhof R. (1990) A gas exchange–wind speed relationship measured on Lake 302N with SF_6. *EOS*, **71**, 1234.

Csanady G.T. (1990) The role of breaking wavelets in air–sea gas transfer. *Journal of Geophysical Research*, **95**, 749–759.

Dacey J.W.H., Wakeman S.G. & Howes B.L. (1984) Henry's law constants for dimthyl-sulfide in freshwater and seawater. *Geophysical Research Letters*, **11**, 991–994.

Danckwerts P.V. (1951) Significance of liquid-film coefficients in gas absorption. *Industrial Engineering Chemistry*, **43**, 1460–1467.

Deacon E.L. (1977) Gas transfer to and across an air–water interface. *Tellus*, **29**, 363–374.

Deardorff J.W. (1968) Dependence of air–sea transfer coefficients on bulk stability. *Journal of Geophysical Research*, **73**, 2549–2557.

Dillon T.M. (1982) Vertical overturns: A comparison of Thorpe and Ozmidov length scales. *Journal of Geophysical Research*, **87**, 9601–9613.

Dillon T.M. & Caldwell D.R. (1980) The Batchelor spectrum and dissipation in the upper ocean. *Journal of Geophysical Research*, **85**, 1910–1916.

Dobbins W.E. (1956) The nature of the oxygen transfer coefficient in aeration systems. In: McCabe J. & Eckenfelder W.W. (eds) *Biological Treatment of Sewage and Industrial Wastes*, vol. I. Reinhold Book Corporation, New York.

Donelan M. (1990) Air–sea interaction. In: *The Sea: Ocean Engineering Science*, vol. 9, pp. 239–292. John Wiley & Sons.

Donelan M.A., Elder F.C. & Hamblin P.F. (1974) Determination of the aerodynamic drag coefficient from wind set-up (IFYGL). *Proceedings of the 17th Conference Great Lakes Research*, pp. 778–788.

Ellins K.K., Wanninkhof R., Roman-Mas A., Quinones-Aponte V. & Davison C. (1990) Monitoring groundwater/surface discharge fluctuations, using ^{222}Rn and SF_6. *EOS*, **71**, 1340.

Elsinger R.J. & Moore W.S. (1983) Gas exchange in the Pee Dee river based on ^{222}Rn evasion. *Geophysical Research Letters*, **10**, 443–446.

Emerson S. (1975a) Chemically enhanced CO_2 gas exchange in an eutrophic lake; a general model. *Limnology and Oceanography*, **20**, 743–753.

Emerson S. (1975b) Gas exchange in small Canadian Shield lakes. *Limnology and Oceanography*, **20**, 754–761.

Erickson D.J. III. (1993) A stability dependent theory for air–sea gas exchange. *Journal of Geophysical Research*, **98**, 8471–8488.

Fortesque G.E. & Pearson J.R.A. (1967) On gas absorption into a turbulent liquid. *Chemical Engineering Science*, **22**, 1163–1176.

Fozdar F.M., Parker G.J. & Imberger J. (1985) Matching temperature and conductivity sensor response characteristics. *Journal of Physical Oceanography*, **15**, 1557–1569.

Frankignoulle M. (1988) Field measurements of air–sea CO_2 exchange. *Limnology and Oceanography*, **33**, 313–322.

Gates D.M. (1966) Spectral distribution of solar radiation at the earth's surface. *Science*, **151**, 523–529.

Goldman J.C., Dennet M.R. & Frew N.M. (1988) Surfactant effects on air–sea gas exchange under turbulent conditions. *Deep-Sea Research*, **35**, 1953–1970.

Gregg M.C. (1987) Diapycnal mixing in the thermocline: A review. *Journal of Geophysical Research*, **92**, 5249–5286.

Hale G.M. & Query M.R. (1973) Optical constants of water in the 200 nm to 200 µm wavelength region. *Applied Optics*, **13**, 555–563.

Happell J.D. (1992) *Stable isotopes as tracers of methane oxidation in the rhizosphere and at the sediment water interface in Florida wetlands*. PhD dissertation, Florida State University.

Happell J.D., Chanton J.P. & Showers W.J. Methane transfer across the water–air interface in stagnant wooded swamps of Florida. *Limnology and Oceanography* (in press).

Harriot P. (1962) A random eddy modification of the penetration theory. *Chemical Engineering Science*, **17**, 149.

Hartman B. & Hammond D.E. (1984) Gas exchange rates across the sediment–water and air–water interfaces in south San Francisco Bay. *Journal of Geophysical Research*, **89**, 3593–3603.

Hartman B. & Hammond D.E. (1985) Gas exchange in San Francisco Bay. *Hydrobiologica*, **129**, 59–68.

Hayduk W. & Laudie H. (1974) Prediction of diffusion coefficients for non-electrolytes in dilute aqueous solutions. *Journal of the American Institute of Chemical Engineers*, **20**, 611–615.

Herczeg A.L. (1987) A stable carbon isotope study of dissolved inorganic carbon cycling in a soft water lake. *Biogeochemistry*, **4**, 231–263.

Hicks B.B. (1972) Some evaluations of drag and bulk transfer coefficients over water bodies of different sizes. *Boundary Layer Meteorology*, **3**, 201–213.

Higbie R. (1935) The rate of absorption of a pure gas into a still liquid during short periods of exposure. *Transactions of the American Institute of Chemical Engineers*, **31**, 365–388.

Holley E.R. & Yotsukura N. (1984) Field techniques for reaeration measurements in rivers. In: Brutsaert W. & Jirka G.H. (eds) *Gas Transfer at Water Surfaces*, pp. 381–401. Reidel, Boston, Massachusetts.

Holmen K. & Liss P.S. (1984) Models for air–water gas transfer: an experimental investigation. *Tellus*, **36B**, 92–100.

Hoover T.E. & Berkshire D.C. (1969) Effects of hydration in carbon dioxide exchange across an air–water interface. *Journal of Geophysical Research*, **74**, 456–464.

Hunt J.C.R. (1984) Turbulence structure and turbulent diffusion near gas–liquid interfaces. In: Brutsaert W. & Jirka G.H. (eds) *Gas Transfer at Water Surfaces*, pp. 67–82. Reidel, Boston, Massachusetts.

Hunt J.C.R. & Graham J.M.R. (1978) Free-stream turbulence near plane boundaries. *Journal of Fluid Mechanics*, **84**, 209–235.

Imberger J. (1985) The diurnal mixed layer. *Limnology and Oceanography*, **30**, 737–770.

Imberger J. & Boashash B. (1986) Application of the Wigner–Ville distribution to temperature gradient microstructure: A new technique to study small-scale variations. *Journal of Physical Oceanography*, **16**, 1997–2012.

Imberger J. & Ivey G. (1991) On the nature of turbulence in a stratified fluid. Part 2: Application to lakes. *Journal of Physical Oceanography*, **21**, 659–680.

Imberger J. & Patterson J.C. (1990) Physical limnology. *Advances in Appied Mechanics*, **27**, 303–475.

Isenogel S.S. (1985) A laboratory study of gas transfer across an air–water interface. MS thesis, University of Southern California.

Itsweire E.C. & Helland K.N. (1989) Spectra and energy transfer in stably stratified turbulence. *Journal of Fluid Mechanics*, **207**, 419–452.

Ivey G.N. & Imberger J. (1991) On the nature of turbulence in a stratified fluid. Part 1: The efficiency of mixing. *Journal of Physical Oceanography*, **21**, 650–658.

Jähne B., Heinz G. & Dietrich W. (1987) Measurement of the diffusion coefficients of sparingly soluble gases in water with a modified Barrer method. *Journal of Geophysical Research*, **92**, 10767–10776.

Jähne B., Huber W., Dutzi A., Wais T. & Imberger J. (1984) Wind/wave-tunnel experiment on the Schmidt number – and wave field dependence of air/water gas exchange. In: Brutsaert W. & Jirka G.H. (eds) *Gas Transfer at Water Surfaces*. Reidel, Boston, Massachusetts.

Jähne B., Munnich K.O., Bosinger R., Dutzi A., Huber W. & Libner P. (1987) On the parameters influencing air–water gas exchange. *Journal of Geophysical Research*, **92**, 1937–1950.

Jähne B., Wais T., Memery L. *et al.* (1985) He and Rn gas exchange experiments in the large wind–wave facility of IMST. *Journal of Geophysical Research*, **90**, 11989–11997.

Jassby A. & Powell T. (1975) Vertical patterns of eddy diffusion during stratification in Castle Lake, California. *Limnology and Oceanography*, **20**, 530–543.

Jellison R.S. & Melack J.M. (1993) Meromixis in hypersaline Mono Lake, California I: and vertical mixing stratification during the onset, persistence, and breakdown of meromixis. *Limnology and Oceanography*, **38**, 1008–1019.

Jirka G.H. & Brutsaert W. (1984) Measurements of wind effects in water-side controlled gas exchange in riverine systems. In: Brutsaert W. & Jirka G.H. (eds) *Gas Transfer at Water Surfaces*, pp. 437–446. Reidel, Boston, Massachusetts.

Khoo B.C. & Sonin A.A. (1992) Augumented gas exchange across wind-sheared and shear-free air–water interfaces. *Journal of Geophysical Research*, **97**, 14413–14416.

Kitaigorodskii S.A. & Donelan M.A. (1984) Wind–wave effects on gas transfer. In: Brutsaert W. & Jirka G.H. (eds) *Gas Transfer at Water Surfaces*, pp. 147–170. Reidel, Boston, Massachusetts.

Kitaigorodskii S.A., Donelan M.A., Lumley J.L. & Terray E.A. (1983) Wave-turbulence interactions in the upper ocean. Part II: Statistical characteristics of wave and turbulent components of the random velocity field in the marine surface layer. *Journal of Physical Oceanography*, **13**, 1988–1999.

Komori S. (1991) Surface-renewal motions and mass transfer across gas–liquid interfaces in open-channel flows. In: Hewitt G.F., Mayinger F. & Riznic J.R. (eds) *Phase-Interface Phenomena in Multiphase Flow*, pp. 31–40. Hemisphere Publishing, New York.

Komori S., Nagaosa R. & Murakami Y. (1990) Mass transfer into a turbulent liquid across the zero-shear gas–liquid interface. *Journal of the American Institute of Chemical Engineers*, **36**, 957–960.

Kwasnik J.M. & Feng T.S. (1979) Development of a modified tracer technique for measuring the stream reaeration rate. *National Technology Information Service*, **PB**, 296317.

Lam K. & Banerjee S. (1992) On the conditions of streak formation in a bounded turbulent flow. *Physics of Fluids,* **A4**, 306–320.

Lamont J.C. & Scott D.S. (1970) An eddy cell model of mass transfer into the surface of a turbulent liquid. *Journal of the American Institute of Chemical Engineers*, **16**, 513–519.

Lange P.A., Jähne B., Tschiersch J. & Ilmberger J. (1982) Comparison between an amplitude-measuring wire and a slope-measuring inserwater wave guage. *Review of Scientific Instrumentation*, **53**, 651–655.

Large W.P. & Pond S. (1981) Open ocean momentum flux measurements in moderate to strong winds. *Journal of Physical Oceanography*, **11**, 324–336.

Large W.P. & Pond S. (1982) Sensible and latent heat flux measurements over the ocean. *Journal of Physical Oceanography*, **12**, 464–482.

Ledwell J.R. (1984) The variation of the gas transfer coefficient with molecular diffusivity. In: Brutsaert W. & Jirka G.H. (eds) *Gas Transfer at Water Surfaces*, pp. 293–302. Reidel, Boston, Massachusetts.

Lewis W.K. & Whitman W.G. (1924) Principles of gas absorption. *Industrial and Engineering Chemistry*, **16**, 1215.

Liss P.S. (1983) Gas transfer: experiments and geochemical implications. In: Liss P.S. & Slinn W.G. (eds) *Air–Sea Exchange of Gases and Particles*, pp. 241–299. Reidel, Boston.

Liss P.S. & Merlivat L. (1986) Air–sea gas exchange rates: Introduction and synthesis. In: Buat-Menard P. (ed) *The Role of Air–Sea Exchange in Geochemical Cycling*, pp. 113–129. Reidel, Boston, Massachusetts.

Liss P.S., Balls P.W., Martinelli F.N. & Coantic M. (1981) The effect of evaporation and condensation on gas transfer across an air–water interface. *Oceanologica Acta*, **4**, 129–138.

Liu P.C. & Schwab D.J. (1987) A comparison of methods for estimating u_* from given u_z and air–sea temperature difference. *Journal of Geophysical Research*, **92**, 6488–6494.

Lovelock J.E. & Ferber G.L. (1982) Exotic tracers for atmospheric studies. *Atmospheric Environment*, **16**, 1467–1471.

Lovelock J.E. & Watson A.J. (1978) The electron capture detector, theory and practice 2. *Journal of Chromatography*, **158**, 123–138.

MacIntyre S. (1984) Current fluctuations in the surface waters of small lakes. In: Brutsaert W. & Jirka G.H. (eds) *Gas Transfer at Water Surfaces*, pp. 125–132. Reidel, Boston, Massachusetts.

MacIntyre S. (1993) Vertical mixing in a shallow, eutrophic lake: Possible consequences for the light climate of phytoplankton. *Limnology and Oceanography*, **38**, 798–817.

MacKenzie B.R. & Leggett W.C. (1991) Quantifying the contribution of small-scale turbulence to the encounter rates between larval fish and their zooplankton prey: effects of wind and tide. *Marine Ecology Progress Series*, **73**, 149–160.

Merlivat L. & Memery L. (1983) Gas exchange across an air–water interface: experimental results and modeling of bubble contribution to transfer. *Journal of Geophysical Research*, **88**, 707–724.

Millero F.J. (1974) Seawater as a multi-component electrolyte solution. In: Goldberg E.D. (ed.) *The Sea*, vol. 5, pp. 3–80. John Wiley and Sons, New York.

Monahan E.C. & O'Muircheartaigh I.G. (1980) Optimal power-law description of oceanic whitecap coverage dependence on wind speed. *Journal of Physical Oceanography*, **10**, 2094–2099.

O'Connor D.J. & Dobbins W.E. (1958) Mechanism of reaeration in natural streams. *ASCE Transactions*, **123**, 641.

Oakey N.S. (1982) Determination of the rate of dissipation of turbulent energy from simultaneous temperature and velocity shear microstructure measurements. *Journal of Physical Oceanography*, **12**, 256–271.

Ocampo-Torres F.J., Donelan M.A., Merzi N. & Jia F. (1994) Laboratory measurements of mass transfer of carbon dioxide and water vapour for smooth and rough flow conditions. *Tellus*, Series B. *Chemical and Physical Meteorology*, **46**, 16–32.

Oldham C. *A Portable Microprofiler with a Fast-response Oxygen Sensor. Limnology & Oceanography* (in press).

Oldham C. & Imberger J. (1992) *The Effects of Mixing on the Distribution of Dissolved Oxygen in a Lake. I. The Surface Waters*. Centre for Water Research Report ED 694 CO. University of Western Australia, Nedlands.

Osborn T.R. (1974) Vertical profiling of velocity microstructure. *Journal of Physical Oceanography*, **4**, 109–115.

Osborn T.R. (1980) Estimates of the rate of vertical diffusion from dissipation measurements. *Journal of Physical Oceanography*, **10**, 83–89.

Osborn T.R. & Cox C.S. (1972) Oceanic fine structure. *Geophysical and Astrophysical Fluid Dynamics*, **3**, 321–345.

Oxburgh R., Broecker W.S. & Wanninkhof R.H. (1991) The carbon budget of Mono Lake. *Global Biogeochemical Cycles*, **5**, 359–372.

Peixoto J.P. & Oort A.H. (1992) *Physics of Climate*. American Institute of Physics, New York.

Peng T.H. & Broecker W. (1980) Gas exchange rates for three closed basin lakes. *Limnology and Oceangraphy*, **25**, 789–796.

Quinn J.A. & Otto N.C. (1971) Carbon dioxide exchange at the air–sea interface: flux augmentation by chemical reaction. *Journal of Geophysical Research*, **76**, 1539–1548.

Rathbun R.E. (1977) Reaeration coefficients of streams – state-of-the-art. *Journal of Hydraulics Division* ASCE, **103**, 409–424.

Rathbun R.E. (1979) Estimating the gas and dye quantities for modified tracer technique measurements of stream reaeration coefficients. *USGS Water Research Investigations*, **15, WRI, 70–27**.

Rathbun R.E. & Grant R.S. (1978) Comparison of the radioactive and modified techniques for measurement of stream reaeration coefficients. *USGS Water Research Investigations and NSTL Station, Miss*, p. 57.

Reid R.C., Prausnitz J.M. & Poling B.E. (1987) *The Properties of Gases and Liquids* McGraw-Hill, New York.

Riley J.P. & Skirrow G. (1975) *Chemical Oceanography*. Academic Press, New York.

Robertson J.E. & Watson A.J. (1992) Thermal skin effect of the surface ocean and its implications for CO_2 uptake. *Nature*, **358**, 738–740.

Ryan P.J. & Harleman D.R.F. (1973) *An Analytical and Experimental Study of Transient Cooling Pond Behavior*. Technical Report 161, R.M. Parsons Laboratory for Water

Resources and Hydrodynamics, MIT, Cambridge.

Saltzman E.S., King D.B., Holmen K. & Leck C. (1993) Experimental determination of the diffusion coefficient of dimethylsulfide in water. *Journal of Geophysical Research*, **98**, 16481–16486.

Sebacher D.I., Harriss R.C. & Bartlett K.B. (1983) Methane flux across the air–water interface: air velocity effects. *Tellus*, **35B**, 103–109.

Shay T.J. & Gregg C.S. (1986) Convectively driven mixing in the upper ocean. *Journal of Physical Oceanography*, **16**, 1777–1798.

Smethie W.M., Takahashi T.T., Chipman D.W. & Ledwell J.R. (1985) Gas exchange and CO_2 flux in the tropical Atlantic Ocean determined from ^{222}Rn and pCO_2 measurements. *Journal of Geophysical Research*, **90**, 7005–7022.

Smith A.V. (1985) Physical, chemical and biological characteristics of CO_2 gas flux across the air-water interface, *Plant, Cell and Environment*, **8**, 387–398.

Soloviev A.V. & Schluessel P. (1994) Parameterization of the temperature difference across the cool skin of the ocean and of the air–ocean gas transfer on the basis of modelling surface renewal. *Journal of Physical Oceanography*, **24**, 1319–1332.

Sonin A.A., Shimko M.A. & Chun J.-H. (1986) Vapor condensation onto a turbulent liquid – 1. The steady condensation rate as a function of liquid-side turbulence. *International Journal of Heat and Mass Transfer*, **29**, 1319–1332.

Strub P.T. (1983) The response of a small lake to atmospheric forcing during fall cooling. PhD dissertation. University of California at Davis.

Stumm W. & Morgan J.J. (1981) *Aquatic Chemistry*. Wiley, New York.

Tang S. & Wu J. (1992) Suppression of wind-generated ripples by natural films: a laboratory study. *Journal of Geophysical Research*, **97**, 5301–5306.

Taylor G.I. (1935) Statistical theory of turbulence. *Proceedings of the Royal Society of London*, **A151**, 421–478.

Tennekes H. & Lumley J.L. (1972) *A First Course in Turbulence*. The MIT Press, Cambridge, Massachusetts.

Theofanous T.G. (1984) Conceptual models of gas exchange. In: Brutsaert W. & Jirka G.H. (eds) *Gas Transfer at Air–Water Surfaces*, pp. 271–282. Reidel, Boston, Massachusetts.

Theofanous T.G., Houze R.N. & Brumfield L.K. (1976) Turbulent mass transfer at free, gas–liquid interfaces, with applications to open-channel, bubble and jet flows. *International Journal of Heat and Mass Transfer*, **19**, 613–624.

Thorpe S.A. (1977) Turbulence and mixing in a Scottish loch. *Philosophical Transactions of the Royal Society of London*, **A286**, 125–181.

Torgersen T., Mathieu G., Hesslein R.H. & Broecker W.S. (1982) Gas exchange dependency on diffusion coefficient: direct ^{222}Rn and ^3He comparisons in a small lake. *Journal of Geophysical Research*, **87**, 546–556.

Tsivoglou E.C. (1967) *Tracer Measurements of Stream Reaeration*. Washington DC Federal Water Pollution Control Administration, US Department of Interior.

T.V.A. (1972) *Heat and Mass Transfer Between a Water Surface and the Atmosphere*. Division of Water Resources Research Laboratory Report, No. 14, Tennessee Valley Authority.

UNESCO (1983) *Algorithms for Computation of Fundamental Properties of Seawater*. UNESCO technical papers in marine science, No. 44.

Upstill-Goddard R.C., Watson A.J., Liss P. & Liddicoat M.I. (1990) Gas transfer in lakes measured with SF_6. *Tellus*, **42B**, 364–377.

Wallace D.W.R. & Wirick C.D. (1992) Dissolved O_2 time-series records large air–sea gas fluxes associated with breaking waves. *Nature*, **356**, 694–696.

Wanninkhof R. (1992) Relationship between wind speed and gas exchange over the ocean. *Journal of Geophysical Research*, **97**, 7373–7382.

Wanninkhof R. & Bliven L. (1991) Relationship between gas exchange, wind speed and radar backscatter in a large wind–wave tank. *Journal of Geophysical Research*, **96**, 2785–2796.

Wanninkhof R., Ledwell J.R. & Broecker W.S. (1985) Gas exchange – wind speed relationship measured with sulfur hexafluoride on a lake. *Science*, **227**, 1224–1226.

Wanninkhof R., Ledwell J.R. & Crucius J. (1991) Gas Transfer Velocities on Lakes Measured with Sulfur Hexafluoride. In: *Air Water Gas Tranfer*. Wilhelms S.C. & Gulliver J.S. (eds), pp. 441–458. American Society of Civil Engineers, New York.

Wanninkhof R., Ledwell J.R., Broecker W.S. & Hamilton M. (1987) Gas exchange on Mono Lake and Crowley Lake, California, *Journal of Geophysical Research*, **92**, 14567–14580.

Wanninkhof R., Mulholland P.J. & Elwood J.W. (1990) Gas exchange rates for a first order stream determined with deliberate and natural tracers. *Water Resources Research*, **26**, 1621–1630.

Weast R.C.W. & Astle M.J. (1980) *CRC Handbook of Chemistry and Physics*. CRC Press, Boca Raton, Florida.

Wei Y. & Wu J. (1992) *In situ* measurements of surface tension, wave damping, and wind properties modified by natural films. *Journal of Geophysical Research*, **97**, 5307–5313.

Weisman R.N. & Brutsaert W. (1973) Evaporation and cooling of a lake under unstable atmospheric conditions. *Water Resources Research*, **9**, 1242–1247.

Weiss R.F. (1974) Carbon dioxide in water and seawater: the solubility of a non-ideal gas. *Marine Chemistry*, **2**, 203–215.

Weiss R.F. & Price B.A. (1980) Nitrous oxide solubility in water and seawater. *Marine Chemistry*, **8**, 347.

Wiesenburg D.A. & Guinasso N.L. (1979) Equilibrium solubilities of methane, carbon monoxide, and hydrogen in water and sea water. *Journal of Chemical Engineering Data*, **24**, 356–360.

Wilcock R.J. (1984) Methyl chloride as a gas-tracer for measuring stream reaeration coefficients: I. Laboratory studies. *Water Research*, **18**, 47–52.

Wilcock R.J. (1988) Study of river reaeration at different flow rates. *Journal of Environmental Engineering*, **114**, 91–105.

Wilhelm E., Battino R. & Wilcock R.J. (1977) Low pressure solubility of gases in liquid water. *Chemistry Review*, **77**, 219–262.

Wilke C.R. & Chang P. (1955) Correlation of diffusion coefficients in dilute solutions. *Journal of the American Institute of Chemical Engineers*, **1**, 264–270.

Woolf D.K. & Thorpe S.A. (1991) Bubbles and the air–sea exchange of gases in near-saturation conditions. *Journal of Marine Research*, **49**, 435–466.

Yu S.L. & Hamrick J.M. (1984) Wind effects on air–water oxygen transfer in a lake. In: Brutsaert W. & Jirka G.H. (eds) *Gas Transfer at Water Surfaces*, pp. 357–368. Reidel, Boston, Massachusetts.

CHAPTER 4

Trace gas exchange in freshwater and coastal marine environments: ebullition and transport by plants

J.P. CHANTON & G.J. WHITING

4.1 Introduction

Gas exchange in freshwater and coastal marine ecosystems proceeds by the mechanisms of bubble ebullition and transport through air-filled hollow spaces/tissues (lacunae/aerenchyma) in emergent aquatic plants in addition to gas flux across the water–air interface (Fig. 4.1). Bubble ebullition is a one-way transport of gases from organic-rich reducing sediments to the atmosphere and to a lesser extent to the water column. In the process of transporting oxygen to their roots, emergent aquatic plants ventilate the sediments or water-logged soils in which they are rooted releasing reduced gases to the atmosphere. Plants play a part in the emission of CH_4, sulphur gases, ammonia and non-methane hydrocarbons (NMHC) such as isoprene and terpene. When plants are present, they often dominate CH_4 emission relative to water–air transport or ebullition. Trace gas emission by plants has recently been reviewed by Sharkey *et al.* (1991). This chapter will focus on gas ebullition and CH_4 and sulphur gas emission from emergent macrophytes and marshes. We will review the importance of and factors controlling these gas transport modes and describe techniques for their measurement.

4.2 The process of ebullition

Ebullition or bubble transport is closely tied to the production of CH_4. Methanogenesis is the terminal process of organic carbon remineralization and occurs when there is a high input of labile organic material in the absence of dissolved oxygen and alternative inorganic electron acceptors such as sulphate (Martens & Berner, 1974; Zehnder, 1978). Methane is only sparingly soluble in water (Yamamoto *et al.*, 1976). Bubbles are formed because the production of CH_4 raises the sum of the partial pressures of the dissolved gases above the hydrostatic pressure in the sediment (Chanton & Dacey, 1991). A common misunderstanding is that dissolved CH_4 must exceed the solubility of the gas in water (roughly 1 mmol at 20°C and $P_{CH_4} = 1$ atm) for bubbles to form. In fact, bubbles can form at CH_4 concentrations well below saturation. Bubble CH_4 content varies between 10 and 90% CH_4 and is at equilibrium with dissolved pore water CH_4 (Chanton *et al.*, 1989). Radon[222], a chemically

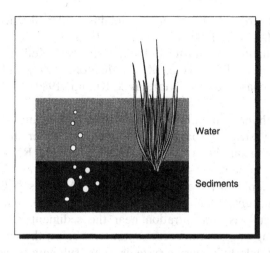

Fig. 4.1 Modes of gas transport from aquatic ecosystems are across the air–water interface, via emergent plants which serve as gas exchange conduits between reducing sediments and the atmosphere, and via bubble ebullition which is a direct transfer between the sediments and the atmosphere allowing reduced gases to bypass the mediating effects of an oxic water column.

inert radioactive gas which has been employed as a tracer of gas exchange processes, also exhibits equilibrium between the bubble and pore water dissolved phase (Martens & Chanton, 1989). Besides CH_4 and Rn, sedimentary bubbles contain 1–10% CO_2, 4–90% N_2 and >1% Ar. Sulphur gases, predominantly H_2S (13.4 µmol l^{-1} gas), but also COS (10 nmol l^{-1}), CH_3SH (0.4 nmol l^{-1}), DMS (0.4 nmol l^{-1}) and CS_2 (1.2 nmol l^{-1}) have been found in bubbles stirred from a salt-marsh tidal creek (M. Hines & J. Chanton, unpublished data). H_2S (2.8 µmol l^{-1}) was found in bubbles from Cape Lookout Bight (Chanton *et al.*, 1987).

The quantity of bubble gas held in a water-saturated sediment or soil can vary widely. Bubble gas volumes as high as 14–15 l m^{-2} have been found in subtidal freshwater sediments (Chanton *et al.*, 1989). Despite its relative insolubility, the dissolved CH_4 inventory is generally greater than the bubble inventory by a factor of 2–8 (Chanton *et al.*, 1989). These quantities shift seasonally due to temperature-induced solubility changes (gases are more soluble in cold water than in warm water).

Usually associated with some sort of perturbation, sedimentary gas bubbles are released from sediments and rise quickly through the water column to the atmosphere. Ebullition transports CH_4 and other gases directly to the atmosphere, bypassing the mediating effects of an oxygenated sediment–water interface, water column or plant rhizome. Environments where ebullition is an important mode of gas transport include lakes (Baker-Blocker *et al.*, 1977; Crill *et al.*, 1988; Keller, 1990;

Keller & Stallard, 1994), river floodplains under flooded conditions (Bartlett *et al.*, 1988, 1990; Devol *et al.*, 1988, 1990), bayous (sluggish tributaries, Chanton & Martens, 1988), estuaries (Kelley *et al.*, 1990), swamps (Happell, 1992; Happell & Chanton, 1993; Pulliam, 1993), harbours and coastal basins (Martens & Klump, 1980, 1984; Martens *et al.*, 1986).

When methane-rich bubbles are formed within and released from sediments they strip or sparge the sediments of other dissolved gases. Bubble stripping has been observed for N_2, Ar and Rn which behave conservatively in methanogenic sediments where nitrate (fueling denitrification) is depleted (Reeburgh, 1969, 1972; Martens & Berner, 1977; Kipphut & Martens, 1982; Martens & Chanton, 1989). While molecular diffusion affects gas concentration near the sediment–water interface, bubble stripping of gases can occur centimetres below the sediment–water interface, particularly in marine systems where sulphate reduction inhibits CH_4 production in surficial layers. The removal of gases from sediments by bubble stripping can be distinguished from molecular diffusion by use of radon, which has served as a particularly useful tracer in sorting out effects of the various gas exchange mechanisms (Kipphut & Martens, 1982; Martens & Chanton, 1989; Fig. 4.2).

Bubbles suffer little dissolution as they rise from the sediments through the water column. Martens & Klump (1980) found only a 15% change in bubble volume as bubbles rose through 3 m of water column. Martens & Chanton (1989) showed that bubbles which had been released naturally and had migrated through the water column were only slightly different from bubbles stirred from the sediment and immediately captured. The differences were mainly in bubble CO_2 content, which was lower by a factor of 5 in the natural bubbles. CO_2 is particularly soluble in water, and is rapidly stripped from bubbles as they migrate through the water column. H_2S can react with water by disassociation, as does CO_2, and behaves similarly (J.P. Chanton, unpublished data). Thus, except for CO_2 and H_2S content, bubbles stirred from the sediment are representative samples of bubbles reaching the atmosphere by natural ebullition.

In addition to directly transporting gases, ebullition can result in enhanced diffusion of gases and dissolved solutes across the sediment–water interface (Klump & Martens, 1981) due to the formation of tubes through which the bubbles exit the sediments. Bubble tubes can enhance diffusion by factors of up to 3 by increasing the surface area of the sediment–water interface contact from a two-dimensional plane to a three-dimensional Swiss cheese-like structure (Martens & Klump, 1980; Martens *et al.*, 1980; Chanton *et al.*, 1987). However, not all methanogenic sediments have the cohesiveness to support and maintain bubble tubes (Kelley *et al.*, 1990).

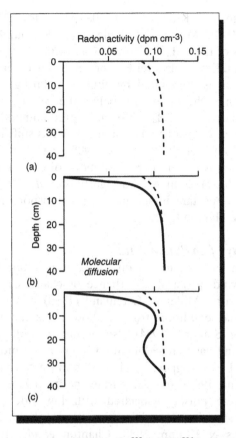

Fig. 4.2 Effects of gas transport processes on Rn^{222} (—)/Ra^{226} (----) equilibrium in sediment pore waters. (a) No transport and resulting secular equilibrium between Rn and Ra; porosity controls the vertical activity variation. (b) Molecular diffusion: radon deficiency caused by diffusion from sediments to overlying waters depletes Rn concentration near the sediment water interface. (c) Rn deficiencies caused by diffusion (surface) and gas bubble stripping due to ebullition, which can deplete gas concentrations at depth, particularly in marine systems where sulphate concentrations inhibit methane production in surficial layers. (From Martens & Chanton, 1989.)

4.2.1 Importance of ebullition in CH_4 transport

Ebullition is the dominant mode of CH_4 emission from the Amazon floodplain, accounting for 50% or more of the total surface emissions from the region (Bartlett *et al.*, 1988, 1990; Devol *et al.*, 1988, 1990). In a floodplain lake which was studied in detail, ebullition accounted for 70% of the total CH_4 flux (Crill *et al.*, 1988). In Central American lakes, ebullition is also the dominant mode of CH_4 transport (Keller, 1990). In the temperate zone, bubbling has been shown to account for 50–80% of the CH_4 transported from subtidal freshwater and marine organic-rich sediments (Martens & Klump, 1980, 1984; Chanton & Martens, 1988; Kelley *et al.*, 1990). Miller & Oremland (1988) found that ebullition transported 98% of the CH_4 flux from a California lake (Searsville Lake). In Italian rice paddies, Holzapfel-Pschorn & Seiler (1986) have observed that CH_4 emission is dominated by ebullition during the first 6 weeks after flooding when the plants are below the water surface. After the plants emerge, however, CH_4 emission is predominantly through plant stems. Schutz *et al.* (1989a,b) have shown that this shift in CH_4 transport is from over 90% via ebullition to over 90% via plant transport as rice matures. Ebullition is also an important mode of gas transport from small organic-rich tundra lakes in Alaska (Martens *et al.*, 1992) and has been suggested to be important in subsurface gas transport in some northern bogs (Fechner & Hemond, 1992).

4.2.2 Variability of ebullition rates
While ebullition may occur at any time, it is often triggered by an event such as wind, physical disturbance or changes in atmospheric or hydrostatic pressure. Miller & Oremland (1988) and Keller & Stallard (1994) found that ebullition from shallow lakes increased with wind speed. Keller & Stallard (1994) also observed ebullition triggered by currents disturbing gas-rich sediments. Changes in atmospheric pressure on the order of 1–3% triggered high ebullition rates from lake sediments (Matson & Likens, 1990). In tidal marine and freshwater areas, ebullition occurs as discrete pulses associated with low tide (Fig. 4.3) when hydrostatic pressure is at a minimum, allowing bubble expansion and release (Martens & Klump, 1980; Chanton *et al.*, 1989; Martens & Chanton, 1989). Total pressure changes over the tidal cycle are on the order of 5–7% at these sites. For ebullition associated with tidal events, one should commence measurements about 2 h before low tide, and continue measurements until bubbling stops.

In temperate areas, ebullition rates are controlled by temperature, and bubbling occurs mainly in warmer months (Martens & Klump, 1980; Chanton & Martens, 1988; Martens & Chanton, 1989). In tropical areas, ebullition rates may vary seasonally with changing water levels. As water levels fall, bubbles stored in the sediments expand and are more frequently released. Additionally, the sediments become more susceptible to perturbation from wind and other such events as water depth decreases. Keller & Stallard (1994) showed direct relationships between water depth and ebullition rate in a Panama Lake. Bartlett *et al.* (1988, 1990) observed that although ebullition was a significant component of CH_4 emissions during both rising and falling water levels on the Amazon river floodplain, the frequency of bubbling and its contribution to the total flux was lower during the period of rising water than during falling water.

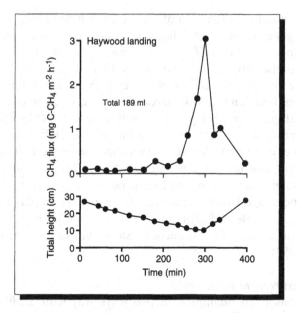

Fig. 4.3 Ebullition rate and tidal height as a function of time in the White Oak River Estuary, NC. Maximal bubble release coincided with the lowest point of the tidal cycle, when hydrostatic pressure was at a minimum. (From Chanton *et al.*, 1989.)

Increasing rates of ebullition on either spatial or temporal scales cause sedimentary bubbles to become more CH_4 rich and nitrogen poor as higher ebullition rates strip more nitrogen from pore waters. Sedimentary N_2 content is controlled by a dynamic balance between removal by bubble stripping and resupply by molecular diffusion from overlying water. Dissolved (pore water) and bubble N_2 concentrations can reflect the history of bubbling over a period of months (Martens & Chanton, 1989). Seasonal measurements have shown bubble N_2 content (Chanton *et al.*, 1989; Martens & Chanton, 1989) and pore water N_2 content (Kipphut & Martens, 1982) to drop dramatically in summer related to bubble stripping of N_2 from the sediments. When bubbling ceases in the late autumn due to reduction in CH_4 production rates (Crill & Martens, 1983) pore water and bubble N_2 concentrations increase as a result of diffusion into sediments from overlying waters. Spatial variations in bubbling rate are also indicated by bubble N_2 content (Chanton *et al.*, 1989). The relationship between bubble N_2 content and ebullition rate derived by Chanton *et al.* (1989) was applied by Martens *et al.* (1992) to estimate roughly ebullition rates from Alaska tundra lakes.

Chanton & Dacey (1991) have suggested that the relative importance of molecular diffusion and ebullition in the export of CH_4 from sediments should be a non-linear function of the CH_4 production rate. At low rates of methanogenesis, diffusion is likely to be the major process, since CH_4 is unlikely to accumulate sufficiently to generate bubbles under hydrostatic pressure. As the rate of methanogenesis increases and begins to exceed the capacity for diffusive loss, bubbles begin to form. Continued increase in the CH_4 production rate will increase the concentration of CH_4 and the rates of both gas transport processes will increase. As the CH_4 concentration approaches its saturation point for solubility in water (cf. Yamamoto *et al.*, 1976), diffusive export approaches a limit, constrained by the steepness of the concentration gradient, while ebullition continues to track the existing rate of methanogenesis. In unvegetated sediments with extremely high CH_4 production rates (e.g. Cape Lookout Bight, NC, $6\,mol\,m^{-2}\,y^{-1}$, Martens & Klump, 1984), the ebullition rate is approximately equal to the CH_4 production rate within the sediments (Crill & Martens, 1983).

4.2.3 Measurement techniques

Ebullition has been quantified in two ways: (a) with air-filled floating chambers; and (b) with submerged water-filled bubble traps or funnels. The former method consists of a chamber which floats on the surface of the water. Samples of chamber head space air are drawn periodically through either a septum or narrow diameter tube and stored as appropriate to the gas of interest for subsequent analysis. Septum-fitted chambers are slightly disadvantageous because one must closely approach the chamber to sample it. Ambient air fills the chamber at the initiation of the experiment, and the emitted gases, particularly CH_4, increase in concentration over time. Some workers have found it helpful to vent chambers after placing them on the water surface with a 'blow-out' fan held over a large diameter hole in the chamber. A second larger hole allows gas flow through the chamber. These holes are then plugged with large rubber stoppers before commencing measurements. Flushing the chamber with ambient air as described above is particularly helpful when collecting samples of chamber head space for isotopic analysis as the experiment starts with known ambient levels of CH_4.

The chamber collects gas diffusing into the chamber from the water under the chamber, in addition to bubbles which enter the chamber from below. Gas flux into the chamber from water–air transfer causes a smooth linear increase in concentration. When bubbles break into the chamber, this smooth increase is punctuated by abrupt increases in concentration as shown in Fig. 4.4 (see also Miller & Oremland, 1988). Obviously, better resolution is obtained when samples are collected frequently, generally every 3–5 min. A continuous CH_4 sampling device has also been employed, which utilizes a closed system recirculating

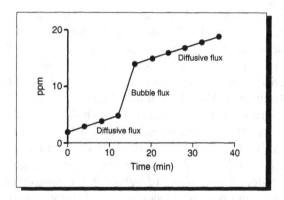

Fig. 4.4 Abrupt changes in the rate of methane increase in a chamber floating upon water indicates the capture of bubbles released from the sediments below. The smooth linear increase of methane within the chamber is supported by diffusion of methane from the water on which the chamber rests.

chamber and a gas filter correlation technique (Sebacher & Harriss, 1982). Studies which have quantified ebullition using floating chambers include Bartlett *et al.* (1988, 1990), Crill *et al.* (1988), Devol *et al.* (1988, 1990) and Wilson *et al.* (1989).

Bubble traps consist of large inverted funnels or pyramids which are suspended over the sediment from which bubble collection measurements are to be made. The top of the funnel or pyramid is fitted with a length of tubing (for remote sampling) attached to a valve. The traps and tubing are filled with water at the initiation of the experiment and as bubbles are trapped they displace the water. The traps may be sampled periodically by a syringe fitted to the other end of the tubing. The syringe can be used to measure bubble volume and retain gas samples for later measurements of gas concentration using chromatographic techniques.

Keller & Stallard (1994) used arrays of 26 cm diameter plastic funnels weighted at the bottom with fishing weights and a foam float glued to the neck for buoyancy. Funnels with long narrow necks were used to minimize CH_4 dissolution. A cannula pierced flush to the bottom of a rubber stopper was fitted into the funnel top. A plastic clip was tied to each funnel and several funnels were attached to guidelines placed in the lake water. Bubble gas was sampled every 2 hours. Strayer & Tiedje (1978) used similar funnels anchored at a depth of 5 m. An inverted graduated cylinder was attached over the funnel stem to collect bubbles. Gas bubble volume was measured and samples were collected at weekly intervals. Baker-Blocker *et al.* (1977) used a similar arrangement and collected samples at intervals which ranged from 3 days to 2 weeks.

Martens & Klump (1980) built a rigid acrylic pyramid with a 1 m² base area which could be lowered into the water column and suspended over the sediments from a boat. The tip of the pyramid was fitted with 3 mm polypropylene tubing which could be sampled via syringes. Samples were collected from the bubble chamber over 10–60 min intervals. These sampling devices were used in subsequent studies by Martens *et al.* (1986), Burke *et al.* (1988), Chanton & Martens (1988), Chanton *et al.* (1989) and Martens & Chanton (1989). In shallow areas, the samplers were floated over the sediments and restrained by lines tied to shore or stakes. Currently Martens and co-workers have constructed an automated ebullition sample collector. This device can be left in the field and programmed to collect a time series of bubble flux measurements, retaining samples for later analysis (W. Ussler, personal communication, 1992).

The frequency at which bubble traps have been sampled has varied from intervals of 10 min (Martens & Klump, 1980) to several weeks (Baker-Blocker *et al.*, 1977). Keller & Stallard (1994) conducted a field experiment to examine sample integrity in bubble traps. After 2 h less than 7% of the total gas volume was lost. CH_4 concentrations actually increased slightly within the traps because other gases, especially CO_2, dissolved more rapidly than CH_4. The authors report that less than 3% of the trapped CH_4 was lost after 2 h. Chanton *et al.* (1989) report a similar experiment conducted in the laboratory. After 4 h CH_4-rich bubbles held in a funnel over stirred water lost $7 \pm 3\%$ of their CH_4 content and $4 \pm 3\%$ of their volume. The upper limit of time which bubbles can be left in bubble traps is undetermined. CH_4 oxidation could become a factor in longer deployments. In addition to volume, it is important to measure the CH_4 concentration in gas trapped in bubble collectors. If the bubble collectors are moored on the surface of the water and become heated by the sun, the heated water within the collector may degas, diluting CH_4 concentrations.

Keller & Stallard (1994) performed a direct comparison between bubble traps and floating chambers as techniques for measuring ebullition. While the two methods agree well, there was more variation associated with the floating chamber technique, primarily because it was used for 15–20 min time intervals while the traps were left out for 2 h.

For measurements of CH_4 or CO_2, plastic syringes fitted with three-way stopcocks are appropriate sample containers. While it is best to analyse samples as soon as possible after collection, Crill *et al.* (1988) found that these syringes could hold 5 ppm CH_4 for over 2 days with less than 1% loss of CH_4. Samples with higher CH_4 concentrations may have a shorter shelf life. When working with CH_4, separate sets of syringes should be used for samples of differing concentrations (e.g. from low to high), ambient air, flux chamber, water column and pore water/bubble. Between use, the syringes should be disassembled, rinsed with water if necessary, and stored apart so that interior plastic and rubber plunger tips can equilibrate with ambient air before reuse (Crill *et al.*, 1988). Care should be taken not to mix the plungers between syringe types (e.g. pore water syringes and flux syringes). We have found it useful to mark syringes and plungers with different coloured label tape to distinguish them. Some investigators have also used evacuated containers to store samples. Evacuated serum vials with butyl rubber stoppers (Bellco, Vineland, New Jersey, part 2048-11800) work well. When transferring a sample to the vial via syringe and needle, pressurize it to 2 atm, if possible. When running the sample, connect a needle directly to the gas chromatograph sample loop, pierce the septum and the sample can be introduced directly to the instrument.

4.3 The role of plants in gas exchange in aquatic ecosystems

Gas transport is important for aquatic macrophytes, for their lower parts live in a water-saturated, often organic-rich anaerobic environment (Hook & Crawford, 1978). Because of these anaerobic conditions, the plants themselves must provide oxygen below ground to fuel root respiration (Armstrong, 1978, 1979). A byproduct of this gas circulation is the transport to the atmosphere of reducing gases from the sediments, most notably CH_4 (Chanton & Dacey, 1991; Schutz *et al.*, 1991). Plants are effective transporters of CH_4 to the atmosphere because they allow CH_4 produced within the anaerobic sediment to bypass the sediment–water interface, which is usually a zone of nearly quantitative aerobic CH_4 oxidation (King, 1990; Oremland & Culbertson, 1992; Happell *et al.*, 1993).

Environments populated by emergent plants include marshes, swamps, lakes and river tributaries. Emergent plants populate water depths that range from 0 to roughly 2 m. Although submersed plants are known to aerate the sediments of their root zone or rhizosphere, which is the area of soil that surrounds and is influenced by the roots (Sand-Jensen *et al.*, 1982; Carpenter *et al.*, 1983), emergent macrophytes are more important for gas transport between the atmosphere and water-logged soils and sediments. The higher proficiency for gas transport of emergent plants relative to submersed ones is seen in the zonation of their habitats: emergent macrophytes can tolerate sediments with higher oxygen demands than submersed macrophytes (Smits *et al.*, 1990).

When present, emergent aquatic plants dominate gas exchange. The effect of aquatic plants as gas conduits is well documented in rice fields, where it has been shown that the plants transport over 90% of the total CH_4 flux (Holzapfel-Pschorn *et al.*, 1985, 1986; Holzapfel-Pschorn & Seiler, 1986; Schutz *et al.*, 1989a,b). Similar results have been found in natural wetlands (Whiting *et al.*, 1991; Bartlett *et al.*, 1992; Chanton *et al.*, 1992a,b; Morrissey & Livingston, 1992; Happell *et al.*, 1993). It has been hypothesized that emergent plants can actually suppress ebullition (Chanton *et al.*, 1989; Chanton & Dacey, 1991).

4.3.1 Mechanisms of gas transport
To facilitate gas transport, aquatic plants have developed extensive intercellular gas spaces called lacunae in their stems and roots. These spaces are interconnected and can account for up to 70% of the total root volume in some plants (Dacey, 1981a; Schutz *et al.*, 1991). Sebacher *et al.* (1985) have shown photographs of typical stem cross-sections. Some stem lacunae are continuous and unrestricted from leaf to rhizome, particularly in the lilies (*Nymphaea*, *Nuphar* and *Nelumbo*). Other stems or rhizomes have diaphragms, septae or membrane partitions which may but do not necessarily impede gas flow (e.g. *Phragmites*, Armstrong & Armstrong, 1988, 1991). Other plants have stems filled with spongy pith (*Peltandra*, *Pontederia*, *Sagittaria*), which impedes gas flow (Frye, 1989; Harden, 1992). Ease of gas flow down plant stems can easily be quantified by attaching the stem via tubing to the top of a water-filled pipette and timing the rate at which water flows out of the pipette as described in Conway (1987), Yamasaki (1984), Harden & Chanton (1994).

Gases may diffuse through these hollow or pith-filled tubular spaces passively following Fick's law (concentration gradient dependence) (Barber *et al.*, 1962; Armstrong, 1978, 1979), or gases may be advected (bulk flow) by several mechanisms (Dacey, 1981b; Schutz *et al.*, 1991). Non-through-flow convection (one-way flow from atmosphere to sediments or vice versa) can be driven by tidal pumping, and below-ground gas consumption or production. Since O_2 is consumed in the roots, producing CO_2, subsequent dissolution of that CO_2 in pore waters can cause a pressure deficit of the order of 20%, the concentration of O_2 in air. This pressure deficit can cause gas to flow towards the roots (non-through-flow convection, Raskin & Kende, 1983, 1985). However, not all pore waters are low in CO_2; some are extremely CO_2 rich, which can actually cause gas flow from the roots to the atmosphere (Koncalova *et al.*, 1988). Beckett *et al.* (1988) have challenged the importance of this type of non-through-flow convection.

Convective through-flow (pressurized ventilation, flow from atmosphere through rhizome and back to the atmosphere) is driven by the process of thermal osmosis (Dacey, 1981a, 1987; Schutz *et al.*, 1991) and the evaporation of water (Leuning, 1983). In water lilies and the reeds *Typha* and *Phragmites*, temperature differences between leaves and air driven by sunlight cause gas to enter young leaves and flow downward to flush the rhizosphere and exit out of older leaky leaves (Dacey, 1981a,b;

Sebacher *et al.*, 1985; Armstrong & Armstrong, 1991; Grosse *et al.*, 1991; Schutz *et al.*, 1991). Plants with this pressurized ventilation system transport gases four to five times faster in the light than in darkness, at which time transport is only via molecular diffusion (Grosse & Schroder, 1984, 1985; Grosse & Mevi-Schutz, 1987; Mevi-Schutz & Grosse, 1988).

4.3.2 CH$_4$ emission from marshes and vegetated wetlands

Plants affect CH$_4$ emissions in several ways: they serve as conduits for gas exchange, they transport O$_2$ to the rhizosphere and foster below-ground aerobic CH$_4$ oxidation, and they stimulate methanogenesis by above and below ground organic carbon production and root exudation. The supply of O$_2$ for below ground CH$_4$ oxidation attenuates CH$_4$ emission while the conduit and production effects facilitate and enhance emission. Estimates of CH$_4$ oxidation in the rhizosphere vary widely, from a maximum of 80 to 90% of CH$_4$ produced below ground (Holzapfel-Pschorn *et al.*, 1985, 1986; Schutz *et al.*, 1989b, 1991; Sass *et al.*, 1990) to less significant amounts (deBont *et al.*, 1978; King *et al.*, 1990; Chanton *et al.*, 1992b; Gerard, 1992; Happell *et al.*, 1993). Evidently the two positive factors (conduit effect and production) often outweight the negative factor (oxidation), for positive linear relationships between CH$_4$ emission and plant biomass and photosynthetic activity have been observed in *Cladium*, *Carex* and rice (Fig. 4.5, Whiting *et al.*, 1991; Whiting & Chanton, 1992; Happell *et al.*, 1993; G.J. Whiting & J.P. Chanton, unpublished data).

An additional manner in which plants may affect below ground gas reservoirs arises from plants' use of water (Chanton & Dacey, 1991). Transpirational water release from wetlands can be as high as several litres per square metre per day (Dacey & Howes, 1984). Root uptake of water may draw overlying water into the sediment, delivering oxidants such as oxygen or nitrate or, if water levels fall, air may enter the sediments. High transpiration rates can lower water levels in wetlands causing CH$_4$-emitting swamps to shift from CH$_4$ production to consumption of atmospheric CH$_4$ (Harriss *et al.*, 1982; Happell & Chanton, 1993; Happell *et al.*, 1993).

4.3.3 Measurement techniques

The most simple technique for the quantification of CH$_4$ emission from plants is to use chambers. A portable temperature and CO$_2$-controlled light-transparent chamber ('phytochamber') has recently been developed by G.J. Whiting (Bartlett *et al.*, 1990; Whiting *et al.*, 1991, 1992; Chanton *et al.*, 1992b; Whiting & Chanton, 1992). The closed-system design is made up of three major parts (Fig. 4.6): (a) sensor; (b) the chamber; and (c) a climate control system. Chamber air temperature, relative humidity, incident photosynthetically active radiation (PAR) and CO$_2$ concentra-

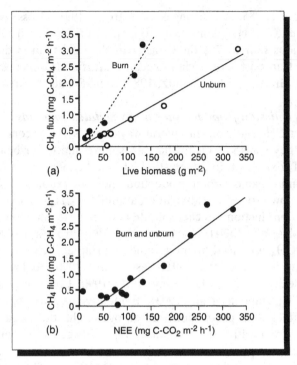

Fig. 4.5 (a) Linear relationships between plant biomass and methane emission at two sites (burned and unburned) in the Everglades. Independent linear relationships were found between methane flux and biomass at each site. (b) Methane data from these two sites converge when plotted versus net ecosystem exchange (NEE), of CO_2 which attest to the strength of this variable as a factor controlling methane emission rates from vegetated wetlands. (From Whiting *et al.*, 1991.)

tion are measured by a LICOR Model 6200 portable photosynthesis system (LI-COR, Inc., Lincoln, Nebraska, USA). Different size chambers with bases ranging from 0.4 to 1 m^2 have been employed; the smaller size is much easier to handle. The chambers with the 0.4 m^2 base consisted of stackable units which were sealed together with closed cell foam and clamps. Plant heights of up to 3 m can be covered. Shorter chambers fitted with Styrofoam bars can be floated on the surface of the swamp flood water or chambers can be set upon the sediment surface placed upon aluminium frames or collars. All sides of the aluminium-framed chamber are clear; three sides and the top are covered with transparent (~90% PAR transmission) Teflon film; the remaining side is composed of clear polycarbonate sheeting (Lexan) upon which external components are mounted. An alternative design is built completely out of plexiglas or polycarbonate (the latter is stronger). This type of chamber works equally

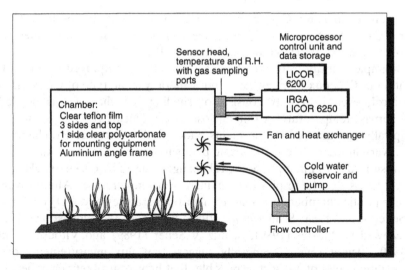

Fig. 4.6 A temperature- and CO_2-controlled light-transparent chamber suitable for measurements of CO_2 and CH_4 exchange from aquatic macrophytes.

well but is heavier and lacks flexible sides which can dampen temperature-induced pressure fluctuations.

Temperature within the closed-system chamber is maintained within 1°C of outside ambient temperature by pumping chilled water to a heat exchanger (automobile heater radiator) mounted within the chamber and blowing air across it with a 12 V brushless fan(s). Air temperature is regulated inside the chamber by controlling the flow of cold water pumped with a 12 V bilge pump from the ice-water reservoir through the heat exchanger. Air within the chambers is mixed by mounting several of these fans at strategic locations within the chamber.

Once the chamber is deployed onto a collar placed into the sediments several days previously, it is vented for periods of about 30 min using a fan mounted onto a 10 cm hole cut into the chamber, or by removing the chamber top. An additional 10 cm hole is present to allow air to escape. After this equilibration period, the chamber is closed, and measurements of CH_4 emission commenced. Gas samples for CH_4 analysis are collected in syringes at 2–5 min intervals for periods of 20–60 min, after which the chambers are opened, vented, closed and measurements commenced again. Smaller chambers without temperature control can be employed to quantify CH_4 release from areas between the emergent plants. Inverted 4 litre paint cans work well for this purpose.

CO_2 concentration is monitored within the chamber with the LICOR apparatus. The chamber system described above was originally developed for making CO_2 exchange measurements (Whiting *et al.*, 1992), but it can be run at constant CO_2 concentration by the addition of tank CO_2, or CO_2 exchange and CH_4 exchange can be measured simultaneously (Whiting *et al.*, 1991; Whiting & Chanton, 1992).

Simpler variations of the above system may be employed for quantification of CH_4 flux in many cases, but generally some type of temperature control, even if it only consists of shading the chamber, is needed. Otherwise temperatures within enclosures can quickly reach extremes not typically encountered by the plants under study. Uncontrolled chambers can vary in their effects. Enclosing plants in opaque chambers may affect emissions due to plant physiological changes caused by darkness, elevated CO_2 levels due to respiration, or increased temperature. Alternatively, clear plastic chambers can lead to high temperatures and CO_2 depletion (Seiler *et al.*, 1984; Schutz *et al.*, 1989a,b). These techniques have been reviewed by Mosier (1989), Schutz & Seiler (1989) and Vitousek *et al.* (1989). These reviews generally suggest that flux measurements over vegetation be kept as short as possible, that light-transparent chambers be used, that fans be used to mix chamber air, that large chamber volumes be used to minimize the impact of increasing concentration on the reduction of natural concentration gradients, and that collars or chamber bases be used to reduce the impact of installing chambers.

For some plants, darkness and CO_2 enrichment or depletion do not appear to affect emissions on a short-term basis. Chanton *et al.* (1992b) contrasted CH_4 emission rates from *Peltandra virginica* in a series of hourly manipulations of light intensity, photosynthetic rate and CO_2 concentration within the phytochamber described above. No systematic variations of CH_4 flux could be attributed to any of these manipulations. The authors concluded that short-term changes in environmental parameters can be tolerated when employing simple uncontrolled chambers over *Peltandra*. They suggested that this conclusion may be extended to other similar plants which do not employ pressurized ventilation (convective through-flow, Chanton & Dacey, 1991) such as *Pontederia*, *Sagittaria* (Sebacher *et al.*, 1985), *Carex* (Koncalova *et al.*, 1988) and rice (Lee *et al.*, 1981; Seiler *et al.*, 1984). However, their results cannot be applied to plants utilizing pressurized ventilation systems. These include many species of the water lily family: *Nymphaeaceae* (Dacey & Klug, 1979, 1982; Dacey 1981a,b; Grosse & Mevi-Schutz, 1987), *Nelumbo* (Mevi-Schutz & Grosse, 1988), *Phragmites* (Armstrong & Armstrong, 1991), *Typha* (Sebacher *et al.*, 1985), *Alnus glutinosa* (Grosse & Schroder, 1984, 1985) and *Nymphoides* (Grosse *et al.* 1991). Since these plants require sunlight to drive their gas circulation systems, it is likely that dark chambers would alter their rates of CH_4 emission. Effects of reduced CO_2 levels which can be caused by uncontrolled clear chambers have also not been investigated in pressurized plants.

Methane efflux from plants employing molecular diffusion may not be affected by changes in CO_2 concentration or illumination because methane release from many plants has been hypothesized to be independent of the stomatal aperture of the plant. Lack of stomatal control of CH_4 flux has been inferred by Seiler *et al.* (1984), who showed that altering CO_2 partial pressures within chambers to near the compensation point, 100 ppm, or raising them to 40 000 ppm did not affect CH_4 efflux from rice plants. Nouchi *et al.* (1990) showed that changes from light to darkness and changes in transpiration rate had only an instantaneous effect on the rate of CH_4 emission from rice, which then returned to the rate of emission occurring before the manipulation.

The reason for lack of stomatal control of CH_4 release is that the location from which plants emit CH_4 is not always associated with the leaves. In rice, CH_4 is emitted at the culm, which is an aggregation of leaf sheaths, but not from the leaf blade. Micropores, which are different from stomata, were found in the epidermis of the rice leaf sheath (Nouchi *et al.*, 1990). Results from tracer experiments suggest that the petioles of *Peltandra* are longitudinally resistant to gas flow, and laterally leaky (Frye, 1989). Frye (1989) suggested that CH_4 release from *Peltandra* occurs through pores in the petiole rather than through the leaves. Harden (1992) has found similar results for *Pontederia* and *Sagittaria*. These pores are important for delivery of O_2 to below ground organs. Armstrong (1978) reports that root systems aeration is little affected by blocking lenticels lying a few centimetres above the waterline. However, if the lenticels at or immediately above the water line are blocked, oxygen levels within roots drop considerably.

A study critical of the use of chambers for measuring CH_4 emissions from *Typha* has recently been published by Knapp & Yavitt (1992). By using bags to cover portions of leaves they concluded that chambers effects on stomatal conductance can affect CH_4 emissions, and that chamber incubation times longer than several minutes underestimate CH_4 emissions from plants because they reduce the concentration gradient between the plant internal gas and the atmosphere. However, in an experiment conducted in the Everglades, the rate of CH_4 increase in closed system phytochambers placed over whole canopies of *Typha* did not flatten or 'tail off' due to reduction in the CH_4 concentration gradient between the plant and the outside air (Fig. 4.7, Chanton *et al.*, 1993). Rates of CH_4 concentration increase within chambers did not show any variation as absolute concentration within the chamber varied over an order of magnitude. Additionally, Knapp & Yavitt (1992) observed reduced stomatal conductance and increasing concentrations of CH_4 within shaded leaves of *Typha*, and suggested that this was due to stomatal control of CH_4 emission. Morrissey *et al.* (1993) have suggested

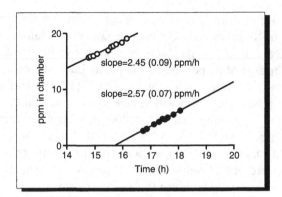

Fig. 4.7 Rates of methane concentration increase within a chamber placed over *Typha* did not show any variation as absolute concentration within the chamber varied over an order of magnitude illustrating that methane emission rates can be measured using closed-system chambers without causing excessive changes in the plant–atmosphere concentration gradient. This experiment was performed by measuring the methane emission rate from a *Typha* plot after the chamber had been closed and in place for several minutes accumulating methane. The chamber was then vented, closed and the flux measurement repeated.

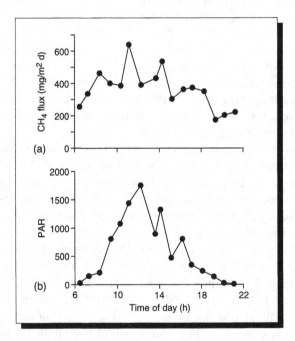

Fig. 4.8 (a) Methane emission rates from *Typha*, a plant utilizing pressurized through-flow convection, are affected by the variations of solar inolation (PAR, photosynthetically active radiation, $\mu mol\,m^{-2}\,s^{-1}$, (b) caused by clouds (J.P. Chanton & J.D. Happell, unpublished data). Measurements were made in a temperature and CO_2-controlled light-transparent chamber. Dacey (1981a) has shown variations of similar magnitude in internal pressurization within the yellow water lily associated with the passage of clouds.

similar effects in *Carex*. An alternative explanation may be that the shading reduced pressurized convective through-flow within *Typha* and CH_4 concentrations within the plants increased due to the inhibition of bulk flow gas transport (Chanton *et al.*, 1993). Whatever the cause, short-term variations in solar insolation are associated with reduced CH_4 emission from *Typha* (Fig. 4.8).

4.3.4 Sulphur gases from marshes and vegetated wetlands

Recent reviews of the biogenic sulphur cycle include Bates *et al.* (1992) and Saltzman & Cooper (1989). Giblin & Wieder (1992) have reviewed sulphur cycling in saline and freshwater wetlands. DMS and H_2S dominate gas emissions from marshes and tidal mudflats, although MeSH, COS, CS_2 and DMDS are also emitted (Jorgensen & Okholm-Hansen, 1985; Cooper *et al.*, 1987a,b,c; Goldan *et al.*, 1987; Lamb *et al.*, 1987; Hines, 1993). Fluxes of S gases from coastal wetlands are 10–100 times higher than sulphur fluxes from the ocean, but their areal extent is limited (Cooper *et al.*, 1987a; Wakeham & Dacey, 1989). DMS flux is related to plant physiological processes, while H_2S flux is related to sediment chemistry (Jorgensen & Okholm-Hansen, 1985; Dacey *et al.*, 1987).

DMS flux varies as a function of plant type and biomass, temperature, season, time of day, water inundation and osmotic stress (Cooper *et al.*, 1987a,c; Lamb *et al.*, 1987; de Mello *et al.*, 1987; Morrison & Hines, 1990). Diel variations in S gas exchange are correlated with temperature and can be as great as an order of magnitude (de Mello *et al.*, 1987; Cooper *et al.*, 1987a,b). Drier sites have been observed to yield higher emission rates than wetter ones. Positive relationships between *Spartina alterniflora* live above ground and root biomass and DMS emission have been observed by Morrison & Hines (1990) and de Mello *et al.* (1987). Cooper *et al.* (1987a) observed a positive relationship between DMS emission and plant biomass and a negative relationship between biomass and H_2S emission in plots of *Disticlis spicata*. H_2S emission is related to seasonality, water depth, temperature and the tidal cycle. Due to tidal pumping, Cooper *et al.* (1987a) found 90% of H_2S emission in a *Spartina alterniflora* marsh to occur only during 3% of the tidal cycle in a narrow region along the water's edge. Oxidation can be an important control of H_2S emission; Ingvorsen & Jorgensen (1982) reported seasonal variations in H_2S flux from an intertidal sediment on the order of 1000 to 10 000, while the microbial sulphide production rate only varied by a factor of 20.

DMS is a degradation product of the osmoregulant dimethylsulphonio-proprionate (DMSP; Dacey & Blough, 1987) which is found in the tissues of *Spartina alterniflora*, *Spartina anglica* and *Zostera marina* (Dacey *et al.*, 1987, and references therein). DMS fluxes have been found to be substantially higher in *Spartina alterniflora* than in plants which do not contain DMSP, e.g. *Spartina patens*, *Juncus roemerianus*, *Distichlis spicata*, *Avicennia germinans*, *Batis maritima* and *Cladium jamaicense* (Cooper *et al.*, 1987b; Morrison & Hines, 1990). Because DMS is a product of plant physiological processes, as opposed to sedimentary processes (like H_2S which is emitted from the sediment surface), it has been argued (Dacey *et al.*, 1987; Wakeham & Dacey, 1989) that DMS emission rates may be particularly sensitive to chamber effects which could stress plants. Since DMSP plays a part in osmotic regulation, its chemistry may be influenced by short-term changes in heat and leaf water balance. DMS appears to be released from the leaves themselves, while the sediments act as a sink for DMS (Dacey *et al.*, 1987). Therefore, it is particularly important to minimize impact to plants when measuring S emissions from wetlands.

Enhancement of DMS emission and production has been observed with plant ingestion by animals. When phytoplankton are consumed by zooplankton, DMS is released into marine waters (Dacey & Wakeham, 1986). In plots of intertidal Alaskan *Carex* which were actively grazed by geese, DMS emissions were doubled when goose faeces were left within the flux chamber (Hines & Morrison, 1992). Hines *et al.* (1993) observed higher emission in Everglades mangrove areas which were associated with fiddler crab burrows. These latter results suggest that some plants may contain small quantities of a DMS precursor other than DMSP.

Information on emissions from freshwater wetlands has been reviewed by Hines (1993). Generally emissions are fairly low (e.g. Alaskan tundra, Hines & Morrison, 1992; Everglades sawgrass, Cooper *et al.*, 1987b; Hines *et al.*, 1993; cattails, Goldan *et al.*, 1987), but values may be as high as those observed in marine wetlands that contain plants which do not produce DMSP (Hines, 1993; Hines *et al.*, 1993). Emission rates from freshwater environments and terrestrial ecosystems may be related to atmospheric inputs. Relatively high rates of DMS emission have been observed in bogs and fens in New Hampshire and Ontario (Nriagu *et al.*, 1987; Hines, 1993). Further results on plants and S emissions may be found in Fall *et al.* (1988), Kesselmeier (1991) and Rennenberg (1991). Richards *et al.* (1991) report S gas loss to the atmosphere from lakes of the Canadian shield.

Methods for the determination of sulphur emissions from wetlands include dynamic (flow-through) and static chambers and micrometeorological techniques. The micrometeorological technique requires measurement of a sulphur gas gradient in the atmosphere and eddy correlation measurement of another parameter (e.g. water vapour), since sulphur detectors do not currently have fast enough response to directly obtain sulphur fluxes. (See Lenschow, Chapter 5, for more information on micrometeorological techniques.) For sulphur gases this technique is subject to extremely high uncertainties (Bates et al., 1992).

The dynamic method consists of a chamber placed over the surface to capture gases of interest. The chamber is swept at rates of roughly $1\,min^{-1}$ with a carrier gas which may consist of tank air free of S gases (Goldan et al., 1987; Morrison & Hines, 1990; Hines et al., 1993), purified ambient air (Adams et al., 1981; Carroll et al., 1986; Cooper et al., 1987a,b,c; Lamb et al., 1987), or ambient air (Stuedler & Peterson, 1985). The flux rate is determined as the difference in gas concentration between the inflow and effluent gas streams:

Flux = (output concentration − input concentration) ∗ flow/area of chamber.

Use of sulphur-free sweep gas increases sensitivity, since the difference between the streams is maximized, and allows greater temporal resolution of flux since only the output gas stream need be sampled. However, atmospheric conditions within dynamic sulphur-free enclosures may alter the natural sulphur gas exchange between the surface and the atmosphere. Castro & Galloway (1991) found significantly different results for COS in comparisons of chambers utilizing sulphur-free sweep air and sweep air containing ambient levels of S gases. The forest soils they studied were a sink for atmospheric COS when ambient levels were passed through the chamber. However, these same soils released COS when S-free air was used as a sweep gas. In Alaskan tundra environments, Hines & Morrison (1992) noted similar effects for COS in comparisons of dynamic chambers utilizing S-free sweep air and static chambers which enclosed ambient air. They suggested that the tundra soil was a COS source while overlying vegetation consumed COS. Vegetation has been postulated to be a major sink for atmospheric COS (Goldan et al., 1988).

de Mello (1992) and de Mello & Hines (1994) compared static (closed) and dynamic chambers of S gas fluxes in *Spahnum* wetlands and found good agreement between these techniques for DMS which is emitted from these sites. However, the dynamic chamber with S-free sweep air was not suitable for measuring COS exchange. de Mello (1992) also found that the removal of vegetation eliminated the COS sink and COS was then emitted.

Detailed descriptions of dynamic chamber systems can be found in Livingston & Hutchinson, Chapter 2 and in Goldan et al. (1987), Morrison & Hines (1990) and Castro & Galloway (1991). A particularly interesting variation on these designs was employed by Stuedler & Peterson (1985). Because of the reactive nature of these gases, all chambers and sampling lines are constructed of FEP or PFA Teflon film and tubing and Teflon must be coated over any aluminium surfaces

(Goldan *et al.*, 1987). Generally, chambers should be swept at a rate sufficient to turn over the chamber atmosphere every 10–15 min. Often the sweep air is introduced through some sort of dispersion device. The presence or absence of CO_2 in the sweep air, which might potentially affect plant stomatal aperture, did not appear to affect short-term DMS emission rates in preliminary experiments by Morrison & Hines (1990). Samples of chamber air are generally passed through detachable cryogenic loop traps after drying the air stream by passing it through a trap in contact with dry ice. To prevent condensation of oxygen in the cryogenic traps immersed in liquid N_2, pressure within the traps can be held under 0.5 atm using mass flow controllers and a vacuum pump (Morrison & Hines). Other investigators have resorted to liquid argon or oxygen for their cryogen to avoid this problem (Cooper *et al.*, 1987a,c; Hines *et al.*, 1993) but these cryogens are expensive and dangerous respectively. Samples of sulphur gases can be held within sample loop traps in the cryogen for at least 8 h (Morrision & Hines, 1990). For analysis, the loops are put into the carrier flow of a gas chromatograph and heated. The sulphur compounds are then separated and quantified on a gas chromatograph with a flame photometric detector.

4.4 Scaling up

The use of chamber technology has often been used to define initially the roles of various ecosystems in gas exchange (Bartlett & Harriss, 1993). However, it is clear that emissions can vary substantially on a variety of scales, both within and among specific source types. In a paper on CH_4 emissions from the Florida Everglades, Bartlett *et al.* (1989) examined the spatial variability within major vegetation associations and examined possible relevant controlling environmental factors. They concluded that by utilization of the mapping capability of high resolution remote sensing data, an inventory of vegetation type could be used as an empirical indicator of flux and thus provide a means of estimating regional emissions. Hines *et al.* (1992) used spectral data from a scene from the Landsat thematic mapper to map habitats in the Everglades. Six vegetation categories were identified and S gas emissions were extrapolated for the entire Everglades National Park. In a study of CH_4 emission, Whalen & Reeburgh (1990) grouped their flux sites into specific habitats based upon topography and plant assemblages. The areal extent of each habitat was estimated from a latitudinal transect, and spatial extrapolations prepared. Because of the limitation of chamber-based measurements in providing large-scale regional estimates, ground-based tower and aircraft micrometeorological techniques have been employed (NASA, GTE-ABLE-3A and -3B). Chamber measurements compare favourably to these more areal extensive techniques (CO_2, Whiting *et al.*,

1992; CH_4, Bartlett, 1989). However, on a global scale even these techniques cannot suffice. Satellite-based remote sensing and other mapping techniques may provide estimates of areal coverage of particular types of wetlands but what is needed is a better understanding of the factors controlling trace gas emissions which act universally across ecosystems. Possibly, the best means of determining these relationships is to manipulate systems experimentally both in the field and laboratory. In such manipulations, chambers can be an important examination tool.

Acknowledgements

We thank Cheryl Kelley, Mark Hines, Rik Wanninkhof and Sally MacIntyre and an anonymous reviewer for comments on the manuscript. Amy Monroe, Pat Klein and Patricia Washington assisted with preparation.

References

Adams D.F., Farwell S.O., Pack M.R., Robinson E. & Bamesberger W.L. (1981) Biogenic sulfur source strengths. *Environmental Science and Technology*, **15**, 1493–1498.

Armstrong J. & Armstrong W. (1988) *Phragmites australis* – A preliminary study of soil oxidizing sites and internal gas transport pathways. *New Phytologist*, **108**, 373–382.

Armstrong J. & Armstrong W. (1991) A convective through-flow of gases in *Phragmites australis* (Cav.). *Trin. Ex. Steud. Aquatic Botany*, **39**, 75–88.

Armstrong W. (1978) Root aeration in the wetland condition. In: Hook D.D. & Crawford R.M.M. (eds) *Plant Life in Anaerobic Environments*, pp. 269–298. Ann Arbor Science, Ann Arbor, Michigan.

Armstrong W. (1979) Aeration in higher plants. In: Woolhouse H.W. (ed) *Advances in Botanical Research*, pp. 226–333. Academic Press, New York.

Baker-Blocker A., Donahue T. & Mancy K.H. (1977) Methane flux from wetland areas. *Tellus*, **29**, 245–250.

Barber D.A., Ebert M. & Evans N.T.S. (1962) The movement of ^{15}O through barley and rice plants. *Journal of Experimental Botany*, **13**, 397–403.

Bartlett D.S. (1989) The Yukon–Kuskokwim delta of Alaska: Characteristics of the regional biogeochemical environment. *EOS (American Geophysical Union Transactions)*, **70**, 284.

Bartlett D.S., Bartlett K.B., Hartmen J.M. *et al.* (1989) Methane emissions from the Florida Everglades: Patterns of variability in a regional wetland ecosystem. *Global Biogeochemical Cycles*, **3**, 363–374.

Bartlett K.B. & Harriss R.C. (1993) Review and assessment of methane emissions from wetlands. *Chemosphere*, **26**, 261–320.

Bartlett K.B., Crill P.M., Bonassi J.A., Richey J.E. & Harriss R.C. (1990) Methane flux from the Amazon River floodplain: Emissions during rising water. *Journal of Geophysical Research*, **95**, 16773–16778.

Bartlett K.B., Crill P.M., Sass R.L., Harriss R.C. & Dise N.B. (1992) Methane emissions from tundra environments in the Yukon–Kuskokwim Delta, Alaska. *Journal of Geophysical Research*, **97**, 16645–16660.

Bartlett K.B., Crill P.M., Sebacher D.I., Harriss R.C., Wilson J.O. & Melack J.M. (1988) Methane flux from the Central Amazonian Floodplain. *Journal of Geophysical Research*, **93**, 1571–1582.

Bates T.S., Lamb B.K., Guenther A., Dignon J. & Stoiber R.E. (1992) Sulfur emissions to the atmosphere from natural sources. *Journal of Atmospheric Chemistry*, **14**, 315–337.

Beckett P.M., Armstrong W., Justin S.H. & Armstrong J. (1988) On the relative importance of convective and diffusive gas flows in plant aeration. *New Phytologist*, **110**, 463–468.

Burke R.A., Martens C.S. & Sackett W.M. (1988) Seasonal variations of D/H and ^{13}C ratios of microbial methane in surface sediments. *Nature (London)*, **332**, 829–831.

Carpenter S.H., Elser J.J. & Olson K.M. (1983) Effects of roots of *myriophyllum verticillatum* on sediment redox conditions. *Aquatic Botany*, **17**, 243–249.

Carroll M., Heidt L., Cicerone R. & Prinn R. (1986) OCS, H_2S, and CS_2 fluxes from a salt water marsh. *Journal of Atmospheric Chemisty*, **4**, 375–395.

Castro M.S. & Galloway J.N. (1991) A comparison of sulfur-free and ambient air enclosure techniques for measuring the exchange of reduced sulfur gases between soils and the atmosphere. *Journal of Geophysical Research*, **96**, 15427–15437.

Chanton J.P. & Dacey J.W.H. (1991) Effects of vegetation on methane flux, reservoirs, and carbon isotopic composition. In: Sharkey T., Holland E. & Mooney H. (eds) *Trace Gas Emissions form Plants*, pp. 65–92. Academic Press, San Diego.

Chanton J.P. & Martens C.S. (1988) Seasonal variations in ebullitive flux and carbon isotopic composition of methane in a tidal freshwater estuary. *Global Biogeochemical Cycles*, **2**, 289–298.

Chanton J.P., Martens C.S. & Goldhaber M.B. (1987) Biogeochemical cycling in an organic-rich coastal marine basin. 7. Sulfur mass balance, oxygen uptake, and sulfide retention. *Geochimica et Cosmochimica Acta*, **51**, 1187–1189.

Chanton J.P., Martens C.S. & Kelley C.A. (1989) Gas transport from methane-saturated tidal freshwater and wetland sediments. *Limnology and Oceanography*, **34**, 807–819.

Chanton J.P., Martens C.S., Kelley C.A., Crill P.M. & Showers W.J. (1992a) Mechanisms of methane transport and isotope fractionation in macrophytes of Alaskan tundra lakes. *Journal of Geophysical Research*, **97**, 16681–16689.

Chanton J.P., Whiting G.J., Happell J.D. & Gerard G. (1993) Contrasting rates and diurnal patterns of methane emission from two Everglades macrophytes. *Aquatic Botany*, in press.

Chanton J.P., Whiting G.J., Showers W.J. & Crill P.M. (1992b) Methane flux from *Peltandra virginica*: stable isotope tracing and chamber effects. *Global Biogeochemical Cycles*, **6**, 15–31.

Conway V.M. (1987) Studies in the autecology of *Cladium mariscus* R.BR. Part III. The aeration of the subterranean parts of the plant. *New Phytologist*, **36**, 64–96.

Cooper D.J., Cooper W.J., de Mello W.Z., Saltzman E.S. & Zika R.G. (1987a) Variability in biogenic sulfur emissions from Florida wetlands. In: Saltzman E.S. & Cooper W.J. (eds) *Biogenic Sulfur in the Environment*, pp. 31–43. ACS Symposium Series 393, Washington, DC.

Cooper W.J., Cooper D.J., de Mello W.Z. *et al.* (1987b) Emissions of biogenic sulphur compounds from several wetland soils in Florida. *Atmospheric Environment*, **21**, 1491–1495.

Cooper D.J., de Mello W.Z., Cooper W.J. *et al.* (1987c) Short-term variability in biogenic sulphur emissions from a Florida *Spartina alterniflora* marsh. *Atmospheric Environment*, **21**, 7–12.

Crill P.M. & Martens C.S. (1983) Spatial and temporal fluctuations of methane production in anoxic coastal marine sediments. *Limnology and Oceanography*, **28**, 1117–1130.

Crill P.M., Bartlett K.B., Wilson J.O. *et al.* (1988) Tropospheric methane from an Amazonian floodplain lake. *Journal of Geophysical Research*, **93**, 1564–1570.

Dacey J.W.H. (1981a) How aquatic plants ventilate. *Oceanus*, **24**, 43–51.

Dacey J.W.H. (1981b) Pressurized ventilation in the yellow waterlily. *Ecology*, **62**, 1137–1147.

Dacey J.W.H. (1987) Knudsen-transitional flow and gas pressurization in leaves of Nelumbo. *Plant Physiology*, **85**, 199–203.

Dacey J.W.H. & Blough N.V. (1987) Hydroxide decomposition of dimethylsulfoniopro-pionate to form dimethyl sulfide. *Geophysical Research Letters*, **14**, 1246–1249.

Dacey J.W.H. & Howes B.L. (1984) Water uptake by roots control water table movement and sediment oxidation in short *Spartina* marsh. *Science*, **224**, 487–489.

Dacey J.W.H. & Klug M.J. (1979) Methane emission from lake sediment through water lilies. *Science*, **203**, 1253–1255.

Dacey J.W.H. & Klug M.J. (1982) Floating leaves and nighttime ventilation in *Nuphar*. Am. *American Journal of Botany*, **69**, 999–1003.

Dacey J.W.H. & Wakeham S.G. (1986) Oceanic dimethyl sulfide: Production during zooplankton grazing on phytoplankton. *Science*, **233**, 1314–1316.

Dacey J.W.H., King G.M. & Wakeham S.G. (1987) Factors controlling emission of dimethylsulfide from salt marshes. *Nature*, **330**, 634–645.

Dacey J.W.H., Wakeham S.G. & Howes B.L. (1984) Henry's law constants for dimethyl-sulfide in freshwater and seawater. *Geophysical Research Letters*, **11**, 991–994.

deBont J.A., Lee K.K. & Bouldin D.F. (1978) Bacterial oxidation of methane in rice paddy soils. *Ecological Bulletin*, **26**, 91–96.

deMello W.Z. (1992) Factors controlling the emissions of sulfur gases from *Spahnum* dominated wetlands. PhD thesis, University of New Hampshire, Durham.

deMello W.Z., Cooper D.J., Cooper W.J. *et al.* (1987) Spatial and diel variability in the emissions of some biogenic sulfur compounds from a Florida *Spartina alterniflora* coastal zone. *Atmospheric Environment*, **21**, 987–990.

de Mello W.Z. & Hines M.E. (1994) Application of static and dynamic enclosures for determining dimethal sulfide and carbonyl sulfide in Sphagnum pentlands: Implications for the magnitude and direction of flux. *Journal of Geophysical Research*, **99**, 14601–14607.

Devol A.H., Richey J.E., Clark W.A., King S.L. & Martinelli L.A. (1988) Methane emissions to the troposphere from the Amazon floodplain. *Journal of Geophysical Research*, **93**, 1583–1592.

Devol A.H., Richey J.E., Forsberg B.R. & Martinelli L. (1990) Seasonal dynamics in methane emissions from the Amazon River floodplain to the troposphere. *Journal of Geophysical Research*, **98**, 16417–16426.

Fall R., Albritton D.L., Fehsenfeld F.C., Kuster W.C. & Goldan P.D. (1988) Laboratory studies of some environmental variables controlling sulfur emission from plants. *Journal of Atmospheric Chemistry*, **6**, 341–362.

Fechner E.J. & Hemond H.F. (1992) Methane transport and oxidation in the unsaturated zone of a *Spagnum* peatland. *Global Biogeochemical Cycles*, **6**, 33–45.

Frye J.P. (1989) Methane movement in Peltandra virginica, MS thesis. University of Virginia, Charlottesville, Virginia.

Gerard G. (1992) Role of aquatic macrophytes in methane dynamics. MS thesis. Florida State University, Tallahassee, Florida.

Giblin A.E. & Wieder (1993) Sulfur cycling in saline and freshwater wetlands – A review. In: Howarth R.W. & Stewart J.W.B. (eds) *Sulfur Cycling in Terrestrial Systems and Wetlands*. John Wiley, New York.

Goldan P.D., Fall R., Kuster W.C. & Fehsenfeld F.C. (1988) Uptake of COS by growing vegetation: A major tropospheric sink. *Journal of Geophysical Research*, **93**, 14186–14192.

Goldan P.D., Kuster W.C., Albritton D.S. & Fehsenfeld F.C. (1987) The measurement of natural sulfur emissions from soils and vegetation: Three sites in the eastern United States revisited. *Journal of Atmospheric Chemistry*, **5**, 439–467.

Grosse W. & Mevi-Schutz J. (1987) A beneficial gas transport system in *Nympnoides pelata*. *American Journal of Botany*, **74**, 947–952.

Grosse W. & Schroder P. (1984) Oxygen supply of roots by gas transport in Alder trees. *Z. Naturforsch*, **39c**, 1186–1188.

Grosse W. & Schroder P. (1985) Aeration of the roots and chloroplast-free tissues of trees. *Ber. Deutsch Bot. Ges. Bd*, **98**, 311–318.

Grosse W., Buchel H. & Tiebel H. (1991) Pressurized ventilation in wetland plants. *Aquatic Botany*, **39**, 89–98.

Happell J.D. (1992) Methane dynamics in Florida wetlands. PhD dissertation. Florida State University, Tallahassee, Florida.

Happell J.D. & Chanton J.P. (1993) Carbon remineralization in a North Florida swamp forest: the effects of water level on the pathways and rates of soil organic matter decomposition. *Global Biogeochemical Cycles*, in press.

Happell J.D., Chanton J.P., Whiting G.J. & Showers W.S. (1993) Stable isotope tracing of methane dynamics in Everglades marshlands with and without active populations of methane oxidizing bacteria. *Journal of Geophysical Research*, in press.

Harden H. (1992) Locus of methane release and mass dependent fractionation from two wetland macrophytes. MS thesis. Florida State University, Tallahassee, Florida.

Harden H. & Chanton J.P. (1994) Locus of methane release and mass dependent fractionation from two wetland macrophytes. *Limnology and Oceanography*, **39**, 148–154.

Harriss R.C., Sebacher D.I. & Day F.P. (1982) Methane flux in the Great Dismal Swamp. *Nature*, **297**, 673–674.

Hines M.E. (1993) Emissions of sulfur gases from wetlands. In: Adams D.D., Seitzinger S.P. & Crill P.M. (eds) *Cycling of Reduced Gases in the Hydrosphere*, in press.

Hines M.E. & Morrison M.C. (1992) Emissions of biogenic sulfur gases from Alaskan tundra. *Journal of Geophysical Research*, **97**, 16703–16707.

Hines M.E., Pelletier R.E. & Crill P.M. (1993) Emissions of sulfur gases from marine and freshwater wetlands of Florida Everglades: Rates and extrapolation using remote sensing. *Journal of Geophysical Research*, **98**, 8991–8999.

Holmen K. & Liss P.S. (1984) Models for air–water gas transfer: an experimental investigation. *Tellus*, **36B**, 92–100.

Holzapfel-Pschorn A. & Seiler W. (1986) Methane emission during a vegetation period from an Italian rice paddy. *Journal of Geophysical Research*, **91**, 11803–11841.

Holzapfel-Pschorn A., Conrad R. & Seiler W. (1985) Production oxidation and emission of methane in rice paddies. *FEMS Microbiology Ecology*, **31**, 343–351.

Holzapfel-Pschorn A., Conrad R. & Seiler W. (1986) Effects of vegetation on the emission of methane from submerged paddy soil. *Plant and Soil*, **92**, 223–233.

Hook D.D. & Crawford R.M.M. (1978) Plant life in anaerobic environments, p. 564. Ann Arbor Science, Ann Arbor, Michigan.

Ingvorsen K. & Jorgensen B.B. (1982) Seasonal variation in H_2S emission to the atmosphere from intertidal sediments in Denmark. *Atmospheric Environment*, **16**, 855–865.

Jorgensen B.B. & Okholm-Hansen B. (1985) Emissions of biogenic sulfur gases from a Danish estuary. *Atmospheric Environment*, **19**, 1737–1749.

Keller M.M. (1990) Biological sources and sinks of methane in tropical habitats and tropical atmospheric chemistry. PhD dissertation. Princeton University, Princeton, New Jersey.

Keller M.M. & Stallard R.F. (1994) Methane emission by bubbling from Gatun Lake, Panama. *Journal of Geophysical Research*, **99**, 8307–8319.

Kelley C.A., Martens C.S. & Chanton J.P. (1990) Variations in sedimentary carbon remineralization rates in the White Oak river estuary, NC. *Limnology and Oceanography*, **35**, 372–383.

Kesselmeier J. (1991) Emission of sulfur compounds from vegetation and global scale extrapolation. In: Sharkey T., Holland E. & Mooney H. (eds) *Trace Gas Emissions from Plants*, pp. 261–266. Academic Press, San Diego.

King G.M. (1990) Regulation by light of methane emissions from a wetland. *Nature*, **345**, 513–515.

King G., Roslev M.P. & Skovgaard H. (1990) Distribution and rate of methane oxidation in sediments of the Florida Everglades. *Applied and Environmental Microbiology*, **56**, 2902–2911.

Kipphut G.W. & Martens C.S. (1982) Biogeochemical cycling in an organic-rich coastal marine basin – 3. Dissolved gas transport in methane-saturated sediments. *Geochimica et Cosmochimica Acta*, **46**, 2049–2060.

Klump J.V. & Martens C.S. (1981) Biogeochemical cycling in an organic-rich coastal marine basin – 2. Nutrient sediment–water exchange processes. *Geochimica et Cosmochimica Acta*, **45**, 257–266.

Knapp A.K. & Yavitt J.B. (1992) Evaluation of a closed-chamber method for estimating methane emission from aquatic plants. *Tellus*, **44b**, 63–71.

Koncalova H., Pokorny J. & Kvet J. (1988) Root ventilation in *Carex gracilis*: diffusion or mass flow? *Aquatic Botany*, **30**, 149–155.

Lamb B.K., Westberg H., Allwine G., Bamesberger L. & Guenther G. (1987) Measurement of biogenic sulfur emissions from soils and vegetation: application of dynamic enclosure methods with Natusch Filter and GC/FPD analysis. *Journal of Atmospheric Chemistry*, **5**, 469–491.

Lee K.K., Holst R.W., Watanabe I. & App A. (1981) Gas transport through rice. *Soil Science and Plant Nutrition*, **27**, 151–158.

Leuning R. (1983) Transport of gases into leaves. *Plant, Cell and Environment*, **6**, 181–194.

Martens C.S. & Berner R.A. (1974) Methane production in the interstitial waters of sulfate depleted marine sediments. *Science*, **185**, 1067–1069.

Martens C.S. & Berner R.A. (1977) Interstitial water chemistry of Long Island Sound sediments – 1. Dissolved gases. *Limnology and Oceanography*, **22**, 10–25.

Martens C.S. & Chanton J.P. (1989) Radon tracing and biogenic gas equilibration and transport from methane saturated sediments. *Journal of Geophysical Research*, **94**, 3451–3459.

Martens C.S. & Klump J.V. (1980) Biogeochemical cycling in an organic-rich coastal basin 1. Methane sediment–water exchange processes. *Geochimica et Cosmochimica Acta*, **44**, 471–490.

Martens C.S. & Klump J.V. (1984) Biogeochemical cycling in an organic-rich coastal basin 4. An organic carbon budget for sediments dominated by sulfate reduction and methanogenesis. *Geochimica et Cosmochimica Acta*, **48**, 1987–2004.

Martens C.S., Kipphut G.W. & Klump J.V. (1980) Coastal sediment–water chemical exchange traced by in situ Rn-222 flux measurements. *Science*, **208**, 285–288.

Martens C.S., Blair N.E., Green C.D. & Des Marais D.J. (1986) Seasonal variations in the stable carbon isotopic signature of biogenic methane in a coastal sediment. *Science*, **233**, 1300–1302.

Martens C.S., Kelley C.A. Chanton J.P. & Showers W. (1992) Carbon and hydrogen isotopic composition of methane from wetlands and lakes of the Yukon–Kuskokwim Delta and the Alaskan tundra. *Journal of Geophysical Research*, **97**, 16689–16703.

Mosier A.R. (1989) Chamber and isotope techniques. In: Anereae M.O. & Schimel D.S. (eds) *Exchange of Trace Gases between Terrestrial Ecosystems and the Atmosphere*, pp. 175–187. John Wiley and Sons, London.

Matson M.D. & Likens G.E. (1990) Air pressure and methane fluxes. *Nature*, **347**, 718–719.

Mevi-Schutz & Grosse W. (1988) A two-way gas transport system in *Nelumbo nucifera*. *Plant, Cell and Environment*, **11**, 27–34.

Miller L.G. & Oremland R.S. (1988) Methane efflux from the pelagic regions of four lakes. *Global Biogeochemical Cycles*, **2**, 269–268.

Morrison M.C. & Hines M.E. (1990) The variability of biogenic sulfur flux from a temperate salt marsh on short time and space scales. *Atmospheric Environment*, **24A**, 1771–1779.

Morrissey L.A. & Livingston G.P. (1992) Methane emissions from Alaska arctic tundra: and assessment of local spatial variability. *Journal of Geophysical Research*, **97**, 16661–16670.

Morrissey L.A., Zobel D.B. & Livingston G.P. (1993) Significance of stomatal control on methane emission from *Carex* dominated wetlands. *Chemosphere*, **26**, 339–356.

Nouchi I., Mariko S. & Aoki K. (1990) Mechanism of methane transport from the rhizosphere to the atmosphere through rice plants. *Plant Physiology*, **94**, 59–66.

Nraigu J.O., Holdway D.A. & Coker R.D. (1987) Biogenic sulfur and the acidity of rainfall in remote areas of Canada. *Science*, **237**, 1189–1191.

Oremland R.S. & Culbertson C.W. (1992) Importance of methane-oxidizing bacteria in the methane budget as revealed by the use of a specific inhibitor. *Nature*, **356**, 421–422.

Pulliam W.M. (1993) Carbon dioxide and methane exports from a southeastern floodplain swamp. *Ecological Monographs*, **63**, 29–54.

Raskin I. & Kende H. (1983) How does deep-water rice solve its aeration problem? *Plant Physiology*, **72**, 447–454.

Raskin I. & Kende H. (1985) Mechansim of aeration in rice. *Science*, **228**, 327–329.

Reeburgh W.S. (1969) Observations of gases in Chesapeake Bay sediments. *Limnology and Oceanography*, **14**, 368–375.

Reeburgh W.S. (1972) Processes effecting gas distributions in estuarine sediments. *Memoirs Geological Society of America*, **133**, 383–389.

Rennenberg H. (1991) The significance of higher plants in the emission of sulfur compounds from terrestrial ecosystems. In: Sharkey T., Holland E. & Mooney H. (eds) *Trace Gas Emissions from Plants*, pp. 217–260. Academic Press, San Diego.

Richards S.R., Kelley C.A. & Rudd J.W.M. (1991) Organic volatile sulfur in lakes of the Canadian shield and its loss to the atmosphere. *Limnology and Oceanography*, **36**, 468–482.

Saltzman E.S. & Cooper W.J. (1989) *Biogenic Sulfur in the Environment*. ACS Symposium Series 393, Washington, DC.

Sand-Jensen K., Prahl C. & Stokholm H. (1982) Oxygen release from roots of submerged aquatic macrophytes. *Oikos*, **38**, 349–354.

Sass R.L., Fisher F.M. & Harcombe P.A. (1990) Methane production and emission in a Texas rice field. *Global Biogeochemical Cycles*, **4**, 47–68.

Schutz H. & Seiler W. (1989) Methane flux measurements: methods and results. In: Anereae M.O. & Schimel D.S. (eds) *Exchange of Trace Gases between Terrestrial Ecosystems and the Atmosphere*, pp. 209–228. John Wiley and Sons, Ltd.

Schutz H., Schroder P. & Rennenberg H. (1991) Role of plants in regulating the methane flux to the atmosphere. In: Sharkey T., Holland E. & Mooney H. (eds) *Trace Gas Emissions from Plants*, pp. 26–64. Academic Press, San Diego.

Schutz H., Seiler W. & Conrad R. (1989b) Processes involved in formation and emission of methane in rice paddies. *Biogeochemistry*, **7**, 33–53.

Schutz H., Holzapfel-Pschorn A., Rennenberg H., Seiler W. & Conrad R. (1989a) A three year continuous record on the influence of daytime, season, and fertilizer treatment on methane emissions rates from an Italian Rice Paddy. *Journal of Geophysical Research*, **94**, 16405–16416.

Sebacher D.I. & Harriss R.C. (1982) A system for measuring fluxes from inland and coastal wetland environments. *Journal of Environmental Quality*, **11**, 34–37.

Sebacher D.I., Harriss R.C. & Bartlett K.B. (1985) Methane emissions to the atmosphere through aquatic plants. *Journal of Environmental Quality*, **14**, 40–46.

Seiler W., Holzapfel-Pschorn A., Conrad R. & Scharffe D. (1984) Methane emission from rice paddies. *Journal of Atmospheric Chemistry*, **1**, 241–268.

Sharkey T.D., Holland E.A. & Mooney H.A. (1991) *Trace Gas Emissions by Plants*. Academic Press, San Diego.

Smits A.J.M., Laan P., Thier R.H. & van der Velde G. (1990) Root aerenchyma, oxygen leakage patterns and alcoholic fermentation ability of the roots of some nymphaeid and isoetid macrophytes in relation to the sediment type of their habitat. *Aquatic Botany*, **38**, 3017.

Strayer R.F. & Tiedje J.M. (1978) *In-situ* methane production in a small hypereutrophic hardwater lake: loss of methane from sediments by vertical diffusion and ebullition. *Limnology and Oceanography*, **23**, 1201–1206.

Stuedler P.A. & Peterson B.J. (1985) Annual cycle of gaseous sulfur emissions from a New England *Spartina alterniflora* marsh. *Atmospheric Environment*, **19**, 1411–1416.

Vitousek P.M., Denmead D.T., Fowler D. *et al.* (1989) Group report: What are the relative roles of biological production, micrometeorology and photochemistry in controlling the flux of trace gases between terrestrial ecosystems and the atmosphere. In: Anereae M.O. & Schimel D.S. (eds) *Exchange of Trace Gases between Terrestrial Ecosystems and the Atmosphere*, pp. 249–261. John Wiley and Sons, London.

Wakeham S.G. & Dacey J.W.H. (1989) Biogeochemical cycling of dimethyl sulfide in marine environments. In: Saltzman E.S. & Cooper W.J. (eds) *Biogenic Sulfur in the Environment*, pp. 152–166. ACS Symposium Series 393, Washington, DC.

Whalen S.C. & Reeburgh W.S. (1990) A methane flux transect along the trans-Alaska pipeline road. *Tellus*, **42b**, 237–249.

Whiting G.J. & Chanton J.P. (1992) Plant-dependent CH_4 emission in a subarctic Canadian Fen. *Global Biogeochemical Cycles*, **6**, 225–231.

Whiting G.J., Bartlett D.S., Fan M., Bakwin P. & Wofsy S. (1992) Biosphere/atmosphere CO_2 exchange in tundra ecosystems: community characteristics and relationships with multispectral surface reflectance. *Journal of Geophysical Research*, **97**, 16671–16681.

Whiting G.J., Chanton J., Bartlett D. & Happell J. (1991) Methane flux, net primary productivity and biomass relationships in a sub-tropical grassland community. *Journal of Geophysical Research*, **96**, 13067–13071.

Wilson J.O., Crill P.M., Bartlett K.B., Sebacher D.I., Harriss R.C. & Sass R.L. (1989) Seasonal variation of methane emissions from a temperate swamp. *Biogeochemistry*, **8**, 55–71.

Yamamoto S., Alcauskas J.B. & Crozier T.F. (1976) Solubility of methane in distilled water and seawater. *Journal of Chemical Engineering Data*, **21**, 78–80.

Yamasaki S. (1984) Role of plant aeration in zonation of *Zizania latifolia* and *Phragmites australis*. *Aquatic Botany*, **18**, 287–297.

Zehnder A.J. (1978) Ecology of methane formation. In: Mitchell R. (ed) *Water Pollution Microbiology*, Vol. 2. Wiley, New York.

CHAPTER 5

Micrometeorological techniques for measuring biosphere–atmosphere trace gas exchange

D.H. LENSCHOW

5.1 Introduction

Trace gases are both emitted and absorbed at the Earth's surface. The fluxes resulting from these exchange processes are of great importance in studying the budgets of trace species – that is, their sources, sinks, transformations and transports. One reason for the importance of exchange processes near the surface is the relative ease with which transport takes place in the lower atmosphere. This is because the atmosphere near the surface is almost always turbulent, which means that trace gases are rapidly diffused to (or from) the surface by irregular or random motions which are generated by wind shear and buoyancy forces. As a result, the lower part of the atmosphere responds quickly to any change in surface exchange or other surface properties. Diffusion by turbulence is many orders of magnitude larger than molecular diffusion. Therefore, molecular diffusion can be neglected except within a few millimetres of the surface.

This turbulent exchange process can be measured in several ways. The preceding chapter discusses measurement of the actual emission of trace gases from test sources collected within enclosed volumes. This chapter discusses techniques that involve measurement of the vertical transport of trace gases through the atmosphere. The advantages of this approach are:
1 it is possible to measure the transport without intruding upon or disturbing the surface being monitored; and
2 it inherently averages over a surface area that increases with the height of the measurement above the surface (for tower measurements) and length of the flight leg (for aircraft measurements).
The main difficulties are, first, the most straightforward way of making direct flux measurements (i.e. without the use of some empirically determined relationship) requires fast-response sensitive trace gas and vertical velocity fluctuation measurements at nearly the same point in space over the entire frequency spectrum of the irregular fluctuations that contribute to the flux. Second, if empirically determined micrometeorological relationships are used for estimating the flux, the accuracy is dependent upon how accurately the relationship has been determined and on how well the assumptions leading to the relationship are fulfilled.

In dealing with a topic such as this that has a central role in a variety of disciplines – from fundamental micrometeorological studies to estimates of trace constituent budgets – there are many complexities and perspectives to consider which cannot all be covered in a single chapter. Fortunately, many aspects of trace gas exchange have been covered elsewhere. Baldocchi *et al.* (1988) and Fowler & Duyzer (1989) have written very similar reviews from somewhat different perspectives that complement the discussion here. Similarly, Denmead (1983) reviews micrometeorological flux-measuring techniques applied to nitrogen losses from the ground. Stull (1988) and Garratt (1992) have written comprehensive micrometeorology textbooks that provide additional details on many topics presented here. Another book that reviews most of the field, but with somewhat more emphasis on turbulence measurements, and flow in canopies and over irregular terrain is Kaimal & Finnigan (1992). Baldocchi (1991) discusses control of trace gas emissions by canopies, and Wyngaard (1990, 1991) discusses problems in applying micrometeorological techniques to measuring fluxes of trace gases. Lenschow & Hicks (1989) assess current technology in measuring trace gas fluxes, and Businger (1986) assesses techniques for measuring dry deposition of trace constituents.

In this chapter, the basic structure of the atmospheric boundary layer (ABL) is discussed, emphasizing its turbulent nature and effects of turbulence on the measurement of fluxes. The similarity relationships are discussed that have an important role in generalizing boundary-layer structure and are the basis for many flux-measuring techniques that are subsequently described. Next various flux-measuring techniques are described, including the eddy correlation technique, and their advantages and limitations discussed. Interspersed among this discussion are more detailed aspects of boundary-layer structure that are necessary to understand and apply the particular technique. This approach was chosen, rather than a single comprehensive discussion of boundary-layer structure (which can already be found in the literature), to focus on the flux-measuring techniques.

5.2 ABL structure

5.2.1 Basic considerations

The ABL is the lower part of the atmosphere that interacts with the biosphere and is closely coupled to the surface by turbulent exchange processes. Typically the time required to mix a constituent released at the surface throughout the ABL is on the order of an hour. This layer varies from a few tens of metres when the air near the surface is stably stratified – i.e. the surface is colder than the air above, as is typically the case over land at night – to several kilometres when the surface is heated by

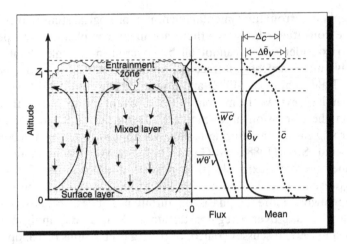

Fig. 5.1 A convective boundary layer. On the left, the sublayers that make up the boundary layer are shown, along with a schematic of flow patterns – upward-moving buoyant thermals forming near the top of the surface layer, extending through the mixed layer, and dissipating in the upper part of the boundary layer. Part of their kinetic energy is dissipated in the entrainment zone by entraining warmer, more buoyant air from above into the boundary layer. The thermals have a smaller total area, and consequently a larger velocity magnitude than the compensating downward-moving air in between the thermals. The flux profiles in the middle panel show the virtual potential temperature flux (which has a universal shape), and the flux of a scalar c which (in this case) has a source at the surface and whose mean concentration decreases with height throughout and immediately above the boundary layer. The virtual potential temperature $[\delta\theta_v \simeq \delta T_v + (g/c_p)\delta z]$ can normally be assumed to be a conserved variable in a well-mixed clear boundary layer. The mean virtual potential temperature and scalar concentration profiles, including their jumps across the top of the boundary layer, are shown on the right.

the sun on a clear summer day and the air is convectively unstable. The lower part of the ABL, where vertical fluxes of conserved quantities can be considered constant (to within about 10%), we call the surface layer. This layer is approximately 10% of the depth of the ABL. As shown in Fig. 5.1, which is a schematic of a convective boundary layer (CBL), although the flux is nearly constant, the concentration can vary considerably near the surface. This is a reflection of the fact that near the surface the size of the turbulent eddies scales with height above the surface; as a result, the ratio of the flux to the gradient increases approximately linearly with height.

Since turbulent fluctuations are by nature irregular and random, we normally deal with them quantitatively by means of statistical averages. We first need to select a time series of sufficient length that the averages calculated over the time series are statistically well behaved. For example, for a variable $y(t)$, we first select a sufficiently long period, then calculate its average over that period. The averaging process itself is denoted by an overbar, so the average of $y(t)$ is $\overline{y(t)}$. I will also, for convenience, sometimes use capitals for the average, i.e. $Y \equiv \overline{y}$.

Although we normally deal with time series, it is actually more straightforward to deal with horizontal displacement x rather than time since the appropriate scaling parameter in the surface layer is the height above the ground; similarly in the remainder of the CBL, which is called the mixed layer, it is the boundary-layer depth. For ground-based measurements, we use the mean wind speed and for aircraft measurements the true airspeed U to convert to the spatial separation $x = Ut$. The assumption that time series from a probe moving through the flow can be used to estimate a spatially varying field of turbulence is known as Taylor's hypothesis. Basically, it assumes that the field of turbulence is frozen as it advects by the observation point. For observations of fluxes in the surface layer from a tower, this assumption is normally valid. Here time and space coordinates are used interchangeably. The fluctuation from the mean $y'(x)$ is then given by $y'(x) = y(x) - \overline{y(x)}$. We see immediately that the mean of the fluctuations $\overline{y'} = 0$.

The density flux of a constituent $\rho_c(x)$ is given by the average of the instantaneous product of $\rho_c(x)$ with vertical velocity w,

$$F_c = \overline{w\rho_c} = \overline{w'\rho_c'} + \overline{w}\,\overline{\rho_c}. \tag{5.1}$$

It has sometimes been assumed that over horizontally homogeneous surfaces $\overline{w} = 0$, and thus $F_c = \overline{w'\rho_c'}$. Webb *et al.* (1980), however, pointed out that the assumption that $\overline{w}\,\overline{\rho_c}$ is 0 and does not contribute to the flux is not justified, even over a horizontally homogeneous surface. In fact, both temperature and water vapour fluxes contribute to the measured density flux of a trace species. This arises from the constraint that the mass flux of dry air must be zero at the surface. Intuitively, we can see that in the case of a heated surface, rising parcels of air will be on average warmer and consequently less dense than their surroundings, while descending parcels will be colder and more dense so that $\overline{w'\rho_c'} < 0$. Therefore, under the assumption of zero mass flux at the surface ($F_{c0} = 0$), Eqn. 5.1 leads to $\overline{w} > 0$. Wyngaard (1990) summarizes the derivation of Webb *et al.* (1980) and obtains the expression

$$F_c \simeq \overline{w'\rho_c'} + a\frac{\overline{\rho_c}}{\rho_a}\overline{w'\rho_v'} + \overline{\rho_c}\left(1 + a\frac{\overline{\rho_v}}{\rho_a}\right)\frac{\overline{w'T'}}{\overline{T}} \tag{5.2}$$

for the total vertical flux of a trace constituent, where $\overline{w'\rho_v'}$ and $\overline{w'T'}$ are the measured fluxes of water vapour density and temperature, $a = m_a/m_v$ is the ratio of the molecular weight of dry air to that of water vapour, and ρ_a is the dry air density. This means that for cases with small values of $\overline{w'\rho_c'}/\overline{\rho_c}$, there may be significant contributions by the water vapour and temperature fluxes to the measured flux which need to be removed to obtain the actual trace species flux. In this context, a value of $\overline{w'\rho_c'}/\overline{\rho_c} < 10^{-2}\,\mathrm{m\,s^{-1}}$ is cause for consideration. For relatively long-lived (in the atmosphere) species such as CO_2, CH_4 and N_2O, the correction terms are often comparable to the measured flux (Wesely *et al.*, 1989). Fowler & Duyzer (1989) present a plot showing the percentage correction for moderate temperature and water vapour fluxes as a function of deposition velocity, which is the ratio of the flux to the mean concentration at some reference height in the surface layer.

If instead of measuring the constituent density, we measure its mixing ratio with respect to dry air $c = \rho_c/\rho_a$, Webb *et al.* (1980) have shown that no correction is necessary; i.e.

$$F_c = \bar{\rho}\,\overline{w'c'}, \tag{5.3}$$

where $\bar{\rho}$ is the moist air density. Two ways that come to mind for measuring constituent flux without having to correct for temperature and water vapour fluxes are:

1 preprocess the air before its constituent density is measured by drying it and keeping its temperature and pressure constant; and

2 measure the ratio of the constituent density to the density of a species that is uniformly mixed – e.g. N_2 or O_2, possibly by absorption measurements.

This second approach has been used (Melfi, 1972; Melfi & Whiteman, 1985) for remote measurements of water vapour mixing ratio by Raman scattering using N_2 as the reference gas for dry air.

Finally, if mixing ratio with respect to total air density is measured, then the measured flux need only be corrected for water vapour flux. Typically the water vapour correction is smaller. For the same latent heat flux λF_q as sensible heat flux $c_p\bar{\rho}F_T$, where λ is the latent heat of vaporization of water vapour and c_p is the specific heat of air at constant pressure, the correction term involving temperature flux is about five times that for water vapour flux.

As Webb *et al.* (1980) point out, these corrections for temperature and water vapour fluxes are also necessary for estimating flux by other techniques such as by differences in concentration with height.

In the following discussion of sampling errors, the flux of a trace constituent is referred to as $\overline{w'c'}$, with the understanding that for sampling errors c can equally be density or mixing ratio.

5.2.2 Turbulence characteristics of the boundary layer

Because of the inherent randomness of the turbulent processes responsible for vertical transport, we develop and use techniques for studying the characteristics of turbulent transport that are based on the statistics of random variables. Ideally, we would like to measure the statistical variables over temporal or spatial scales long enough that enough samples are obtained to ensure sufficient accuracy. In practice, this may be difficult to achieve because in the time (or distance) required to obtain sufficient samples, there may be systematic changes in the conditions or processes that control the variable. Therefore, it is important to have quantitative measures for estimating the accuracy of flux measurements as a function of averaging time (or length). A basic assumption in this discussion is that the turbulent field is stationary (for tower measurements) or horizontally homogeneous (for airplane measurements) – that is, the turbulent statistics are independent of time (stationarity) or horizontal distance (homogeneity).

We define the autocovariance function e.g. for w as

$$R_w(\xi) \equiv \overline{w'(x)w'(x + \xi)}, \tag{5.4}$$

which is a measure of how well w is correlated with itself as a function of separation distance ξ. The autocovariance is used in determining the integral scale, which is defined as

$$l_w \equiv \frac{1}{\sigma_w^2} \int_0^\infty R_w(\xi)d\xi, \tag{5.5}$$

where $\sigma_w^2 \equiv \overline{w'^2}$ is the variance of w. Of course, in practice, the limit of integration cannot be infinite; it is sufficient to integrate over a separation distance that extends far enough that the autocovariance shows no correlation and fluctuates about 0. To estimate the integral scale requires a separation distance of at least several integral scales as shown in Fig. 5.2, and a time series of the order of at least a hundred times the integral scale.

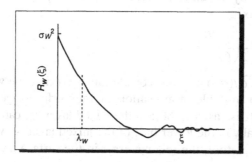

Fig. 5.2 Schematic diagram of the autocovariance function for w as a function of the separation distance ξ showing the integral scale l_w, which is the integral of the normalized autocovariance function $R_w(\xi)/R_w(0)$, also known as the autocorrelation function. The variance of w, $\sigma_w^2 \equiv R_w(0)$. For values of ξ several times l_w, values of w are independent of each other, and thus $R_w(\xi)$ becomes small.

The integral scale is useful in estimating the random error in flux measurement due to the limited length of any real set of measurements in the ABL. The error variance of a flux measurement is given by

$$\tilde{\sigma}_f^2(L) \equiv \{[F(L) - \langle F(L)\rangle]^2\}, \tag{5.6}$$

where L is the averaging length of the flux. The braces $\{\ \}$ denote an ensemble average, which is an average over an infinite number of identical realizations of the same experiment. Lenschow & Kristensen (1985) showed that

$$\tilde{\sigma}_f^2 \leq 4\sigma_w^2\sigma_c^2\frac{\min(l_w l_c)}{L}, \tag{5.7}$$

where $\sigma_c^2 \equiv \overline{c'^2}$ and $\min(l_w l_c)$ is the smaller of the integral scales of w and c. Normally, in the ABL, $l_w \leq l_c$, so that

$$\tilde{\sigma}_f^2 \leq 4\sigma_w^2\sigma_c^2\frac{l_w}{L}. \tag{5.8}$$

From the spectral results of Kaimal *et al.* (1972), obtained from tower measurements in the surface layer, the value of l_w/z, where z is the height above the ground, ranges from ~ 0.3 to ~ 1 as the hydrodynamic stability ranges from neutral to very unstable. In the mixed layer above the surface layer, Lenschow & Stankov (1986) found from aircraft measurements that

$$l_w/z_i \simeq 0.24\left(\frac{z}{z_i}\right)^{1/2}, \tag{5.9}$$

where z_i is the depth of the mixed layer. These results are plotted in Fig. 5.3. For practical applications, we consider the relative flux measurement error obtained by dividing the square root of Eqn. 5.8 by the absolute value of the flux,

$$\frac{\tilde{\sigma}_f}{|\overline{w'c'}|} \leq \frac{2}{|r_{wc}|}\left(\frac{l_w}{L}\right)^{1/2}, \tag{5.10}$$

where $r_{wc} \equiv \overline{w'c'}/(\sigma_w\sigma_c)$ is the correlation coefficient between w and c. For neutral to unstable stratification in the surface layer, $|r_{wc}| \geq 0.4$ (Lenschow & Kristensen, 1986); in the mixed layer r_{wc} can vary considerably, but typically in the lower part of the mixed layer we would expect it to be $\simeq 0.4$. Therefore, the relative error in the surface layer is

$$\frac{\tilde{\sigma}_f}{|\overline{w'c'}|} \leq 5\left(\frac{z}{L}\right)^{1/2}, \tag{5.11}$$

and in the lower part of the mixed layer,

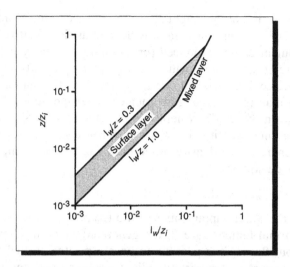

Fig. 5.3 Plot of the normalized integral scale of vertical velocity l_w/z_i vs. normalized height z/z_i through both the surface layer and the mixed layer. In the surface layer, the integral scale is proportional to height, while in the mixed layer, $l_w/z_i \simeq 0.24(z/z_i)^{1/2}$. The line $l_w/z = 0.3$ represents the neutral surface layer and $l_w/z = 1$ a very unstable surface layer.

$$\frac{\tilde{\sigma}_f}{|\overline{w'c'}|} \leqslant 2.4\left(\frac{z}{L}\right)^{1/2}\left(\frac{z_i}{z}\right)^{1/4}. \tag{5.12}$$

This says that, for example, at a measurement height of 10 m in the surface layer, the required averaging length for a relative error of <20% is $L \geqslant 6.2$ km, or at a wind speed of $5\,\mathrm{m\,s^{-1}}$, the required averaging time is $\geqslant 20$ min. At a height of 50 m in a 1000 m deep ABL, the mixed-layer measuring length will be $L \geqslant 32.2$ km. If the airplane is flying at $100\,\mathrm{m\,s^{-1}}$, the averaging time will be $\geqslant 5.4$ min.

As mentioned earlier, these considerations apply if the turbulent field is stationary or horizontally homogeneous. If this is not the case, it may become difficult to define an integral scale; for example, if significant variability exists on scales that are a significant fraction of the record length, the integral scale will be a significant fraction of the length of the time series, and thus cannot be accurately estimated. Generally, the effect of inhomogeneity, even if it occurs on scales a small fraction of the record length (e.g. a patchwork of different crops), is likely to increase the integral scale, and thus longer measurement times would be necessary. This is, however, an area in need of further research.

The commonly used measure of hydrodynamic stability in the surface layer is the Monin–Obukhov length. Monin–Obukhov similarity theory is based on expressing variables in the surface layer as functions of the dimensionless ratio of the height above the surface to the Monin-Obukhov length $\mathscr{L}$. This dimensionless ratio is defined as the negative of the ratio of the turbulent energy produced (or consumed) by buoyancy to the turbulent energy generated by wind shear (assuming a neutral wind profile) near the surface. The buoyant energy source (or sink) is a surface warmer (or cooler) than the overlying air (neglecting evaporation). The buoyant energy is given by $gF_{b0}/\overline{T}$, where $(\;)_0$ denotes a surface-layer value, g is the acceleration of gravity, and F_{b0} is the surface virtual temperature flux, which is generated by density differences due to both temperature and humidity fluctuations:

$$F_{b0} = (\overline{w'T'})_0 + 0.61\overline{T}(\overline{w'q'})_0, \tag{5.13}$$

where q is the water vapour mixing ratio (ρ_v/ρ_a).

Since virtual temperature flux is generated by both temperature and water vapour fluxes, it is possible to have a positive virtual temperature flux with the surface temperature slightly cooler than the atmosphere if the evaporation rate is sufficient to overcompensate for a negative temperature flux. This occurs frequently over the ocean where typically the atmosphere and ocean are at nearly the same temperature.

The turbulent energy generated by wind shear near the surface, assuming a neutral wind profile, is given by u_*^3/kz, where u_* is the friction velocity and $k \simeq 0.4$ is von Kármán's constant. This surface-layer scaling velocity is defined by $u_*^2 \equiv -(\overline{u'w'})_0$, which, when multiplied by air density, is the turbulent momentum transport at the surface, or equivalently, the drag that the atmospheric flow exerts on the Earth's surface. The Monin–Obukhov length then is

$$\mathscr{L} \equiv -\frac{\overline{T}u_*^3}{kgF_{b0}}, \tag{5.14}$$

and variables in the surface layer such as the gradient and variance of a constituent c can be written as normalized functions of $z/\mathscr{L}$. These normalized functions have been determined empirically for many surface-layer variables, with sufficient confidence to confirm that the hypothesis of surface-layer similarity is an extremely useful tool for quantitative calculations of surface-layer structure. Examples are (Kaimal & Finnigan, 1992)

$$\frac{kz}{c_*}\frac{\partial C}{\partial z} = \phi_c(z/\mathscr{L}), \tag{5.15}$$

where

$$\phi_c(z/\mathscr{L}) \simeq (1 - 16z/\mathscr{L})^{-1/2}; \qquad -2 \leqslant z/\mathscr{L} \leqslant 0 \tag{5.16}$$

$$= 1 + 5z/\mathscr{L}; \qquad 0 \leqslant z/\mathscr{L} \leqslant 1$$

and

$$\frac{\sigma_c}{|c_*|} = \phi_s(z/\mathscr{L}), \tag{5.17}$$

where

$$\phi_s(z/\mathscr{L}) = 2(1 - 9.5z/\mathscr{L})^{-1/3}; \qquad -2 \leqslant z/\mathscr{L} \leqslant 0, \tag{5.18}$$

$$= 2(1 + 0.5z/\mathscr{L})^{-1}; \qquad 0 \leqslant z/\mathscr{L} \leqslant 1, \tag{5.19}$$

and $c_* = -F_{c0}/u_*$ is a scaling parameter for c. The absolute value is used in Eqn. 5.17 because variance is generated equally by either a positive or negative surface flux of c.

The above formulations are specified only for the range $-2 < z/\mathscr{L} < 1$. For $z/\mathscr{L} < -2$, free convection scaling (Wyngaard & Coté, 1971) applies and u_* is no longer a scaling parameter. The free convection region is a transition between surface-layer and mixed-layer scaling. For $z/\mathscr{L} > 1$, the flow is only intermittently coupled to the surface by turbulence, and therefore no longer follows surface-layer similarity theory.

In the mixed layer a different similarity structure applies. Here the relevant height scale is no longer z, but the depth of the CBL z_i. This reflects the fact that the turbulent eddies that predominantly effect vertical transport scale with z_i. Further, wind shear near the surface is no longer the predominant source of turbulent energy; rather buoyancy dominates in the mixed layer. Therefore, the relevant velocity scale becomes the convective velocity

$$w_* \equiv \left(\frac{g}{T}F_{b0}z_i\right)^{1/3}. \tag{5.20}$$

Similarly, the mixed-layer scaling parameter used for variables involving c becomes $C_* = -(\overline{w'c'})_0/w_* = -F_{c0}/w_*$. One additional complication for mixed-layer scaling is that the behaviour of mixed-layer variables depends not only on the surface fluxes, but also the fluxes through the top; i.e. the entrainment fluxes. The turbulent eddies impinging on the capping inversion entrain air from above the ABL and mix it with boundary-layer air. Therefore, the mean normalized gradient is a function of both fluxes.

One way of representing this is to consider that the scalar mixing ratio field can be decomposed into two fields whose sum equals the observed field (Moeng & Wyngaard, 1984; Wyngaard & Brost, 1984). One field is denoted as 'bottom-up' diffusion ( )$_0$, the other as 'top-down' diffusion ( )$_i$. The bottom-up field is that which exists due to a scalar flux that decreases linearly from F_{c0} at the surface to zero at the top; the top-down field is that due to a scalar flux which is F_{ci} at the top and decreases linearly to zero at the surface. Thus, $C(z/z_i) = C_0(z/z_i) + C_i(z/z_i)$. The normalized gradient functions for these two fields are

$$-\frac{w_* z_i}{F_{c0}}\frac{\partial C_0}{\partial z} = g_0(z/z_i), \tag{5.21}$$

where

$$g_0(z/z_i) \simeq 0.7(1 - z/z_i)^{-2}, \tag{5.22}$$

and

$$-\frac{w_* z_i}{F_{ci}}\frac{\partial C_i}{\partial z} = g_i(z/z_i), \tag{5.23}$$

where

$$g_i(z/z_i) \simeq 0.4(z/z_i)^{-3/2}. \tag{5.24}$$

The normalized gradient functions (Eqns 5.22 & 5.24) have been estimated from numerical calculations using large eddy simulations (LES) of the mixed layer (Moeng & Wyngaard, 1984, 1989).

Moeng & Wyngaard also obtained normalized variance functions from LES results for both top-down and bottom-up diffusion. For bottom-up diffusion in the lower part of the ABL ($z/z_i < 0.1$), they obtained results that agree well with observations (Kaimal et al., 1976). These results follow the curve which can be obtained from Eqn. 5.17 for $-z/\mathscr{L} \gg 1$:

$$\sigma_c/|C_*| = 1.3(z/z_i)^{-1/3}. \tag{5.25}$$

Compared with the surface layer, the mixing process in the mixed layer is so efficient that vertical concentration gradients of conserved species are small (see Fig. 5.1). Here the vertical transport is dominated by thermals, which are buoyant plumes generated within the surface layer by surface heating. They extend through the mixed layer, mixing with the surrounding air, and eventually lose their momentum near the top of the ABL as they impinge on the overlying inversion. The turbulent energy given off by impingement of these turbulent eddies at the top entrains overlying stable air, warmer than the boundary-layer air, into the CBL. The net result is a growing turbulent CBL which rapidly diffuses any trace constituent, whether introduced at the surface, through the top, or somewhere in between.

The rate at which this diffusion occurs is specified by a characteristic time scale τ which is estimated in the surface layer as follows: the turbulent eddy diffusivity for c, K_c, defined by the relation

$$F_c \equiv -K_c\frac{\partial C}{\partial z}, \tag{5.26}$$

can be specified in terms of boundary-layer parameters as the product of a characteristic transfer velocity times a characteristic length scale, where the transfer velocity is given by the ratio of the length scale to τ. In the surface layer, from Eqn. 5.15,

$$K_c(z) = \frac{z^2}{\tau} = u_* kz\phi_c^{-1}(z/\mathscr{L}).$$ (5.27)

Therefore,

$$\tau \simeq \frac{z\phi_c}{ku_*}.$$ (5.28)

In the mixed layer, the concept of a vertical eddy diffusivity is not as well defined since it depends upon the fluxes both at the bottom and at the top, and the eddy diffusivities associated with each of these processes are different (Moeng & Wyngaard, 1984; see also Holtslag & Moeng, 1991). Instead, we use the characteristic turbulence time scale for scalar diffusion in the mixed layer estimated by Weil (1985): $0.7z_i/w_*$. Based on calculations by Weil (1990), a trace constituent introduced into the mixed layer will be mixed throughout the mixed layer in about five to six times this time scale, or

$$\tau \simeq 4z_i/w_*.$$ (5.29)

A typical value of τ in the mixed layer is, from Eqn. 5.29, a few tens of minutes to an hour. From Eqn. 5.28, a few metres above the surface this time scale will be typically more than an order of magnitude less; i.e. a couple of minutes.

This time scale can also be used to estimate the chemical reaction time scale within which the mean concentration profile for chemically reactive species may be affected. If the lifetime of a trace species is, for example, of the same order of magnitude as τ, its diffusion time will be comparable to its chemical lifetime and its mean structure will differ from that predicted for conserved species. This means that a reactive species with a lifetime of a few tens of minutes or less cannot be assumed to be conserved in the mixed layer; if its lifetime is a couple of minutes or less it cannot be assumed to be conserved in the surface layer, and consequently its flux and gradient may depart from those predicted for conserved species (Fitzjarrald & Lenschow, 1983; Lenschow & Delany, 1987; Kramm *et al.*, 1991).

In the stable boundary layer (SBL), which forms over land in the evening as the ground cools, mixing is much reduced compared to the CBL and concentrations of contaminants released at the surface are likely to be larger. In the classical case, turbulence is generated by shear near the surface, and damped by the static stability so that the turbulence energy decreases with height. As a result, the SBL is shallower; furthermore, turbulence becomes increasingly intermittent with height, with no pronounced jump in stability or reduction in turbulence across the top as is the case with the CBL. The depth of the SBL can be specified as the height at which the turbulence energy is reduced to a fraction (e.g. 5%) of its surface layer value (Caughey et al., 1979). Typically this is the order of several tens to several hundred metres. In the upper part of the SBL, the stable stratification gives rise to gravity waves, which result in periodic oscillations in u, v, w, T and c. However, T and c are not in phase with w, so no flux occurs.

In very stable cases, i.e. with light winds and strong surface cooling (clear skies), the SBL is often quite complex due to the interacting processes of clear air radiative cooling, separation of the flow into discrete layers, intermittency of turbulence, and frequent occurrence of a low level jet above the surface inversion layer (Stull, 1988). In this situation, boundary-layer structure cannot be predicted from the usual similarity theory (Kim & Mahrt, 1992). For example, the main source of turbulence energy may be at the top of the surface inversion layer near the low level jet, which leads to an upside-down boundary layer structure.

In the course of a typical diurnal cycle of the ABL over land, the CBL (shown schematically in Fig. 5.1) grows rapidly through the morning and slows down early in the afternoon, eventually reaching a plateau in mid-afternoon, typically one to several kilometres deep. By late afternoon, the turbulence decreases as the difference between the ground and air temperature decreases, and mixing is reduced. In the evening, the ground radiatively cools below the air temperature, and a SBL develops which may grow slowly through the night. After sunrise, the CBL again forms when the ground becomes warmer than the overlying air and warms the stable layer that developed through the night. It may eventually grow through the nearly neutrally stabilized layer left over from the previous day. Often fair weather cumulus clouds form at the top of the CBL. If the clouds continue to grow to several kilometres depth, or cover a significant fraction of the surface area, they are likely to have a significant impact on flux profiles both through the effects of the air circulations induced by the latent heat release (e.g. venting air out through the top of the CBL) as well as by modifying the radiation budget at the surface (Stull, 1988).

Over the ocean, the diurnal cycle is quite different, and often scarcely noticeable. Since the ocean has a much larger effective heat capacity and conductivity, its surface temperature is hardly perturbed by the daily solar cycle. More importantly, in clear daytime conditions the boundary-layer air is heated directly by the sun, which typically more than compensates for long-wave (infrared) radiational cooling. In this case, the ABL is more stable (i.e. smaller virtual temperature flux and less cloudiness) in the daytime than at night.

5.3 Techniques for measuring fluxes

A variety of techniques exist for measuring fluxes by micrometeorological techniques, each with its own advantages and disadvantages. This section discusses the most important of these techniques, and indicates advantages and disadvantages of each. Overall, however, there is no single method that is right for all applications; rather, the techniques discussed here are complementary. Businger (1986) is a useful reference describing limitations and accuracy of many of the techniques that are relevant for measuring deposition of trace constituents at the surface.

5.3.1 Eddy correlation

The most straightforward and basic approach for flux measurement is the eddy correlation technique, where concurrent instantaneous measurements of the vertical velocity w and a scalar quantity c are multiplied together to obtain the flux. A variation of this, which is still a fundamental measurement that does not require some empirical relationship to estimate the flux is the eddy accumulation technique.

Although the flux can be calculated directly as discussed earlier, we can obtain more insight into the size of the eddies that contribute to the flux by looking at the contributions to the flux as a function of frequency f (Hz) or wavenumber κ (rad per unit length). This can be done by a Fourier decomposition and subsequent calculation of the cospectrum of w and c. One way of characterizing this procedure is first to define the cross-covariance between w and c (analogous to the definition of the auto-covariance function of Eqn. 5.3),

$$R_{wc}(\xi) \equiv \overline{w'(x)c'(x + \xi)}. \tag{5.30}$$

The cross-spectrum can then be defined as the Fourier transform of the cross-covariance,

$$Cr_{wc}(\kappa) = C_{wc}(\kappa) - iQ_{wc}(\kappa) = \frac{1}{2\pi} \int_{-\infty}^{\infty} R_{wc}(\xi) e^{-i\kappa\xi} d\xi. \tag{5.31}$$

The real part of the cross-spectrum $C_{wc}(\kappa)$ is the cospectrum and the imaginary part $Q_{wc}(\kappa)$ is the quadrature spectrum. The advantage of this procedure is that the integral of the cospectrum,

$$\int_{-\infty}^{\infty} C_{wc}(\kappa) d\kappa = 2 \int_{0}^{\infty} C_{wc}(\kappa) d\kappa = \overline{w'c'} = F_c. \tag{5.32}$$

Thus, $C_{wc}(\kappa)$ is the contribution to the total flux per wave number interval (or frequency interval). The quadrature spectrum is that part of the cross-spectrum that is out of phase (i.e. one variable leads or lags the other by $\pi/2$ rad). Thus the integral of the quadrature spectrum is identically zero, and does not contribute to the flux. The ratio of the two is a measure of the phase angle between the two variables,

$$\phi(\kappa) = \tan^{-1}\frac{Q(\kappa)}{C(k)}. \tag{5.33}$$

This cross-spectrum approach is useful in specifying the wave number interval, and thus the frequency response and sampling rates required for estimating the flux. Using surface-layer and mixed-layer scaling, we can estimate 'universal' cospectra (actually $n \times C_{wc}$, where n is a normalized frequency), as shown in Figs 5.4 and 5.5 for use in determining the appropriate wave number interval for measuring fluxes throughout the CBL. The surface-layer data for the curve in Fig. 5.4 were obtained from Schmitt *et al.* (1979), while the mixed-layer data used for Fig. 5.5 are from Lenschow & Stankov (1986). These curves are generalizations for design purposes, and not meant to provide accurate representations for specific cases. For example, the flux-transport eddies tend to be elongated along the direction of the wind, and the surface-layer cospectra are not entirely independent of z_i, particularly in very unstable cases. We can relate these results approximately to the integral scale discussion earlier

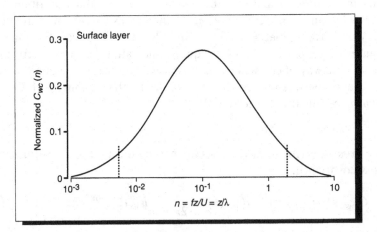

Fig. 5.4 Schematic diagram of a typical normalized (such that the area under the curve is unity) cospectrum of vertical velocity and a scalar variable in a convective surface layer, plotted as a function of normalized frequency, where U is the mean wind speed (for tower measurements) or true airspeed (for airplane measurements) and λ is wavelength. The cospectrum is multiplied by frequency, so that the area under the curve is proportional to the flux. The two dashed vertical lines at $n = 0.005$ and $n = 2$ are the frequencies at which the curve of frequency times the cospectrum is ~20% of its maximum; the flux contained within these two limits is ~93% of the total flux.

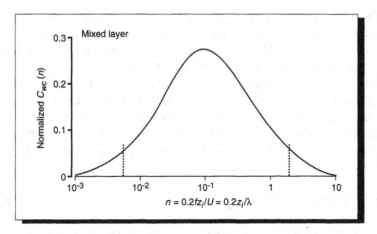

Fig. 5.5 As Fig. 5.4, except that mixed-layer scaling is used, so that the curve is a normalized cospectrum multiplied by frequency. In this case, the length scale used to normalize the frequency is the depth of the boundary layer.

by noting that l_{wc} is related to the peak in the cospectrum κ_{wc} in the following way: the wavelength of the cospectral peak is $\lambda_{wc} = 2\pi\kappa_{wc}^{-1}$; $l_{wc} \simeq \lambda_{wc}/(2\pi)$ (Kristensen *et al.*, 1989). Therefore, $l_{wc} \simeq \kappa_{wc}^{-1}$. Mann & Lenschow (1994) discuss errors in flux measurement in more detail.

These curves show that, for example, at 4 m height above the ground for a wind speed of $5\,\mathrm{m\,s^{-1}}$ in order to measure 93% of the flux, the required band width for flux measurement is $0.006 \leqslant f \leqslant 2.5\,\mathrm{Hz}$. For an airplane flight leg in the surface layer 30 m above the ground and flying at $100\,\mathrm{m\,s^{-1}}$, the required band width for flux measurement is $0.017 \leqslant f \leqslant 7\,\mathrm{Hz}$; in the middle part of a 1000 m deep mixed layer the band width is $0.0025 \leqslant f \leqslant 1\,\mathrm{Hz}$.

A straightforward way of evaluating the instrument response and time lag between two variables is to look at $\phi(\kappa)$. In particular, if we wish to evaluate the time response of a species concentration measurement, its correlation with another scalar such as temperature or another species is typically much greater than with a velocity component, and the variables are either in phase or out of phase. This is because the source of scalar–scalar covariance is the sum of each of the fluxes times its vertical gradient, while the dissipation is by molecular processes on very small scales (Wyngaard *et al.*, 1978). There are no pressure terms, as in the velocity–scalar covariance equations, which can reduce correlations and generate phase shift between the variables. A time lag τ_c on a phase-angle vs. frequency plot manifests itself as a straight line from the origin with slope $\pm 2\pi\tau_c$.

Often a trace species is measured by a sensor which is located some distance away from the sampling point, for example, because of bulkiness (and consequent possible flow distortion effects) or fragility of the sensor. The ducting that transports the sample to the sensor can introduce both a time lag and high frequency attenuation of the concentration fluctuations. Lenschow & Raupach (1991) show how this attenuation can be estimated. For a straight tube with turbulent flow* and an equilibrium velocity profile (i.e. neglecting the transient effects of the intake, which tend to reduce attenuation), they found that the frequency at which the signal variance at the input is reduced by half is

$$f_0 \simeq 0.09 Re^{1/16} \frac{U}{(DL_t)^{1/2}}, \tag{5.34}$$

where L_t is the length of the tube. If the flow in the tube is laminar, they found that near the critical Reynolds number f_0 is roughly a quarter of the value for turbulent flow. Thus, it is important to ensure that the flow in the intake tube is fast enough that the signal is not attenuated in the band width where flux transport is significant; usually this means that the flow should be fast enough to be turbulent. An example of application of these results to a closed-path CO_2 analyser and subsequent correction of the measured flux is presented by Leuning & Moncrieff (1990) and Leuning & King (1992). They also discuss application of the Webb correction to both open-and closed-path sensors.

Another factor to consider is the amount of noise that can be tolerated in the measurement before the noise begins to affect the accuracy of the flux measurement. There are, of course many sources of noise. One of the most common and easiest to deal with is so-called white noise. This random noise may result, for example, from measuring an output proportional to a finite number of photons or electrons, or may be contributed by amplifiers or other electronic components. Here it is assumed that the vertical velocity is noise-free; the noise is present only in the trace species. The resulting error variance for flux measurement, which includes both the error due to the finite length of the sample L, as given by Eqn. 5.8, as well as the contribution due to a white noise variance σ_{nc}^2, is given by (Ritter et al., 1990)

$$\tilde{\sigma}_f^2 = \frac{\sigma_w^2}{L}(4\sigma_c^2 l_w + \sigma_{nc}^2 \Delta x), \tag{5.35}$$

where Δx is the distance between samples. Thus, if we limit the white-noise contribution to the flux to be $\leq$ the sampling-length contribution,

---

* The critical Reynolds number ($Re \equiv DU/\nu$, where D is the diameter and ν the kinematic viscosity) at which the flow becomes turbulent in a straight cylindrical tube is $Re_c \simeq 2300$.

$$\sigma_{nc}^2 \leqslant \frac{4\sigma_c^2 l_w}{\Delta x}.$$ (5.36)

A more detailed discussion of the effect of random noise on turbulence measurements is given by Lenschow & Kristensen (1985).

Although eddy correlation is the most direct flux-measuring technique, it can also be difficult to implement because it requires concurrent and contiguous measurements of w and c with high-frequency response (for tower measurements, typically $\geqslant 1\,Hz$; for airplane measurements, typically $\geqslant 10\,Hz$). Furthermore, measurement of w requires careful consideration of the alignment of the vertical velocity sensor relative to the local Earth vertical and, for a moving measurement platform, the platform motions. Misalignment of the vertical velocity sensor can contaminate the vertical flux measurement with the horizontal turbulent flux. Similarly, the vertical flux can also be contaminated by the horizontal flux through the distortion of flow around the sensors and supporting structures (Wyngaard, 1990, 1991).

A wide variety of sensor technologies have been utilized or proposed for eddy flux measurements. Water vapour flux has been measured with an assortment of sensors, including wet-bulb thermometers, infrared and ultraviolet absorption devices, and dew-point hygrometers; in addition, thin-film capacitative devices have potential for eddy correlation. Lenschow & Hicks (1989) discuss trace gas sensors that have been used or have potential for eddy correlation measurements. These include chemiluminescence systems for O_3, and possibly for NO, NO_2 and the sum of all odd nitrogen species; infrared absorption systems for CO_2, CO and CH_4; flame ionization detectors for total hydrocarbon concentration, flame photometric detection for total sulphur, and electron capture detection for sulphur hexafluoride (SF_6).

The sampling requirements outlined above for standard eddy correlation flux measurement can be relaxed somewhat without significant reduction in the required sampling length. This can be done by making flux measurements only about once per integral scale without increasing the sampling length requirements for a given level of flux measurement accuracy (Lenschow et al., 1994). This technique, which he calls intermittent sampling, is based on collecting an air sample quickly so that the fluctuations in concentration are not averaged over sampling lengths important for flux measurement, then storing the sample and measuring its concentration with an instrument that may not respond fast enough to use for direct flux measurement. In this way, when the concentration is multiplied by w at the time the sample is collected, the short-wavelength contribution to the flux is still retained. Since the integral scale of the flux is roughly an order of magnitude larger than the shortest significant wavelength contribution to the flux, it is possible to use instruments with considerably slower response. If samples are collected at a rate less than the integral scale, the flux can still be measured, but a longer sampling time is required for the same level of accuracy.

5.3.2 Eddy accumulation

The eddy correlation technique is the most direct and fundamental way to measure fluxes. However, for a variety of reasons it is not always feasible or practical to use this technique. Therefore, we also need to consider other techniques for flux measurement. In order to do this systematically, we will order them in roughly the degree to which they rely on empirically determined results or assumptions.

One variation on eddy correlation that does not involve an empirically determined relationship is the eddy accumulation technique (Desjardins, 1972). This technique is representative of what are known as conditional sampling techniques – i.e. collecting samples in more than one reservoir – with the sample collection process determined by some property of the vertical velocity. In the eddy accumulation technique, the air is collected in two reservoirs – one is used when $w' > 0$ and the other when $w' < 0$. Furthermore, the volume rate at which the sample is collected is proportional to the magnitude of w. That is,

$$\frac{dV^{\pm}}{dt} = \kappa |w'| \delta^{\pm},$$
(5.37)

where V^+ and V^- are the volumes of the two samples collected, $\delta^+ = 1$ when $w' > 0$ and 0 otherwise, $\delta^- = 1$ when $w' < 0$ and 0 otherwise, and κ is the constant of proportionality [in units of (length)^2/time] which relates the volume flow rate to w'. Therefore, the total accumulated mass in each of the reservoirs $M_c^{\pm}$ collected over over a time interval T is

$$\frac{1}{T} \int_0^T \rho_c^{\pm} \frac{dV^{\pm}}{dt} dt = \frac{M_c^{\pm}}{T} = \kappa \overline{\rho_c^{\pm} w' \delta^{\pm}}.$$
(5.38)

Neglecting air density fluctuations,

$$\overline{w' \rho_c} = \overline{w'(\overline{\rho_c} + \rho'_c)} = \overline{w' \rho'_c} = \overline{\delta^+ w' \rho_c^+} + \overline{\delta^- w' \rho_c^-}.$$
(5.39)

Therefore, combining Eqns 5.38 and 5.39,

$$\overline{w' \rho'_c} = \frac{M^+ - M^-}{\kappa T}.$$
(5.40)

That is, the flux of c is estimated from the difference in mass collected in each reservoir divided by the collection time. Thus, fast response measurements of c are not required for eddy correlation – in effect, fast flow control has been substituted for fast species measurement.

Since

$$\overline{\kappa|w'|\delta^+} - \overline{\kappa|w'|\delta^-} = \kappa\overline{w'} = 0, \tag{5.41}$$

$V^+ = V^- = V$ (if the pressure and temperature in each reservoir are the same). Thus,

$$\overline{w'\rho_c'} = \frac{(\bar{\rho}_c^+ - \bar{\rho}_c^-)V}{\kappa T}. \tag{5.42}$$

Hicks & McMillan (1984) note that if the separate volumes can be measured, a correction can be made when $V^+ \neq V^-$ to improve the accuracy.

We also note that from Eqn. 5.37 (since the total volume collected is $2V$),

$$V/T = \tfrac{1}{2}\kappa\overline{|w'|}. \tag{5.43}$$

For a particular velocity distribution, $\overline{|w'|}$ is related to σ_w by a proportionality constant. Therefore,

$$\overline{w'\rho_c'} \simeq b\sigma_w(\bar{\rho}_c^+ - \bar{\rho}_c^-). \tag{5.44}$$

Businger & Oncley (1990) have proposed this simplification of the eddy accumulation technique, where the samples are collected at the same rate in each reservoir. Through calculations with surface-layer data sets, they found that the parameter b is about 0.6 over a range of both positive and negative stabilities. Wyngaard & Moeng (1992) show that $b = 0.63$ for a joint Gaussian probability distribution for w and c. Furthermore, they found good agreement with the value $b \simeq 0.6$ for bottom-up diffusion, which normally dominates near the surface, but a smaller value $b \simeq 0.47$ for top-down diffusion, which may be relevant for measurements in the mixed layer.

The merits and problems of the eddy accumulation technique have been discussed by Hicks & McMillan (1984). Some problems are:

1 the difference in mean concentration between the two reservoirs is typically very small – less than a couple of per cent – which could present storage and handling problems;

2 controlling the flow rate with the required speed, accuracy and dynamic range is difficult; and

3 it is necessary to remove any mean offset in the vertical velocity in real time.

As a result, eddy accumulation is viewed with some scepticism by many in the field. In many cases, however, species sensors may not have the required frequency response for direct eddy correlation so that eddy accumulation may be the only viable alternative. This opens the door to many additional sensing technologies such as gas chromatography and collection of samples for later analysis in the laboratory.

5.3.3 Gradient and difference techniques

Direct measurement of fluxes of trace species requires fast-response concurrent measurements of both the vertical air velocity and the trace species. However, there are other ways of measuring flux. They generally require some empirically determined relationship to estimate the flux. The most common derived technique is the gradient method. Here, the flux is estimated from the difference in concentration between two or more levels. This difference can be calculated from the integral of the flux–gradient similarity relationship (Eqn. 5.15). In the surface layer, this takes the form

$$F_{c0} = -\frac{u_* k(C_2 - C_1)}{\int_{z_1}^{z_2} \phi_c(z/\mathcal{L})d\ln z}. \tag{5.45}$$

Close to the surface or in conditions close to neutral, $\phi_c \simeq 1$, so that Eqn. 5.45 reduces to

$$F_{c0} = -\frac{u_* k(C_2 - C_1)}{\ln(z_2/z_1)}. \tag{5.46}$$

Furthermore, close to the surface (i.e. neutral stability) $K_c \simeq K_m \simeq u_* kz$, where K_m is the aerodynamic transfer coefficient for momentum,

$$-(\overline{u'w'})_0 = u_*^2 = K_m\frac{\partial U}{\partial z}. \tag{5.47}$$

Therefore, the mean wind gradient close to the surface is given by

$$\frac{\partial U}{\partial z} = \frac{u_*}{kz}. \tag{5.48}$$

Integrating Eqn. 5.48 between two measurement levels z_1 and z_2,

$$U_2 - U_1 = \frac{u_*}{k}\ln(z_2/z_1). \tag{5.49}$$

Substituting Eqn. 5.49 into 5.45,

$$F_{c0} = -\frac{(U_2 - U_1)k^2(C_2 - C_1)}{[\ln(z_2/z_1)]^2}. \tag{5.50}$$

Thus the flux of c can be estimated solely by differences in wind speed and concentration at two levels.[*]

In order to obtain the mean wind profile, we integrate Eqn. 5.48 from a level z down to a level z_0, known as the roughness length, where $U(z) \rightarrow 0$,

$$U(z) = \frac{u_*}{k}\ln(z/z_0). \tag{5.51}$$

The roughness length is approximately 1/30 the height of individual roughness elements. (Typical values are 3×10^{-5} to 3×10^{-4} m over large expanses of water, 0.004–0.02 m over fairly level grass-covered surfaces, 0.02–0.1 m over farmland, and 0.2–0.5 m over forest; Stull, 1988.) This logarithmic law holds only for the limiting case of a neutral surface layer. Furthermore, over a plant canopy or some other surface containing closely spaced roughness elements, Eqn. 5.51 also needs to be modified by replacing z in the numerator by $z - d$, where d is known as the displacement height, to obtain

$$U(z) = \frac{u_*}{k}\ln\left(\frac{z - d}{z_0}\right). \tag{5.52}$$

The value of d, which results from the canopy acting as a displaced lower boundary, is typically 70–80% of the canopy height (Kaimal & Finnigan, 1992). Values of z_0 and d can be estimated from data by plotting $\ln(z - d)$ versus $U(z)$; the appropriate value of d is that which results in a straight line, the zero intercept is z_0, and the slope of the line is u_*/k.

Similar relations are obtained for species concentrations by integrating Eqn. 5.15. Typically, however, the transfer lengths for species concentrations z_c are not equal to z_0; for temperature and water vapour the transfer lengths are generally a fraction of z_0, reflecting the greater efficiency of a canopy to absorb momentum than scalars (Kaimal & Finnigan, 1992). Substituting $z_1 = z_0$ into Eqn. 5.49 and $z_1 = z_c$ into Eqn. 5.46, and substituting Eqn. 5.49 into 5.46,

$$F_{c0} = -\frac{k^2}{\ln(z_2/z_0)\ln(z_2/z_c)}U_2(C_2 - C_0). \tag{5.53}$$

---

[*] More generally, if we had not assumed neutral stability,

$$F_{c0} = \frac{(U_2 - U_1)k^2(C_2 - C_1)}{-\int_{z_1}^{z_2}\phi_c d\ln z \int_{z_1}^{z_2}\phi_m d\ln z},$$

which requires also an estimate of the stability parameter $\mathscr{L}$.

This formulation is particularly useful for estimating flux over the ocean. In this case the standard measurement height z_2 is commonly selected to be $10 \, \text{m}$, so that

$$F_{c0} = \mathscr{C}_c U_{10}(C_0 - C_{10}), \tag{5.54}$$

where $\mathscr{C}_c$, which is a slowly varying function of $z/\mathscr{L}$, is known as the aerodynamic transfer coefficient. At neutral stability, the aerodynamic transfer coefficient for momentum (the drag coefficient) increases from about 0.001 for U from 2 to $5 \, \text{m s}^{-1}$ to 0.002 at $24 \, \text{m s}^{-1}$. The transfer coefficients for temperature and water vapour (commonly assumed to be independent of wind speed) are $\mathscr{C}_{TN} \simeq 0.0010$ and $\mathscr{C}_{EN} \simeq 0.0012$, respectively (Smith, 1988); thus the fluxes of T and q can be conveniently estimated by measuring the surface temperature, and the mean wind, air temperature and humidity at $10 \, \text{m}$.

Fluxes of other trace species across the air–sea interface are often estimated from the difference between the air and water concentrations of the trace species times a transfer velocity (Liss, 1983). The transfer velocity includes contributions from transfer through both the water and the atmosphere. Following Liss (1983), this transfer velocity may be expressed either in terms of the water concentration or the air concentration, which are related through the Henry's law constant H, which is the dimensionless ratio of the equilibrium trace species concentration in air (kg m^{-3} of air) to the equilibrium concentration of non-ionized dissolved gas in water (kg m^{-3} of water). Expressing the transfer velocity in terms of the concentration in air,*

$$\frac{1}{K_a} = \frac{1}{k_a} + \frac{H}{\alpha k_w}, \tag{5.55}$$

where k_w and k_a are the transfer velocities in water and air, respectively, and α is an enhancement factor that accounts for chemical reactions between the trace species and water. The order of magnitude of k_w is $\sim 0.001 \, \text{cm s}^{-1}$ at $U \sim 5 \, \text{m s}^{-1}$ to $\sim 0.01 \, \text{cm s}^{-1}$ at $U \sim 15 \, \text{m s}^{-1}$; the magnitude of k_a is $\sim 1 \, \text{cm s}^{-1}$ at $3 \, \text{m s}^{-1}$ to $\sim 3 \, \text{cm s}^{-1}$ at $U \sim 10 \, \text{m s}^{-1}$ (Liss & Merlivat, 1986). Therefore, neglecting chemical reactivity effects, for species with $H \ll \alpha k_w/k_a$; i.e. for species with a small Henry's law constant or that are chemically active in water, the flux is controlled by the transfer velocity through the air. Examples of this include H_2O, HCl, SO_2, NH_3 and HNO_3. On the other hand, for $H \gg \alpha k_w/k_a$ the flux is controlled by the transfer through the water; examples include CO_2, CO, CH_4, N_2O, methyl iodide and DMS (Liss & Slater, 1974).

---

* Alternatively, the transfer velocity can be expressed in terms of the concentration in water, $K_w^{-1} = (\alpha k_w)^{-1} + (H k_a)^{-1}$.

Equation 5.55 is often written in the form of resistances in analogy to electrical circuitry,

$$R_a = r_a + r_w, \tag{5.56}$$

where $R_a = K_a^{-1}$, $r_a = k_a^{-1}$, and $r_w = H(\alpha k_w)^{-1}$. This analogy is also frequently used in the context of estimating atmospheric deposition to surfaces for species such as O_3, SO_2 and HNO_3 that react with and are absorbed on vegetation where, in this case, the resistance to deposition is expressed as

$$r = r_a + r_s + r_c. \tag{5.57}$$

In this context, r_a is the aerodynamic resistance of the atmosphere from a reference height z down to the roughness length z_0; r_s is the resistance contributed by the particular molecular or particulate properties of the species and the difference between z_0 and the transfer length of the species; and r_c is the remaining (residual) resistance due to the less-than-perfect absorption of molecules or particles that strike the surface (Wesely, 1983). Often, the deposition velocity, $v_d \equiv r^{-1}$ is used in the literature to describe the absorption of trace gases or particles at the surface, so that $F_c = -v_d C$.

In the mixed layer, as discussed earlier, both the entrainment flux and the surface flux affect the mixed layer profile. Therefore, two equations, and a minimum of three levels are needed to obtain the surface flux. From Eqns 5.21 and 5.23, the relationship between the concentration differences and the top and bottom fluxes is given by

$$C_j - C_k = (z_i w_*)^{-1} \left[-F_{c0} \int_{z_k}^{z_j} g_0 dz - F_{ci} \int_{z_k}^{z_j} g_i dz \right]. \tag{5.58}$$

The surface flux can then be obtained in principle by solving Eqn. 5.58 for two sets of concentration differences. This has been done by Davis (1992) for isoprene measurements at several levels in the mixed layer from a tethered balloon. This technique promises to be particularly useful for estimating fluxes in remote areas or over forest canopies using aircraft or tethered balloons in situations where surface-layer measurements are difficult.

Measurement of fluxes in the vicinity of a plant canopy presents added complications compared with measurements well above the canopy. The structure within the foliage does not follow the similarity laws previously discussed, since exchanges of mass, momentum and heat are occurring within the canopy. Here, simple gradient diffusion models are not adequate for describing vertical transfer. Raupach (1988) has shown, however, that vertical profiles of normalized mean and turbulent quantities within and just above canopies tend to collapse onto universal curves when plotted as functions of height normalized by the canopy height. Raupach *et al.* (1989) also found that canopy turbulence is much more intermittent than above, and exhibits characteristic coherent structures.

Apparent turbulent diffusivities just above the canopy can be as much as two to three times those predicted from classical theory in this so-called 'roughness sublayer', whose depth is estimated to be about three to four times the scale of the horizontal inhomogeneities in the canopy (Shuttleworth, 1989; Cellier & Brunet, 1992). One approach, of course, is always to measure above this roughness layer. This may often be inconvenient or undesirable since higher up the concentration differences are smaller and the fetch requirements are greater. For this reason, Cellier & Brunet (1992) have devised generalized flux–profile relationships that can be used within this roughness sublayer.

Another complication is the differences in diffusivities for different species that may occur just above the canopy depending upon their sources or sinks. For example, during daytime, isoprene is emitted by leaves while N_2O is emitted at the ground. Quantifying these diffusivity differences is an area that requires further research.

5.3.4 Other indirect techniques

Bowen ratio (surface energy budget)
The Bowen ratio is defined as the ratio of sensible ($c_p\bar{\rho}F_T$) to latent (λF_q) surface heat fluxes. It was originally used, together with the other terms in the surface energy budget, to estimate these fluxes at the surface (Denmead, 1983; Stull, 1988; Fowler & Duyzer, 1989). Advantages of this approach are that eddy flux measurements and stability corrections are not required; a drawback is that it requires measurements of the incoming net radiation at the surface R_n and the soil heat flux G. The surface energy budget is

$$R_n - G = c_p\bar{\rho}F_T + \lambda F_q. \tag{5.59}$$

From Eqn. 5.26, if the transfer characteristics for temperature, humidity and a trace species are assumed to be identical,

$$F_c = \frac{R_n - G}{c_p\bar{\rho}\Delta\bar{T} + \lambda\Delta\bar{q}}\Delta C, \tag{5.60}$$

where $\Delta Y \equiv Y_2 - Y_1$, the difference in the variable Y between two levels in the surface layer.

Alternatively, a trace species flux can be estimated from

$$F_c = F_T\frac{\Delta C}{\Delta\bar{T}} = F_q\frac{\Delta C}{\Delta\bar{q}}. \tag{5.61}$$

In this way, flux of a trace species can be obtained from concentration difference measurements of that species plus concurrent flux and difference measurements of either temperature, humidity or another trace species. Similarly, the magnitude (but not the sign) of flux could, in principle, also be obtained from the standard deviation σ_c, and the flux and standard deviation of temperature or humidity (Wesely, 1988),

$$|F_{c0}| = \frac{\sigma_c}{\sigma_T}|F_{T0}| = \frac{\sigma_c}{\sigma_q}|F_{q0}|. \tag{5.62}$$

This technique may be particularly useful in situations such as estimating fluxes over a forest or in other cases where surface–layer similarity relationships are difficult to apply. Again, with all these techniques that use a surrogate variable to characterize the transport process for species concentration we assume that the transport process is identical for both.

Variance and dissipation
Another approach to estimating fluxes is to use turbulence moments which can be related to surface fluxes through the similarity relationships discussed previously. A straightforward example is solving Eqn. 5.17 – which is the normalized standard deviation of fluctuations in species concentration in the surface layer – for the species flux (Wesely, 1988; De Bruin *et al.*, 1993),

$$|F_{c0}| = \sigma_c u_* \phi_s^{-1}, \tag{5.63}$$

where $\phi_s(z/\mathcal{L})$ is given by Eqns 5.18 and 5.19. Similarly, from Eqn. 5.25, in the lower part of the mixed layer, the surface flux can be estimated from

$$|F_{c0}| = 0.77\sigma_c w_*(z/z_i)^{1/3}. \tag{5.64}$$

Of course, the standard deviation is not sensitive to the sign of the flux, since fluctuations can be equally generated by a flux of either sign. Therefore, if the sign of the flux is not obvious, other information is necessary to determine it. One limitation of this technique is the effect of horizontal advection over terrain with horizontally varying surface characteristics which can also generate fluctuations in species concentration. Furthermore, this effect will be different for different quantities if their fluxes have different horizontal distributions (Wesely, 1988; De Bruin *et al.*, 1993).

A related approach is the dissipation technique. This technique utilizes the hypothesis that the dissipation of turbulent kinetic energy, temperature and species concentration variances can be estimated from the spectrum of fluctuations of these quantities in the so-called inertial subrange, which is the frequency range between the large-scale turbulent eddies that are responsible for most of the energy and the smallest eddies which are responsible for the dissipation of the variance through molecular diffusion. In this intermediate range, the variance spectrum is proportional to the $-5/3$ power of wave number (or frequency), so in principle one needs only to measure at a particular frequency in this range to obtain the variance dissipation. By making concurrent measurements of the velocity, temperature and species concentration fluctuations in the inertial subrange it is possible, through consideration of the variance budgets of these variables, to estimate simultaneously the fluxes of these quantities. The relationships needed for this are presented and discussed by Edson et al. (1991) and Fairall & Larsen (1986). This technique is particularly suited for ship measurements because it is not necessary to make corrections for ship motions (since the ship motions are at lower frequencies), and the measurements are less affected than covariance measurements by flow distortion caused by the ship and supporting structures.

Mass balance

A different approach for estimating surface flux is through the use of mass balance or budget techniques. In contrast to variance and profile techniques, this is an absolute technique in the sense that no empirical relationships are necessary to estimate the flux. The budget for the mean concentration of a species is given by (e.g. Stull, 1988)

$$\frac{dC}{dt} = \frac{\partial C}{\partial t} + U(z)\frac{\partial C}{\partial x} + W(z)\frac{\partial C}{\partial z} = -\frac{\partial F_c}{\partial z} + Q_c, \tag{5.65}$$

where Q_c is the sum of the sources and sinks of C due to chemical reactions or phase changes. For simplicity, I have oriented the x axis along the mean horizontal flow direction. Integrating Eqn. 5.65 from the surface up to some height z, assuming a well-mixed layer so that $\partial C/\partial z = 0$, and solving for the surface flux,

$$F_{c0} = z\frac{\partial\{C\}}{\partial t} + z\{U\}\frac{\partial\{C\}}{\partial x} + (\overline{w'c'})_z - z\{Q_c\}, \tag{5.66}$$

where the curly braces denote a vertical average. The application of this technique depends on the particular experimental conditions. Williams (1982) discusses limitations in using it to estimate surface deposition.

We can also integrate Eqn. 5.65 through the entire ABL from the surface to a level just above z_i (where the turbulence, and thus the turbulent flux of c vanishes), and use the relation for the flux just below the top of the ABL (Lilly, 1968), $F_{ci} = -w_e\Delta C$, where ΔC is the jump in C across the thin inversion layer capping a well-mixed CBL and w_e is the entrainment velocity – i.e. the rate at which air from the overlying free atmosphere is mixed by turbulence into the ABL (Wesely *et al.*, 1989), to obtain

$$F_{c0} = z_i \frac{\partial\{C\}}{\partial t} + z_i\{U\}\frac{\partial\{C\}}{\partial x} - \{U\}\Delta C \frac{\partial z_i}{\partial x} - w_e \Delta C - z_i\{Q_c\}. \quad (5.67)$$

The third term results from a slope in the capping inversion layer. It may be possible to evaluate surface flux by this technique from an aircraft for a species which can be measured only with slow response (but very accurate) instruments. The entrainment velocity w_e may be estimated by using another variable for which eddy flux measurements are possible by extrapolating flux measurements at several levels well below the ABL top (to avoid flying through the corrugated inversion layer) and extrapolating the flux profile through the top.

This mass balance approach has also been applied to estimating surface flux from treated (e.g. fertilized) plots of limited along-wind extent (e.g. Denmead, 1983; Wilson *et al.*, 1983; Fowler & Duyzer, 1989). In this case, assuming steady-state, no internal sources or sinks or background concentration of the trace species, no horizontal variation in the mean horizontal velocity field, and integration at some point downwind of the plot from the surface to a depth z that encompasses the entire plume of trace species emitted by the plot, Eqn. 5.65 becomes

$$F_{c0} = \int_0^z U \frac{\partial C}{\partial x} dz = \int_0^z \frac{\partial UC}{\partial x} dz = \frac{1}{X}\int_0^z UC dz. \quad (5.68)$$

This approach permits evaluation of the surface flux from mean measurements of concentration and horizontal wind as a function of height downwind of a test plot, independent of stability. However, stability will affect the height to which the emission plume extends.

5.4 Spatial variability of fluxes

Thus far, flux measurement has been considered only over horizontally homogeneous surfaces. The history of observational micrometeorology is one of experiments carried out over horizontally homogeneous sites, as demanded by similarity theory. Unfortunately, in most practical situations (an obvious exception is the open sea) surfaces are not horizontally homogeneous. Over land, hills and scattered trees are obvious sources of inhomogeneity. However, other more subtle effects may contribute. Varying soil fertility or tilth, varying density of plant cover, or even footpaths used to service ground-based instruments may induce horizontally varying patterns in surface roughness, evaporation, plant growth and trace gas emissions and deposition.

This issue is receiving increasing attention because meso- and large-scale numerical models of the atmosphere implicitly predict grid averaged fluxes, and fluxes inferred from satellite data are area averages. Unfortunately, area-averaged fluxes do not necessarily obey existing flux formulations based on measurements over homogeneous terrain. For example, in transition regions generated by changes at the surface, an internal boundary layer may form (Garratt, 1992) which deepens downwind. Within the internal boundary layer, the flow reaches equilibrium with the new surface conditions. Above the internal boundary layer, the flow is influenced by conditions upstream from the change of surface conditions.

If the surface heterogeneity is not organized into well-defined regions of uniform surface conditions, the concept of an internal boundary layer and existing models of the transition zones are no longer applicable. However, if the surface heterogeneity occurs on a sufficiently small scale and is randomly distributed, then the concept of a blending height might be applicable (Mason, 1988). Above the blending height, the fluxes are in statistical equilibrium with the random surface heterogeneity and the influence of individual elements is no longer obvious due to spatial integration by the turbulent eddies.

If the heterogeneity occurs on scales of tens of kilometres and mean winds are weak, the surface variations may directly generate mesoscale circulations which lead to vertical transport (Yan & Anthes, 1988; Segal & Arritt, 1992). This process, however, is usually of secondary importance. Rather, modulation of the turbulent flux by surface heterogeneity is likely to be relatively more important than the direct mesoscale flux. Since the turbulent flux and vertical gradient are related in a non-linear way, the relationship between the spatially averaged flux and the spatially averaged gradient is different from that predicted by the usual relationships for surface fluxes so that correction terms may be necessary (Mahrt, 1987; Claussen, 1990).

Finally, area-averaged momentum fluxes are often enhanced by bluff roughness effects associated with obstacles such as forest edges and isolated trees (Beljaars & Holtslag, 1991). This effect is due to local pressure drag and does not enhance the area-averaged fluxes of heat or trace gases. As a result, modelling momentum flux is quite different from modelling fluxes of other quantities.

One way of dealing with the issue of heterogeneity, particularly for fixed-point measurements, is to define a flux footprint, which is a weighting function that describes the contribution per unit emission of each element of a surface area source to the vertical species flux measured at a particular height above the surface. Thus, the vertical flux at a point at height z is equal to the area integral of the product of the surface flux distribution and the flux footprint. The flux footprint in the surface layer has been estimated from a simple analytic dispersion model by Schuepp *et al.* (1990), and from a more sophisticated analytical model and a

Lagrangian stochastic dispersion model by Horst & Weil (1992). Weil & Horst (1992) have also applied the Lagrangian model to footprints in the mixed layer.

As an example, the results of Horst & Weil (1994) are applied to the case of a horizontally homogeneous distribution of surface flux, a roughness length $z_0 = 0.01$ m, and a measurement height $z = 2$ m. For neutral stratification (i.e. $z_0/\mathcal{L} = 0$) 90% of the cross-wind integrated flux measured at a height of 2 m is emitted within an upwind distance of about 400 m and 50% within about 35 m. The corresponding distances for $z_0/\mathcal{L} = -10^{-3}$ are 70 m and 20 m, respectively; and for $z_0/\mathcal{L} = 10^{-3}$ they are 750 m and 60 m, respectively. For neutral stratification the traditional rule of thumb that the upwind fetch for horizontally homogeneous conditions should be at least $100z$ is marginally sufficient, while for stable stratification (i.e. $z_0/\mathcal{L} > 0$), it is definitely insufficient. That is, if flux variations occur even somewhat beyond $100z$ in the SBL, the measured flux would be affected. Horst & Weil (1994) further discuss fetch requirements and also note that decreasing the surface roughness increases the required fetch.

In the mixed layer, Weil & Horst (1992) define a length $\Delta x_{0.5}$, which is the width of the region at the surface where the cross-wind integrated flux footprint measured at a height z_m is $\geq$ half its peak value. They obtained the expression

$$\Delta x_{0.5} = 0.9 \frac{U z_i}{w_*} \left(\frac{z_m}{z_i} \right)^{2/3}, \tag{5.69}$$

for $z_m/z_i \leq 0.4$. For $z_m/z_i > 0.4$, the width no longer increases. This is a measure of how small a scale of horizontal flux variability can be resolved from aircraft measurements. As an example, for $U = 5$ m s^{-1}, $z_i = 1000$ m and $z_m = 300$ m, $\Delta x_{0.5} \simeq 2$ km.

5.5 Platform-specific techniques and instruments

Until now, most of what has been discussed applies to both ground-based and airplane measurements. One difference is that *in situ* tower measurements are mostly in the surface layer, while airplane measurements are mostly in the mixed layer. This leads to differences in measurement techniques because of the different scaling laws that apply. In the case of aircraft, the velocity of the air relative to the Earth must be measured from a relatively fast-moving platform. For measurements from ships, the sensors are in the surface layer, but the platform is moving about. Platform motions are of major importance for eddy flux measurements since the vertical air velocity used for flux measurements is the difference between the platform velocity and the air velocity measured by the platform. In order to calculate this velocity difference, the air velocity measurement must be rotated to an Earth-based coordinate system, which requires measurement of the attitude angles of the platform.

5.5.1 Ground-based techniques

Technology for measuring air–surface exchange from ground-based sites has become increasingly reliable and easy to use. The most commonly used instrument for measuring w for eddy correlation is the sonic anemometer. This instrument measures the transit time of sound waves passing between two transducers typically spaced 10 cm to a few tens of centimetres apart. Since there are no moving parts, the frequency response is usually limited by the line averaging over the path between the transducers, and the velocity resolution is more than sufficient for flux measurement. For flux measurements over a surface in an unstably stratified surface layer, Kristensen & Fitzjarrald (1984) have shown that the ratio of the height above the ground to the path length between the transducers should be >5.

One source of error for vertical velocity sensors is misalignment with respect to the local Earth vertical to avoid contamination of the vertical flux by the streamwise flux, which can be as much as three times greater than the magnitude of the vertical flux (Wyngaard et al., 1971). This can best be dealt with by measuring both the horizontal and vertical components of the flux with a three-component anemometer and then analytically rotating the measured fluxes to a coordinate system with axes parallel and normal to the streamlines. Physically aligning the velocity sensor to the local Earth vertical is adequate by itself only over flat terrain.

Contamination of the measured vertical flux by the streamwise flux can also be caused by flow distortion around the sensors and supporting structures. To avoid this flux crosstalk, Wyngaard (1988a) recommends that the sensor array be symmetric about the horizontal plane passing through the point at which the vertical velocity is measured. Wyngaard also notes that even in a symmetric array the vertical velocity is amplified and the flux is thus overestimated due to flow blockage by the measurement array. Thus the bulk of the array should be minimized to reduce its impact on the vertical velocity measurement. Wyngaard & Zhang (1985) discuss the impact of these 'transducer shadow' effects for various sonic anemometer designs.

The sonic anemometer is useful not only for measuring w and $\overline{w'c'}$, but also for measuring $\overline{u'w'}$, which is necessary for calculating $\mathcal{L}$. However, if only w is measured by a single-axis sonic, u_* can be estimated, for example, from the similarity relations (Kaimal & Finnigan, 1992)

$$\sigma_w/u_* = \phi_w \simeq 1.25(1 - 3z/\mathcal{L})^{1/3}; \quad -2 \leqslant z/\mathcal{L} \leqslant 0 \tag{5.70}$$

$$\simeq 1.25(1 + 0.2z/\mathcal{L}); \quad 0 \leqslant z/\mathcal{L} \leqslant 1, \tag{5.71}$$

keeping in mind that u_* is not only explicitly on the left side of Eqns 5.70 and 5.71, but is also contained in $\mathscr{L}$.

Other sensors, such as bivanes and propellers, have also been used to measure w (Kaimal, 1986; Kaimal & Finnigan, 1992). They are simpler and less expensive than sonic anemometers, but have the disadvantages of mechanical bearings and non-linear response (Kristensen, 1992).

5.5.2 Aircraft techniques

It is also feasible to estimate surface fluxes from low level traverses flown by instrumented aircraft. In this way, true spatially averaged fluxes and measurements over surfaces that are difficult to access from the ground, such as forest and ocean can be measured. However, the mobility of the aircraft also means that aircraft motions must be removed from the measured air velocity before velocity fluctuations relative to the Earth can be obtained. The instrumentation and techniques for doing this are described by Lenschow (1986). The velocity and attitude angles of the airplane relative to the Earth are commonly obtained from an inertial navigation system (INS), which contains an orthogonally mounted triad of accelerometers whose outputs are integrated to obtain the airplane velocity, and integrated again to obtain position. The accelerometers may be mounted either on a platform that is stabilized by means of gyroscopes to the local Earth vertical and true north, or mounted directly to the aircraft, with the local Earth vertical and true north updated continuously by means of computer calculations.

Air velocity relative to the aircraft is commonly measured by a combination of instruments. A Pitot tube mounted at the front of the aircraft is used, along with a static pressure port, for a pressure difference measurement which is called the dynamic pressure. This measurement is combined with static pressure and air temperature to obtain the true airspeed U_a, which is the magnitude of the air velocity vector. This measurement is then combined with a measurement of the air flow angles of the velocity vector (using vanes or differential pressure measurements) relative to the longitudinal axis of the airplane. As an illustrative example, for small angular displacements, the vertical air velocity can be approximated by

$$w \simeq U_a(\alpha - \theta) + w_p, \tag{5.72}$$

where α is the angle of the airstream in the vertical plane of the aircraft (in aircraft parlance, the angle of attack), θ is the angle of the longitudinal axis of the airplane measured from the Earth's horizontal plane (the pitch angle), and w_p is the vertical velocity of the airplane relative to the Earth. Both θ and w_p are variables that can be obtained from an INS. Similar equations can be used to estimate the horizontal components. This relationship is very much simplified over what is normally used in practice to obtain w because it does not include the second-order cross terms that result from the aircraft not flying straight and level.

Since the aircraft speed is typically about 10 times the mean wind speed, air compressibility effects, which are not important for ground-based measurements, often need to be considered. This is in addition to whatever effects flow distortion may have on the measured w. Wyngaard (1988b) shows, for example, that at aircraft speeds, species density can be appreciatively altered. Without corrections, this can lead to serious contamination of density flux by temperature and momentum fluxes. On the other hand, species mixing ratio is still conserved. Thus, for airborne flux measurements both compressibility effects, as well as the Webb effect (Eqn. 5.2) argue for the desirability of sensing species mixing ratio; or if that is not possible, making accurate concurrent measurements of the temperature and pressure of the sampled air to calculate its density, and thus the species mixing ratio. Both these errors become more serious for small values of $\overline{w' \rho_c'}/\overline{\rho_c}$.

Although aircraft have been used to measure fluxes at levels $<30\,\mathrm{m}$ above the surface (Desjardins & MacPherson, 1989), they often, for safety reasons, are not flown close enough to the surface to measure the surface flux directly. This can be corrected for, however, by flying at several levels and linearly extrapolating the flux profile to the surface. Justification for a linear extrapolation is based on the assumption that vertical transport in the CBL is dominated by turbulence. The budget equation for a conserved species can be written as

$$\frac{\partial C}{\partial t} = -\frac{\partial F_c}{\partial z} + A, \tag{5.73}$$

where A represents the horizontal advection terms. Differentiating Eqn. 5.73 with height and interchanging the order of differentiation,

$$\frac{\partial}{\partial t}\left(\frac{\partial C}{\partial z}\right) = -\frac{\partial^2 F_c}{\partial z^2} + \frac{\partial A}{\partial z}. \tag{5.74}$$

The time scale for changes in the vertical gradient of the mean concentration is typically an order of magnitude larger than the turbulence mixing time scale (Eqn. 5.29), $4z_i/w_*$. Furthermore, vertical variations in the horizontal advection terms are often negligible (Davis, 1992). Therefore,

$$\frac{\partial^2 F_c}{\partial z^2} \simeq 0. \tag{5.75}$$

Integrating Eqn. 5.75 twice and using the fluxes at the surface and top as boundary conditions,

$$F_c \simeq Az + B = F_{c0} + \frac{F_{ci} - F_{c0}}{z_i}z. \tag{5.76}$$

Mann and Lenschow (1994) discuss errors in fluxes and flux gradients measured by aircraft in the CBL.

5.5.3 Remote techniques

Thus far, only *in situ* techniques have been discussed for flux measurement. There are, however, several potential techniques for measuring fluxes remotely. One possibility is by means of the convective layer gradient technique discussed earlier. LIDAR systems have been developed that can measure concentrations of trace gases remotely by both the Raman technique (Melfi, 1972, describes a ground-based water vapour system) or by the differential absorption LIDAR (DIAL) technique (Browell, 1989, for example describes an airborne ozone system; several water vapour systems have also been developed). If concentration differences through the mixed layer can be measured with sufficient accuracy, it may be possible to use these systems either from the ground or an aircraft (along with an estimate of the convective velocity w_*) to estimate the surface flux (Davis, 1992).

Another possibility is to combine a remote wind measurement with LIDAR-based concentration measurements directly to measure the eddy flux of concentration. A Doppler LIDAR or radio acoustic sounding system (RASS) can measure the along-beam velocity component as a function of height. The concurrent concentration and vertical velocity measurements can, in principle, be multiplied together and averaged to obtain the flux as a function of height. Bösenberg *et al.* (1991) have attempted to measure water vapour flux using a ground-based DIAL concurrently with a RASS. Using this technique, it may be possible to obtain vertical flux profiles either from the ground or from aircraft which can be used to estimate vertical flux divergence. This approach offers a whole new dimension in applications of flux measurements to addressing sources, sinks and distributions of trace gases in the atmosphere.

Acknowledgements

I am grateful for the many helpful comments from K. Davis, T. Horst, L. Kristensen, L. Mahrt and J. Weil. I especially appreciate the thorough review and helpful suggestions of D. Baldocchi.

References

Baldocchi D.D. (1991) Canopy control of trace gas emissions. In: Sharkey T.D., Holland E.A. & Mooney H.A. (eds) *Trace Gas Emissions by Plants*, pp. 293–333. Academic Press, New York.

Baldocchi D.D., Hicks B.B. & Meyers T.P. (1988) Measuring biosphere–atmosphere exchanges of biologically related gases with micrometeorological methods. *Ecology*, **69**(5), 1331–1340.

Beljaars A.C.M. & Holtslag A.A.M. (1991) Flux parameterization over land surfaces for atmospheric models. *Journal of Applied Meteorology*, **30**, 327–341.

Bösenberg J., Senff C. & Peters G. (1991) Measurement of water vapour flux profiles in the PBL with lidar and radar-RASS. In: *Extended Abstracts, Lower Tropospheric Profiling: Needs and Technologies*. The National Center for Atmospheric Research, PO Box 3000, Boulder, CO 80307.

Browell E.V. (1989) Differential absorption lidar sensing of ozone. In: Proceedings, IEEE, **77**.

Businger J.A. (1986) Evaluation of the accuracy with which dry deposition can be measured with current micrometeorological techniques. *Journal of Climate and Applied Meteorology*, **25**, 1100–1124.

Businger J.A. & Oncley S.P. (1990) Flux measurement with conditional sampling. *Journal of Atmospheric and Oceanic Technology*, **7**, 349–352.

Caughey S.J., Wyngaard J.C. & Kaimal J.C. (1979) Turbulence in the evolving stable layer. *Journal of the Atmopheric Sciences*, **6**, 1041–1052.

Cellier, P. & Brunet Y. (1992) Flux gradient relationships above tall plant canopies. *Journal of Agricultural and Forest Meteorology*, **58**, 93–117.

Claussen M. (1990) Area-averaging of surface fluxes in a neutrally stratified, horizontally inhomogeneous atmospheric boundary layer. *Atmospheric Environment*, **24a**, 1349–1360.

Davis K.J. (1992) *Surface fluxes of trace gases derived from convective-layer profiles*. PhD dissertation, University of Colorado. Available as NCAR/CT-139 from NCAR, PO Box 3000, Boulder, CO 80307.

De Bruin H.A., Kohsiek W. & Van Den Hurk B.J.J.M. (1993) A verification of some methods to determine the fluxes of momentum, sensible heat, and water vapour using standard deviation and structure parameter of scalar meteorological quantities. *Boundary-Layer Meteorology*, **63**, 231–257.

Denmead O.T. (1983) Micrometeorological methods for measuring gaseous losses of nitrogen in the field. In: Freney J.R. & Simpson J.R. (eds) *Gaseous Loss of Nitrogen from Plant–Soil Systems*, pp. 133–157. Martinus Nijhoff/W. Junk, The Hague.

Desjardins R.L. (1972) *A study of carbon dioxide and sensible heat fluxes using the eddy correlation technique*. PhD dissertation, Cornell University.

Desjardins R.L. & MacPherson J.I. (1989) Aircraft-based measurements of trace gas fluxes. In: Andreae M.O. & Schimel D.S. (eds) *Exchange of Trace Gases between Terrestrial Ecosystems and the Atmosphere*, pp. 135–172. John Wiley & Sons, New York.

Edson J.B., Fairall C.W., Mestayer P.G. & Larsen S.E. (1991) *An Experimental and Theoretical Investigation of the Inertial-Dissipation Method for Computing Air–Sea Fluxes*. NOAA Technical Memorandum ERL WPL-199. Available from the National Technical Information Service, 5285 Port Royal Road, Springfield, VA 22161.

Fairall C.W. & Larsen S.E. (1986) Inertial dissipation methods and turbulent fluxes at the air–ocean interface. *Boundary-Layer Meteorology*, **34**, 287–301.

Fitzjarrald D.R. & Lenschow D.H. (1983) Mean concentration and flux profiles for chemically reactive species in the atmospheric surface layer. *Atmospheric Environment*, **17**, 2505–2512.

Fowler D. & Duyzer J.H. (1989) Micrometeorological techniques for the measurement of trace gas exchange. In: Andreae M.O. & Schimel D.S. (eds) *Exchange of Trace Gases between Terrestrial Ecosystems and the Atmosphere*, pp. 189–207. John Wiley & Sons, New York.

Garratt J.R. (1992) *The Atmospheric Boundary Layer*. Cambridge University Press, Cambridge.

Hicks B.B. & McMillan R.T. (1984) A simulation of the eddy accumulation method for measuring pollutant fluxes. *Journal of Climate and Applied Meteorology*, **23**, 637–643.

Holtslag A.A.M. & Moeng C.-H. (1991) Eddy diffusivity and countergradient transport in the convective atmospheric boundary layer. *Journal of the Atmospheric Sciences*, **48**, 1690–1698.

Horst T.W. & Weil J.C. (1992) Footprint estimation for scalar flux measurements in the atmospheric surface layer. *Boundary-Layer Meteorolology*, **59**, 279–296.

Horst T.W. & Weil J.C. (1994) How far is far enough? The fetch requirements for micrometeorological measurements of surface fluxes. *Journal of Atmospheric and Oceanic Technology*, **11**, 1018–1025.

Kaimal J.C. (1986) Flux and profile measurements from towers in the boundary layer. In: *Probing the Atmospheric Boundary Layer*, pp. 19–28. American Meteorological Society, Boston, Massachusetts.

Kaimal J.C. & Finnigan J.J. (1992) *Atmospheric Boundary Layer Flows – Their Structure and Measurement*. Oxford University Press, Oxford.

Kaimal J.C., Wyngaard J.C., Haugen D.A., Coté O.R. & Izumi Y. (1976) Turbulence structure in the convective boundary layer. *Journal of the Atmospheric Sciences*, **33**, 2152–2169.

Kaimal J.C., Wyngaard J.C., Izumi Y. & Coté O.R. (1972) Spectral characteristics of surface-layer turbulence. *Quarterly Journal of the Royal Meteorological Society*, **98**, 563–589.

Kim J. & Mahrt L. (1992) Simple formulation of turbulent mixing in the stable free atmosphere and nocturnal boundary layer. *Tellus*, **44A**, 381–394.

Kramm G., Müller H., Fowler D., Höfken K.D., Meixner F.X. & Schaller E. (1991) A modified profile method for determining the vertical fluxes of NO, NO_2, ozone, and HNO_3 in the atmospheric surface layer. *Journal of Atmospheric Chemistry*, **13**, 265–288.

Kristensen L. (1992) *The Cup Anemometer and Other Exciting Instruments. Risø-R-615(EN)*. Risø Library, Risø National Laboratory, PO Box 49, DK-4000, Roskilde, Denmark.

Kristensen L. & Fitzjarrald D.R. (1984) The effect of line averaging on scalar flux measurements with a sonic anemometer near the surface. *Journal of Atmospheric and Oceanic Technology*, **1**, 138–146.

Kristensen L., Lenschow D.H., Kirkegaard P. & Courtney M. (1989) The spectral velocity tensor for homogeneous boundary layer turbulence. *Boundary-Layer Meteorology*, **47**, 149–193.

Lenschow D.H. (1986) Aircraft measurements in the boundary layer. In: Lenschow D.H. (ed) *Probing the Atmospheric Boundary Layer*, pp. 29–55. American Meteorological Society, Boston, Massachusetts.

Lenschow D.H. & Delany A.C. (1987) An analytic formulation for NO and NO_2 flux profiles in the atmospheric surface layer. *Journal of Atmospheric Chemistry*, **5**, 301–309.

Lenschow D.H. & Hicks B.B. (eds) (1989) *Global Tropospheric Chemistry – Chemical Fluxes in the Global Atmosphere*. National Center for Atmospheric Research, Boulder, Colorado.

Lenschow D.H. & Kristensen L. (1985) Uncorrelated noise in turbulence measurements. *Journal of Atmospheric and Oceanic Technology*, **2**, 68–81.

Lenschow D.H. & Kristensen L. (1986) Comments on 'Sampling errors in flux measurements of slowly depositing pollutants.' *Journal of Climate and Applied Meteorology*, **25**, 1785–1787.

Lenschow D.H. & Raupach M.R. (1991) The attenuation of fluctuations in scalar concentrations through sampling tubes. *Journal of Geophysical Research*, **96**, 15259–15268.

Lenschow D.H. & Stankov B.B. (1986) Length scales in the convective atmospheric boundary layer. *Journal of the Atmospheric Sciences*, **43**, 1198–1209.

Lenschow D.H., Mann J. & Kristensen L. (1994) How long is long enough when measuring fluxes and other turbulence statistics? *Journal of Atmospheric and Oceanic Technology*, **11**, 661–673.

Leuning R. & King K.M. (1992) Comparison of eddy-covariance measurements of CO_2 fluxes by open- and closed-path CO_2 analysers. *Boundary-Layer Meteorology*, **59**, 297–311.

Leuning R. & Moncrieff J. (1990) Eddy-covariance CO_2 flux measurements using open- and closed-path CO_2 analysers: corrections for analyser water vapour sensitivity and damping of fluctuations in air sampling tubes. *Boundary-Layer Meteorology*, **53**, 63–76.

Lilly D.K. (1968) Models of cloud-topped mixed layers under a strong inversion. *Quarterly Journal of the Royal Meteorology Society*, **94**, 292–309.

Liss P.S. (1983) Gas transfer: experiments and geochemical implications. In: Liss P.S. & Slinn W.G.N. (eds) *Air–Sea Exchange of Gases and Particle*, pp. 241–298. Reidel, Dordrecht, The Netherlands.

Liss P.S. & Merlivat L. (1986) Air–sea gas exchange rates: Introduction and synthesis. In: Buat-Ménard P. (ed) *The Role of Air–Sea Exchange in Geochemical Cycling*, pp. 113–127. Reidel, Dordrecht, The Netherlands.

Liss P.S. & Slater P.G. (1974) Flux of gases across the air–sea interface. *Nature*, **247**, 181–184.

Mahrt L. (1987) Grid-averaged surface fluxes. *Monthly Weather Review*, **115**, 1550–1560.

Mann J. & Lenschow D.H. (1994) Errors in airborne flux measurement. *Journal of Geophysical Research*, **99**, No. D7, 14519–14526.

Mason P.J. (1988) The formation of areally-averaged roughness lengths. *Quarterly Journal of the Royal Meteorological Society*, **114**, 399–420.

Melfi S.H. (1972) Remote measurements of the atmosphere using Raman scattering. *Applied Optics*, **11**, 1605–1610.

Melfi S.H. & Whiteman D.H. (1985) Observations of lower-atmosphere moisture structure and its evolution using a Raman lidar. *Bulletin of the American Meteorological Society*, **66**, 1288–1292.

Moeng C.-H. & Wyngaard J.C. (1984) Statistics of conservative scalars in the convective boundary layer. *Journal of the Atmospheric Sciences*, **41**, 3161–3169.

Moeng C.-H. & Wyngaard J.C. (1989) Evaluation of turbulent transport and dissipation closures in second-order modeling. *Journal of the Atmospheric Sciences*, **46**, 2311–2330.

Raupach M.R. (1988) Canopy transport processes. In: Steffen W.L. & Denmead O.T. (eds) *Flow and Transport in the Natural Environment: Advances and Applications*, pp. 95–127. Springer-Verlag, Berlin.

Raupach M.R., Finnigan J.J. & Brunet Y. (1989) Coherent eddies in vegetation canopies. In *Proceedings, Fourth Australasian Conference on Heat and Mass Transfer*, pp. 9–12. Christchurch, New Zealand.

Ritter J.A., Lenschow D.H., Barrick J.D.W. *et al.* (1990) Airborne flux measurements and budget estimates of trace species over the Amazon Basin during the GTE/ABLE-2B Expedition. *Quarterly Journal of Geophysical Research*, **95**(D10), 16875–16886.

Schmitt K.F., Friehe C.A. & Gibson C.H. (1979) Structure of marine surface layer turbulence. *Journal of the Atmospheric Sciences*, **36**, 602–618.

Schuepp P.H., Leclerc M.Y., MacPherson J.I. & Desjardins R.L. (1990) Footprint prediction of scalar fluxes from analytical solutions of the diffusion equation. *Boundary-Layer Meteorology*, **50**, 355–373.

Segal M. & Arritt R.W. (1992) Non-classical mesoscale circulations caused by surface sensible heat flux gradients. *Bulletin of the American Meteorological Society*, **73**, 1593–1604.

Shuttleworth W.J. (1989) Micrometeorology of temperate and tropical forest. *Philosophical Transcripts of the Royal Society of London B*, **324**, 299–334.

Smith S.D. (1988) Coefficients for sea surface wind stress, heat flux, and wind profiles as a function of wind speed and temperature. *Journal of Geophysical Research*, **93**, 15467–15472.

Stull R.B. (1988) *An Introduction to Boundary Layer Meteorology*. Kluwer Academic Publishers, Dordrecht, The Netherlands.

Webb E.K., Pearman G.I. & Leuning R. (1980) Correction of flux measurements for density effects due to heat and water vapor transfer. *Quarterly Journal of the Royal Meteorological Society*, **106**, 85–100.

Weil J.C. (1985) Updating applied diffusion models. *Journal of Climate and Applied Meteorology*, **24**, 1111–1130.

Weil J.C. (1990) A diagnosis of the asymmetry in top-down and bottom-up diffusion using a Lagrangian stochastic model. *Journal of the Atmospheric Sciences*, **47**, 501–515.

Weil J.C. & Horst T.W. (1992) Footprint estimates for atmospheric flux measurements in the convective boundary layer. In: Schwartz S.E. & Slinn W.G.N. (eds) *Precipitation Scavenging and Atmosphere–Surface Exchange*, **2**, pp. 717–728.

Wesely M.L. (1983) Turbulent transport of ozone to surfaces common in the eastern half of the United States. In: Schwart S.E. (ed) *Trace Atmospheric Constituents: Properties, Transformations, and Fates*, pp. 346–370. Wiley & Sons, New York.

Wesely M.L. (1988) Use of variance techniques to measure dry air–surface exchange rates. *Boundary-Layer Meteorology*, **44**, 13–31.

Wesely M.L., Lenschow D.H. & Denmead O.T. (1989) Flux measurement techniques. In: Lenschow D.H. & Hicks B.B. (eds) *Global Tropospheric Chemistry: Chemical Fluxes in the Global Atmosphere*, pp. 31–46. National Center for Atmospheric Research, Boulder, Colorado.

Williams R.M. (1982) Uncertainties in the use of box models for estimating dry deposition velocity. *Atmospheric Environment*, **16**, 2707–2708.

Wilson J.D., Catchpole V.R., Denmead O.T. & Thurtell G.W. (1983) Verification of a simple micrometeorological method for estimating ammonia losses after fertilizer application. *Agricultural Meteorology*, **29**, 283–290.

Wyngaard J.C. (1988a) Flow-distortion effects on scalar flux measurements in the surface layer: Implications for sensor design. *Boundary-Layer Meteorology*, **42**, 19–26.

Wyngaard J.C. (1988b) The effects of probe-induced flow distortion on atmospheric turbulence measurements: Extension to scalars. *Journal of the Atmospheric Sciences*, **45**, 3400–3412.

Wyngaard J.C. (1990) Scalar fluxes in the planetary boundary layer – theory, modeling and measurement. *Boundary-Layer Meteorology*, **50**, 49–75.

Wyngaard J.C. (1991) On the maintenance and measurement of scalar fluxes. In: Schmugge T.J. & André J.-C. (eds) *Land Surface Evaporation: Measurement and Parameterization*, pp. 199–229. Springer-Verlag, New York.

Wyngaard J.C. & Brost R.A. (1984) Toward convective boundary layer parameterization: a scalar transport module. *Journal of the Atmospheric Sciences*, **41**, 102–112.

Wyngaard J.C. & Coté O.R. (1971) The budgets of turbulent kinetic energy and temperature variance in the atmospheric surface layer. *Boundary-Layer Meteorology*, **9**, 441–460.

Wyngaard J.C. & Moeng C.-H. (1992) Parameterizing turbulent diffusion through the joint probability density. *Boundary-Layer Meteorology*, **60**, 1–13.

Wyngaard J.C. & Zhang S.-F. (1985) Transducer-shadow effects on turbulence spectra measured by sonic anemometers. *Journal of Atmospheric and Oceanic Technology*, **2**, 548–558.

Wyngaard J.C., Coté O.R. & Izumi Y. (1971) Local free convection, similarity, and the budgets of shear stress and heat flux. *Journal of the Atmospheric Sciences*, **28**, 1171–1182.

Wyngaard J.C., Pennell W.T., Lenschow D.H. & Lemone M.A. (1978) The temperature–humidity covariance budget in the convective boundary layer. *Journal of the Atmospheric Sciences*, **35**, 47–58.

Yan H. & Anthes R.A. (1988) The effect of variations in surface on mesoscale circulations. *Monthly Weather Review*, **116**, 192–208.

CHAPTER 6

Standard analytical methods for measuring trace gases in the environment

P.M. CRILL, J.H. BUTLER, D.J. COOPER & P.C. NOVELLI

6.1 Introduction

Low molecular-weight volatile compounds are ubiquitous in nature and are the means of most of the mass exchange of biogeochemical cycles between the various reservoirs of the planetary biosphere. These molecules play central roles in the oxidation chemistry of the troposphere, the radiation balance of the planet and even the water supply to the stratosphere. In order to understand topics as diverse as net ecosystem mass flows, the health of terrestrial or oceanic biological systems, anthropogenic effects, production or decay of organic reservoirs such as forests, peats or sediments, one has to measure rates of exchange of low molecular weight trace gases.

This chapter will explore some of the standard methods of measurement of trace gases. We will concentrate on carbon, nitrogen and sulphur molecules which have geochemical cycles that are strongly biologically mediated such as CH_4, N_2O, CO, CO_2 and DMS ($(CH_3)_2S$). Gas chromatography (GC) will be the main technique discussed in some detail with only passing mention of other methods. Readers are referred to other chapters of this book on some of the newer methods that are rapidly becoming standard fare in modern research laboratories.

According to Dampier (1971), it was the sixteenth century iatrochemist von Hohenheim, also known as Paracelsus, who first recognized the chaotic or complex nature of the planetary atmosphere and referred to this mixture as 'καοζ' from which van Helmont derived our word 'gas' in the following century. It was not until the rise of the scientific method and the development of chemical techniques in the seventeenth and eighteenth centuries that opened the way to the discoveries of Scheele, Black, Lavoisier, Priestly and others (and the observations of flammable 'marsh gas' or CH_4 by Volta). Another leap in the analysis and discovery of trace volatiles was taken in the nineteenth and early twentieth centuries when spectroscopic methods were developed, allowing the unambiguous identification of gases (e.g. Lockyer's discovery of He in the sun in 1865, a discovery that preceded by 27 years He's isolation and characterization in the laboratory of the great noble-gas chemist William Ramsay).

GC, probably the most common and powerful modern technique for trace gas analysis, is a relative newcomer. Modern GC is the result of the work of A.J.P Martin and his colleagues in the mid-twentieth century (e.g. James & Martin, 1952). Chromatography, the separation of compounds by exploiting selective adsorption–desorption characteristics of compounds with a mobile phase moving across a solid (or immobile) support, was first described in 1906 by M.S. Tswett. It was then promptly forgotten until 1931 when liquid–solid chromatography was resurrected in the laboratory of R. Kuhn, and then modified by A.J.P. Martin and his colleagues to use gas as the mobile phase (cf. Ettre, 1992). Prior to this, the separation of volatiles required one of the many variants of distillation such as that used 50 years previously in the laboratory of W. Ramsay in the isolation and description of the noble gases. With GC, a standard and convenient technique for the separation and quantification of the components of complex mixtures was suddenly available for widespread use. The first commercial machines were available within a decade after James and Martin's first description of gas–liquid partition chromatography in 1952.

As a result, modern environmental scientists have an almost dizzying array of tools available for the study of trace gases in ecosystem processes. Some of the most common methods of analysis of low molecular weight trace gases are listed in Tables 6.1–6.3 together with the approximate detection limit of the method. Reflecting the history of trace gas discoveries, the methods fall into three broad categories: chemical, optical and chromatographic. The appropriateness of a particular method is determined largely by the problem at hand. Many variables must be considered when deciding what method to use such as which gases are to be measured, what precision is required, how accessible is the field site, sampling requirements and potential sampling artefacts.

6.2 Generalities
In this section, basic notions about standardization, sampling and analysis will be introduced. More detailed discussion of the particular issues as they pertain to the analysis at hand will be extended in the sections that follow.

6.2.1 Standardization
Standardization of any analysis is a basic problem. Any technique used in a laboratory should be well worked out and the precision and accuracy of the analysis should be understood. 'Quick and dirty' is simply a waste of time. A discussion of assessment of precision and accuracy is beyond the scope of this chapter but just about any quantitative analytical chemistry textbook will provide guidance. Of all the methods listed in the tables, only with the optical methods are unambiguous and quantitative assess-

Table 6.1 Techniques and references for standard methods of analysis of trace carbon gases

Species	Technique	Sample method	Detection limit	Ambient concentration	Reference
Carbon					
CH$_4$	GC-FID	Batch	10 ppbv	1.75 ppmv	Steele *et al.*, 1987
	GFC with IR absorption	Continuous	10 ppbv		Sebacher & Harriss, 1982
CO	GC with HgO detector	Batch	1 ppbv	0.05–1 ppmv	Robbins *et al.*, 1968
	GC-FID with methanizer	Batch	10 ppb		Rasmussen & Khalil, 1981
	HgO detector	Continuous	3 ppbv		Seiler & Junge, 1970
	GFC with IR absorption	Continuous			Dickerson & Delany, 1988
CO$_2$	Non-dispersive IR adsorption	Continuous	3 ppmv	355 ppmv	Komhyr *et al.*, 1983
	GC-FID with methanizer	Batch	1 ppbv		Rasmussen & Khalil, 1981
	GC-TCD	Batch			Tacket, 1968
	Absorption and titration	Batch			Witkamp & van der Drift, 1961
NMHC	GC-FID after concentration	Batch	50 pptv	0.1–50 ppbv	Sexton & Westberg, 1979
	GC-PID	Batch			Kerfoot *et al.*, 1990

FID, flame ionization dectector; GC, gas chromatography; GFC, gas filter correlation; IR, infrared; PID, photoionization detector; TCD, thermal conductivity detector.

ments of a trace gas possible and even in those cases there are frequently interferences and sampling problems that have to be solved. The non-optical methods require the comparison of the unknown to a known standard. In all cases, the analysis conditions should be explicitly defined and the working standards should be traceable to common standards that are widely available to the research community such as the standard reference materials from the National Institute of Standards and Technology (NIST, formerly NBS). Intercomparisons of techniques and standards between laboratories should also be encouraged (see Fehsenfeld, Chapter 7). Often this is the only way to identify problems with a particular analysis. Collaboration will get a lot more accomplished than competition.

Table 6.2 Techniques and references for standard methods of analysis of trace nitrogen gases

Species	Technique	Sample method	Detection limit (pptv)	Ambient concentration (ppbv)	Reference
Nitrogen					
N_2O	GC-ECD	Batch	40	280–320	Wentworth & Freeman, 1973
NO	NO/O_3 chemiluminescence	Continuous	5	0.001–10	Fontijin *et al.*, 1970
NO_2, NO_x, NO_y	NO/O_3 chemiluminescence	Continuous	10	0.01–100	Fehsenfeld *et al.*, 1987
NH_3	Denuder tubes		20	0.1–50	LeBel *et al.*, 1985
	Condensation collection, IC		20		Gorzelska *et al.*, 1992

ECD, electron capture detector; GC, gas chromatography; IC, ion chromatography.

Table 6.3 Techniques and references for standard methods of analysis of trace sulphur gases

Species	Technique	Sample method	Detection limit	Ambient concentration	Reference
Sulphur					
SO_2	GC with S-doped FPD	Batch	3 pg S/sample	0.005–100 ppbv	Farwell & Barinaga, 1986
	Condensation collection/IC	Batch	10 pptv		Klemm & Talbot, 1991
	Gas/liquid exchange coil, HPLC	Batch	10 pptv		Saltzman *et al.*, 1992
H_2S	$AgNO_3$ filters photometry	Batch	0.2 ppb	1–500 pptv	Natusch *et al.*, 1972
	GC-PID	Batch	100 pptv		Cutter & Oatts, 1987
Organo-S and H_2S	GC with S-doped FPD	Batch	3 pg S/sample		Farwell & Barinaga, 1986
	O_3 chemiluminescence	Batch	200 pptv		
	GC-ECD after fluorination	Batch	0.03 pg S/sample		Johnson & Lovelock, 1988

ECD, electron capture detector; FPD, flame photometric detector; GC, gas chromatography; IC, ion chromatography; PID, photoionization detector.

Quantitative analysis of an unknown can only be performed after sampling errors have been resolved or minimized, the analyte has been isolated or resolved and the analysis technique is reproducibly responding in a known fashion, ideally in a linear way. Then the sample can be quantified using one of three common techniques; (a) normalization; (b) external standardization; or (c) internal standardization/standard addition. For additional details, the reader is referred to a textbook of quantitative analytical chemistry such as Siggia (1968) or Harris (1991).

Normalization assumes that all the compounds in the sample give the same instrument response. The magnitude of the response of the unknown compound is compared to the sum of the responses of the sample and a percentage of the total is computed. For example, this might be useful in the analysis of air samples with a universal detector such as the thermal conductivity detector discussed below.

External standardization involves the comparison of the response of the sample compound to the instrument response of one or more standards of known concentration of the same compound. This method requires separate analysis runs so the analysis conditions between standard runs and sample runs have to remain constant. The standard matrix should also resemble the sample matrix as closely as possible. The concentrations of the standard should bracket the concentration of the analyte or at least be very similar to the concentration of the unknown.

Internal standardization and standard addition techniques require the addition of a pure standard in a known amount to the sample matrix. In standard addition, known amounts of the same compound as the analyte are added and the instrument responses to the different concentrations are analysed. Internal standardization uses known additions of a pure compound that has a similar retention time (in chromatographic cases), a similar instrument response for a given concentration and is added at a concentration similar to that of the unknown compound. Both techniques have the advantage of carrying the standard through the same preparation and delivery as the sample. With both techniques the standard should be unreactive with the sample, be pure, easily and accurately added, and form homogeneous solution with the sample.

6.2.2 Sampling

The evaluation of the method must be considered within the scope of the experiment. In general, sampling exploits either the physical (e.g. melting or boiling points) and/or the chemical properties (e.g. the acid–base chemistry of H_2S or CO_2) of the trace gas of interest in order to isolate or concentrate the analyte before delivering it to the detector required for its analysis. This is true of all the 'batch' methods, especially the chromatographic methods, listed in Tables 6.1–6.3. The sampling could be as simple as isolating a sample in a can, flask or syringe or as complex as trapping samples on metal foils before flash desorption, cryofocusing and subsequent analysis. Only most of the 'continuous' methods listed, especially the optical methods, are capable of analysing the species of interest directly in its ambient matrix. Particular sampling issues will be discussed in the following sections.

There are unique sets of problems with each approach for sampling trace gases. The sampling technique that is used will depend on the environmental question being asked, the sensitivity required and the chemistry of the particular gas, especially of the reactive species. The potential losses of reactive gases during sampling and analysis could result from several different processes depending upon the gas. These include:

1 physical interaction with surfaces in the sampling equipment;
2 oxidative losses during preconcentration; and
3 physical/chemical interactions in the analytical system after desorption of the preconcentrated sample.

Moisture also presents problems in the preconcentration of large volumes of air due to the physical blocking of cryogenic traps and interference or damage if introduced into the analytical system. In order to overcome these environmental factors, design of a suitable sampling and analysis approach involves choosing suitable materials for construction, removal of interfering species and removal of sufficient atmospheric moisture without removal of the gases of interest. Luckily, not all of these problems will be evident for all the gases in every circumstance. A variety of sampling schemes are discussed in the sections that follow about analysis of particular compounds. The reader should remember that aspects of each of these schemes can be generalized for many of the trace gases of interest. For example, cryogenic preconcentration of sulphur gases in air will also be applicable to analysis of non-methane hydrocarbons or the head space equilibration techniques discussed for N_2O in water samples will also work for CH_4.

All sampling schemes must be tested and verified. Verification of a sampling/analysis technique in field operation can be simply achieved in most cases through a standard addition. This involves collecting two simultaneous, or possibly sequential, samples with and without the gas of interest added upstream of the sampling inlet. The addition should be an amount similar to that measured in routine samples and within the range of a normal calibration. This approach tests the system under natural sampling conditions. With perfect collection efficiency, the difference between the two measurements should correspond to the amount of the addition.

6.2.3 Analysis

Chemical methods

There are only a few purely wet chemical methods for the analysis of low-molecular-weight trace gases in common use today mainly because of the much lower sensitivity of the methods compared to optical and chromatographic methods. These methods do have the advantages of being relatively simple and inexpensive and they particularly lend themselves to laboratory radioactive tracer experiments. CO_2 and H_2S are the two gases most commonly analysed with chemical methods that exploit their acid–base chemistry and, in the case of H_2S, its reactivity. CO_2 can be trapped on basic absorbants (e.g. soda lime, Ascarite) or in basic solutions. Analysis can be as simple as taring the CO_2-free absorbant then weighing it after exposure for a given amount of time to the CO_2-containing sample atmosphere (e.g. Edwards & Ross-Todd, 1983) or the absorbed CO_2 can be titrated with HCl (Witkamp & van der Drift, 1961). Recently, two comparisons of field methods for evaluating CO_2 evolution from soils have been published (Raich et al., 1990; Freijer & Bouten, 1991).

H_2S can also be absorbed into basic solutions but, more frequently, it is trapped and stabilized by reaction with Zn or Cd in solution (e.g. Hansen et al., 1978) or on $AgNO_3$-impregnated filters (Natusch et al., 1972) to form the metal sulphide. The amount of sulphide is then determined by titration (Golterman, 1971), spectrophotometrically (e.g. Cline, 1969) or fluorometrically (Jaeschke & Herrmann, 1981). Similarly, NH_3 gas can be absorbed and analysed. However, an acidic absorbant must be used. Ferm (1979) describes a method for atmospheric NH_3 that uses an oxalic acid trap and an ion-specific electrode for analysis. A spectrophotometric method (Solorzano, 1969) can also be used for analysis. Ferm's warning that organic amines may interfere with the analysis can be generalized for all the chemical methods. Other volatiles and particulates which are normally present in trace concentrations, such as the organic nitrogen compounds in the case of NH_3, may become major interferants given the large volumes of air that have to be sampled in order to use chemical methods in natural environments.

Optical methods

Optical methods of analysis exploit the interaction of radiant energy with matter. Measurement techniques based upon radiation absorption (especially infrared (IR) absorption) have proven useful for quantification of atmospheric trace gases because these gases absorb radiation at very specific and unique wavelengths. Absorption features of small molecules appear at discreet, well-defined wavelengths (often referred to as 'lines' or 'bands') and are diagnostic for a particular trace gas. The amount of energy in a beam of radiation of a particular wavelength that passes through a sample matrix that contains an absorbing gas is quantitatively reduced in a fashion that depends directly upon the path length and the concentration of the absorbent. This is known as the Beer–Lambert law,

$$\frac{dI}{dn} = -\kappa I$$

which states that the change in intensity, I, of the incident radiation with respect to the number, n, of absorbing molecules is proportional to the extinction coefficient, κ. In an integrated and rearranged form, the Beer–Lambert law can be stated as,

$$\ln\frac{I}{I_0} = -\kappa b C$$

where the log of the ratio of the measured radiation intensity, I, to the initial intensity, I_0, is equal to the product of the constant, the path length, b, and the concentration of the absorbent, C.

Visible and ultraviolet methods are usually considered separately from those methods that use the absorption of radiative energy at the IR wavelengths. This is because physical interactions of matter and energy in the different spectral regions are due to different mechanisms. Energy in the near-ultraviolet and visible spectral regions is absorbed by the excitation of valence electrons whereas infrared absorption is the result of resonant interactions of IR with the vibrational and rotational transitions of covalent bonds. The result is that IR methods tend to be more useful for trace gas identification and quantification.

The principle exceptions to this generality are the chemiluminescent and fluorescent techniques used for analysis of NO, NO_2 and sulphur gases. In these techniques the energy to excite outer shell electrons is supplied by either chemical reactions (chemiluminescence methods) or a H_2-rich reducing flame (flame photometric detectors). The fluorescent energy emitted as photons when the electrons return to their ground state is then measured.

Non-dispersive IR (NDIR) techniques use a gas filter to absorb selectively IR radiation passing through a gas mixture containing the gas of interest. Figure 6.1 is the basic outline of two common types of NDIR analysers. The first is a dual beam arrangement which has evolved from a negative-filter technique described by Wright & Herscher (1946) and the second is a single beam method called gas filter correlation (GFC) based on a description by Sebacher (1978). In both cases, the absorption of IR by a gas mixture in the sample cell is compared to radiation that is allowed to pass through either the reference cell in the NDIR or through the reference gas half cell (the gas filter) in the GFC. The radiation

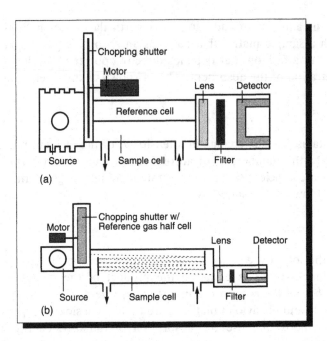

Fig. 6.1 Schematics of two types of non-dispersive infrared gas analysers. (a) Non-dispersive infrared gas analyser. (b) Gas filter correlation analyser.

reaching the detector is differentially modulated only at the wavelengths absorbed by the filter gas. This also has the advantage of using the contributions of a large number of absorption lines of the spectrum of the measured gas thereby increasing the amount of radiation energy absorbed above that when tuning one spectral feature using conventional spectroscopic instruments. Recent advances in spectroscopic techniques for measuring trace gases are covered in more detail in Chapter 8 of this volume.

Chromatographic methods

Probably the most common instrumental methods for the separation and analysis of low-molecular-weight volatile compounds are chromatographic. Of these, GC is the most widely used particularly in the analysis of trace gases. Chromatographic methods all rely on the individual partitioning characteristics of different compounds between a mobile phase (usually gas or liquid) and an immobile or stationary phase (a solid or a liquid phase bound to a solid support). We will concern ourselves with only a general discussion of the basics of GC and the detectors and techniques in common use for the quantification of trace gases found in the environment. The reader should refer to any modern textbook on instrumental

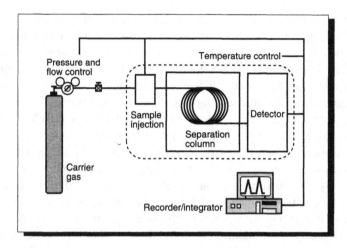

Fig. 6.2 Basic elements of a gas chromatograph. These elements are features common to every gas chromatographic system.

analysis (e.g. Skoog, 1985; Strobel & Heineman, 1989) or GC (e.g. Lee *et al.*, 1984) for a discussion of theoretical concepts of chromatographic separation.

Figure 6.2 shows the principle elements of a GC system. The carrier gas is the mobile phase. The pressure and flow of the carrier is regulated as it moves a compound from a sample introduction system across the stationary phase of the separation column where the compound is isolated from other potentially interfering compounds in the sample matrix and then to the detector. Because temperature affects vapour pressure, viscosity and solubility, the separation column and the detector (and frequently the sample injection system) of a gas chromatograph are always temperature controlled. Temperature programming of the separation column oven, i.e. changing temperatures in a controlled and regular fashion, is often used to facilitate and speed up chromatographic analyses. As the sample components elute from the end of the column they interact with the detector causing a change in electrical signal which is detected and recorded and now, commonly, integrated by a data system. The electrical signals that are caused by the individual components of a sample as they pass through the detector are recorded as a series of peaks above a baseline. This is referred to as the chromatogram. The area under the recorded peak to the baseline, peak area, is directly dependent on the concentration of the component as is the maximum height of the peak, if the peak's shape is constant. The time required for the component to elute from the column, the retention time, serves as a means to identify the component. For identical analysis conditions (i.e. carrier gas, carrier flow rate, column, column length and temperature, detector, detector temperature), the retention time for a given compound will be the same. Absolute identification of a compound is only accomplished when GC is combined with a technique such as mass spectroscopy. So careful attention to run conditions and calibration with known standards are required to gain precise quantification of a compound.

Carrier gases, and fuel gases for the flame detectors, should be as pure as practicable. The most common impurities in gas cylinders will be trace hydrocarbons, O_2 and moisture. The higher the purity the better because impurities will increase detector noise (e.g. hydrocarbons in fuel gas for flames; O_2 will absorb electrons in electron capture detectors) or even burn-out detectors (e.g. O_2 attacking the filaments in a thermal conductivity detector) and destroy columns or at least affect column performance (e.g. moisture affecting the partitioning between mobile and stationary phases). Use traps to remove impurities. First remove hydrocarbons (e.g. with charcoal, molecular sieves or catalytic combustion filters) then moisture (e.g. with molecular sieves or $CaSO_4$) and finally use an O_2 trap. Each cylinder will have different levels of impurities so care will have to be taken to ensure the efficacy of the traps. Always change the tank when the pressure drops below 200 psig. The partial pressure of water and impurities adsorbed to the walls of the cylinder will increase as the pressure in the cylinder decreases so there will be higher levels of impurities delivered as the cylinder empties. Use the most appropriate carrier gas for the detector and the particular analysis. For example, He, because of its very high thermal conductivity, is most appropriate for use with a thermal conductivity detector but, because it is a noble gas, He is not a very good carrier for an electron capture detector.

Cylinder pressure is reduced to operating pressures with a two-stage regulator. The diaphragm of the regulator should be stainless steel to avoid the problem of diffusion of air, water and organics from and across plastic or rubber diaphragms. Outlet pressures at the second stage are frequently set at 40–60 psig. Gas flow to the head of the column is then controlled with either a needle valve or a diaphragm valve. Gas flows for packed column chromatography are usually on the order of 30 ml min^{-1} and are much lower ($\sim$5 ml min^{-1}) when capillary columns are used.

A major advantage to GC is the small sample size required for analysis but a major difficulty is delivering small samples into the GC in a routine and reproducible manner. There are generally two approaches:
1 syringe injection; and
2 valve injection.
Syringe injection involves introducing a known volume of sample into the carrier gas stream through a hypodermic needle that has pierced a septum

in the injection port. Septa can be almost any resealable rubber or plastic though silicone rubber is common. However, careful consideration of the septum material is necessary because the material may be a source of unwanted contamination. Syringe injection of samples is not the preferred method for trace gas analyses. Septa leak and they must be changed frequently. It is very difficult to dry the sample and deliver the exact same volume of sample gas with a syringe. Even the insertion and injection technique can affect the peak shape and thus the reproducibility of the analysis. The precision of the analysis is somewhat technique dependent with only the most experienced chromatographers able to obtain good precisions. Personal experience has shown that precisions of less than 1% and accuracies of less than 5% are very difficult to obtain with direct injection of gas samples. Most of these problems are solved with the use of sample injection valves. Highly reproducible volumes of sample can be injected into the carrier gas stream and precisions on the order of 0.2% for the analysis of 1 ppm CH_4 with a flame ionization detector can be obtained by fairly inexperienced analysts. A six-port injection valve allows a sample to be dried as excess sample is loaded onto a sample loop of, for example 0.5–3 ml. The sample loop can be any volume up to the capacity of the separation column. Then, by switching the valve, the loop is placed directly in the carrier gas stream and the sample is swept onto the column. Water should be removed from the sample before filling the sample loop because the contribution of water vapour to sample volume can be significant. This can be accomplished with a Nafion drier or a tube of desiccant such as Aquasorb (P_2O_5 on an inert support) installed in line.

Any number of valves and ports may be configured to provide flow and column switching depending upon the analysis problem at hand. For example, a 10-port valve might be configured to backflush higher molecular weight gases off a short preliminary or 'guard' column when it is switched back to the load position. In this case, the slower eluting species are contaminants that would interfere with the analysis of a trace gas. The contaminants are retained on the guard column for a certain amount of time that allows the gas of interest to elute from the guard column onto the analytical column. After the sample gas has passed onto the analytical column, the gas flow is reversed on the guard column and the contaminants are backflushed to a vent while the analyte continues to the detector. Valving arrangements are becoming more complex as sampling and analysis is automated and because of the widespread use of microprocessors to control valves.

The separation column is the heart of the gas chromatograph. Packed 1/8″ (3.2 mm) o.d. columns are the most widely used for the analysis of low molecular weight trace gases mainly because they are robust, reproducible and inexpensive. They can also handle the fairly large samples required for the direct analysis of trace components in ambient air grab samples. The choice of packing is dependent upon the separation that is needed. Generally, solid adsorbants such as molecular sieve 5A, silica gel or chromatography grade charcoal and porous polymers such as the Chromosorbs, Porapaks and HayeSeps are used to separate the permanent gases, sulphur gases, N_2O and CH_4. Porous polymers have higher surface areas and higher load capacities than the standard diatomaceous earth which makes them particularly suitable for trace gas analysis. A very good source of information about the specifics of packings are the sample chromatograms published in the catalogues of chromatography suppliers. The column tubing is often stainless steel (which should be thoroughly rinsed with a polarity sequence of solvents before use) though glass and nickel are frequently used and, occasionally, Teflon is used for sulphur gas separation.

Every new column should be conditioned before use to remove adsorbed impurities, moisture, and/or polymerization solvents. This involves heating the column well above its normal operating temperature, but below its maximum temperature, for 12–24 h with carrier gas flowing through the column. The column should not be connected to the detector during conditioning. Also the column should never be heated without a gas flow. Porous polymers tend to shrink upon heating so, in order to limit dead volume, additional packing should be added to the column after its initial conditioning followed by an additional 1–1.5 h conditioning before use. If the GC is in fairly constant use, the column temperature should be raised above 100°C or at least 25–50° above the column's normal operating temperature with reduced carrier flow between analysis runs in order to keep the column dry, clean and conditioned.

There are a variety of detectors available to analyse the column effluent in GC. But of the 40 or so detectors available only five are in common use in the chromatographic analysis of environmental trace gases: thermal conductivity detector (TCD), flame ionization detector (FID), electron capture detector (ECD), flame photometric detector (FPD) and mercuric oxide reduction. Only one of the detectors, FPD, is specific for a particular element (in this case for sulphur) and one is essentially a universal detector (TCD). Yet each of these detectors is particularly suited for a subset of the trace gases in which we are interested and each will be discussed in some detail in the sections that follow. In general, all detectors should be kept clean, should never be heated without carrier gas flow and should always be kept at temperatures that are elevated above the temperatures of the analytical column in order to prevent condensation of effluent onto the detector.

The TCD theoretically will detect any substance that has a thermal conductivity different from that of the carrier gas. It usually consists of a

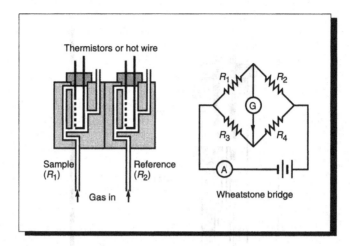

Fig. 6.3 Schematic of a thermal conductivity detector (TCD) and a diagram of a Wheatstone bridge.

pair of thermistors or hot-wire filaments (Fig. 6.3) that form the resistive elements (R_1 and R_2) of a Wheatstone bridge (R_3 and R_4 ratio resisitors to balance the bridge). The current, measured with the ammeter (A), flowing through the elements is dependent upon their temperature. If the bridge is balanced, then the heat removed from the resistors by the carrier gas flow is the same on both sides of the detector. The current will then be affected similarly across R_1 and R_2 and the bridge will remain balanced. If a sample with a different thermal conductivity flows across R_1 then a different amount of heat will be removed and the current through R_1 will change. The Wheatstone bridge will then be unbalanced and a change in the voltage at G will be detected. The ionization detectors are generally much more sensitive than TCDs but TCDs are particularly useful in measuring compounds that are not easily ionized such as N_2, O_2-Ar and CO_2. They are also useful in measuring samples in which normally trace compounds make up a large portion of the sample matrix without having to dilute the sample (such as CH_4 in bubble samples).

Detection of organic compounds by FID is initiated when the compounds are pyrolysed in a H_2/air flame in the detector. The resulting ions and electrons set up a current which can be measured across an electrode and an ion collector (Fig. 6.4). Since the method was described by Harley *et al.* (1958) and McWilliam & Dewar (1958) it has become one of the most popular detectors in use. It responds to all organic compounds except formic acid though its sensitivity decreases with increasing substitution of heteroatoms (e.g. halogens, oxygen or sulphur). It has a large dynamic linear range (10^7) and is very sensitive because of an inherently

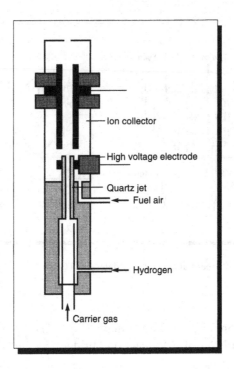

Fig. 6.4 Schematic of a flame ionization detector (FID).

low noise level and excellent baseline stability. It is insensitive to inorganic elements except those in periodic groups I and II which ionize in flames. The principal drawbacks are that it is a destructive detector and it requires fuel gases (H_2 and air) which make it a little less amenable to field use in remote locations.

A variation on flame ionization is the FPD (Fig. 6.5) which uses a much more H_2-rich flame and was described for chromatography by Brody & Chaney (1966). An in-depth review of the theory and applications of the FPD has been written by Farwell & Barinaga (1986). The basic principle of operation is that the GC column effluent is combusted in a H_2-rich flame, forming an excited S_2^* radical as one of the reaction intermediates. As the radical returns to its ground state it releases a photon at a wavelength 394 nm which is detected by a photomultiplier tube (PMT) and electrometer. The selectivity to sulphur stems from the use of a narrow bandpass optical filter at this precise wavelength. A similar selective sensitivity to phosphorus-containing compounds can also be achieved with the use of the appropriate optical filter. The major complicating factors in the use of a FPD are its log-linearity and the potential to have the flame chemistry altered by co-eluting non-sulphur compounds,

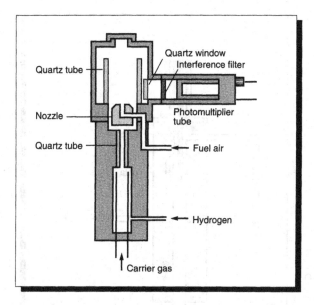

Fig. 6.5 Schematic of a flame photometric detector (FPD).

displacing the S_2^* equilibrium. There is an added practical consideration that, in practice, FPDs take several hours (or even days) to achieve a stable maximum response. This means that frequent calibration is required if the detector is to be used soon after ignition. Most FPDs have a detection limit on the order of a few picomoles of sulphur.

The basic components of an ECD are illustrated in Fig. 6.6. ECDs today are of two types: (a) those with a tritiated-scandium (tritium) foil; and (b) those with a ^{63}Ni foil. Tritium detectors originally were preferred because they were more sensitive, although less linear, than the ^{63}Ni detectors. However, as it became possible to heat the ^{63}Ni detectors to temperatures as high as 375°C, the difference in sensitivity became less pronounced. Today, the ^{63}Ni detector is preferred because it requires less activity (8–15 vs. 500 mCi) and because it is less likely to vent radionuclides during operation. Tritium detectors must be operated at lower temperatures and, under certain conditions, can vent tritiated water.

Stated simply, an ECD detects electrophilic (electron capturing) compounds by monitoring or correcting a drop in current produced in their presence. During normal operation, carrier gas molecules or atoms, either N_2 or Ar, are ionized by low-energy β particles emitted from the radioactive foil source. Each β particle passing through the cell ionizes hundreds of carrier gas molecules,

$$\beta + M \rightarrow \beta + M^+ + e^-$$

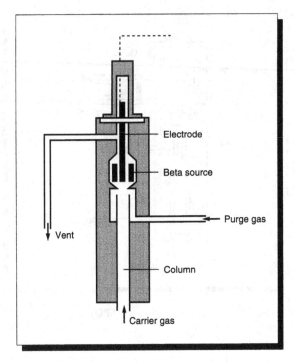

Fig. 6.6 Schematic of an electron capture detector (ECD).

establishing an electron (e^-) density which is a function of cell geometry, carrier gas, temperature, pressure and gas flow rate (Lovelock, 1974; Lovelock & Watson, 1978). Direct ionization of the carrier gas is the most probable process when N_2 is the carrier gas. However, outer shell electrons of noble gases (such as He or Ar) will absorb energy from β particles and form metastable atoms with relatively long lifetimes.

$$\beta + Ar \rightarrow \beta + Ar^*.$$

This results in reduced sensitivity because the deactivation of the metastable atoms by trace sample compounds and impurities in the carrier gas produces electrons increasing the current, the reverse of the desired effect (Connor, 1981). Addition of $\approx 5\%$ CH_4 to the carrier gas will ensure that all the metastable atoms from both the carrier gas as well as from impurities are deactivated. The result is a very sensitive and stable signal. Modern ECDs collect these electrons by pulsing the cell at a controlled frequency to match a reference current. In the presence of an electrophilic compound, which captures some of the electrons in the cell, the pulse rate must be increased. Typically, this frequency is converted to a voltage $(0-1\,V)$ output. The frequency required to match the reference current, and hence the voltage output, is more or less proportional to the amount of electrophilic compounds in the cell. This response varies, as different compounds have different electron affinities, and not all compounds respond linearly. Halogens are excellent candidates for detection by electron capture, as are certain oxygenated compounds or, to a much lesser extent, sulphur atoms. ECDs typically are not sensitive to hydrocarbons.

6.3 Carbon gases

6.3.1 CO₂ analysis

There has been considerable attention paid to the analysis of CO_2 in natural systems since at least the days of de Saussure's quantitative experiments on plant exchange at the turn of the nineteenth century. As with all the trace gases we will consider, there are several analytical methods by which CO_2 can be measured. The evaluation of which method to use must be considered within the scope of the experiment. For example, a chemical absorption and gravimetry or titration method may be appropriate for measuring the integrated release of CO_2 from soils into a head space trapped under a chamber. Such a method is not sensitive enough to measure the seasonal changes in atmospheric CO_2 or fast enough to measure photosynthesis rates. Reference to chemical methods of CO_2 analysis has been made above. The two other principal means of analysis are optical and gas chromatographic.

NDIR absorption

The details of this method are discussed above (see Fig. 6.1). This method operates on the principle of absorption by the 4.3 μm band of CO_2 in a multiple pass optical cell (e.g. Komhyr *et al.*, 1983). Radiation from an IR source first passes through a beam chopper and is then split, in a dual beam instrument, and passed through a reference cell containing CO_2-free N_2 or air and a sample cell. The CO_2-free cell transmits a reference beam to the detector to compare against the radiation absorbed by the sample in the reference side. Some commercial instruments include single-cell models in which the reference is supplied by periodically switching the air flow to the cell through a CO_2 adsorbant such as soda lime. The instruments are stable and accurate. NDIR is the method used by Keeling *et al.* (1982) at Mauna Loa since 1958 and by the National Oceanic and Atmospheric Administration (NOAA) and Climate Monitoring Diagnostics Laboratory (CMDL), previously called the Geophysical Monitoring for Climate Change (GMCC) to monitor ambient CO_2 concentrations in the global troposphere (Komhyr *et al.*, 1985). Komhyr *et al.* (1985) report a precision of 0.1 ppm CO_2 at 350 ppm. The method is also fast enough (time response $\approx 0.1\,s$) to be used in photo-synthesis studies using climate and light controlled chambers (e.g. Bartlett *et al.*, 1990) or micrometeorological exchange methods (e.g. Baldocchi & Meyers, 1991).

GC

CO_2 analysis can also be accomplished by GC using either TCD (Tacket, 1968) or FID analysers (Rasmussen & Khalil, 1981). In both cases, analysis begins with the chromatographic separation of CO_2 from other trace species. For TCD analysis, a porous polymer-packed column (e.g. Porapak Q, HayeSep Q) is commonly used at fairly low temperatures ($\approx 40°$) with He carrier gas ($\approx 30\,ml\,min^{-1}$). He is a better choice than N_2 for use with a TCD because He conducts heat about six times more efficiently than N_2 and about nine times more efficiently than CO_2. This optimizes the analysis for CO_2. The permanent gases (O_2, Ar and N_2) and CH_4 (at trace levels) are not resolved and all elute from the column as one large peak early in the run. Molecular Sieve 5A columns, higher column temperatures and longer run times are required to separate O_2/Ar, N_2 and CH_4. Si gel, carbosieve and alumina are also used as packings in GC-TCD. The range of linear response of TCDs is different with different detector manufacturers so it is important to characterize the detector response across the concentration range of the samples one is analysing.

CO_2 and the other two major carbon species in the atmosphere (CO and CH_4) can also be quantified using GC with FID detection. In this case CO and CH_4 are separated from CO_2 on a heated pre-column, usually of Si gel. The CO and CH_4 are then separated on a molecular sieve column (usually MS5A or MS4A). The gases then pass into a catalytic convertor. The catalytic agent is a metal oxide (usually Ni) on an inert support (Chromasorb or fire brick). In the presence of H_2 at high temperatures (about 400°C), the oxidized carbon gases (CO and CO_2) are reduced to CH_4 which is then detected by flame ionization. Run times are about 15 min (e.g. Rasmussen & Khalil, 1981).

$$4H_2 + CO_2 \rightarrow CH_4 + 2H_2O$$

$$3H_2 + CO \rightarrow CH_4 + H_2O$$

It is necessary to define the efficiency of the methanizer; typically the Ni catalysed conversion is 95–100% efficient. Efficiencies must be determined for each instrument by comparing detector response for a known amount of CO_2 (or CO) to that for a known amount of CH_4. Gas chromatographs designed for this application are available from most GC manufacturers at moderate costs.

6.3.2 CO analysis

There are several analytical methods by which CO can be measured. Methods best suited for measurement of a wide range of CO levels may not necessarily be the best for determining mixing ratios in a more narrow range. For example, in the determination of CO production rates or fluxes in natural environments, one may encounter mixing ratios over several orders of magnitude, while measurements of background remote tropospheric concentrations typically fall in a narrow range (50–250 ppbv; e.g. Seiler, 1974). Below we review the analytical methods available for determining CO mixing ratios and the types of experiments for which they are best suited.

Measurement techniques for CO based upon IR absorption have proven useful because this gas, like many other low molecular weight trace gases, exhibits moderately strong IR absorption.

Solar spectroscopy

Migeotte & Neven (1952) first measured CO in the atmosphere using solar spectroscopy. As the name implies, the method uses the sun as the light source and measures the IR absorption spectra of the entire atmospheric column in the spectral range of the CO 4.7 μm absorption band. The primary limitations of this method are:

1 reliance on solar radiation;
2 the results are total column averaged mixing ratios;
3 the precision of the method is, at best, of the order of 6–10%.

While these constraints preclude its use for CO measurements in many experimental applications, solar spectroscopic measurements have proven useful for estimating the rate of change of CO in the atmosphere (e.g. Dianov-Klokov et al., 1989; Zander et al., 1989).

Tunable diode laser (TDL) spectroscopy

This also measures CO based upon absorption of IR radiation (Ku et al., 1975). However, rather than using the sun as the radiation source, this method uses a semiconductive diode laser which is current or temperature tuned to specific absorption features of CO in the 4.7 μm wavelength region.

The lead salt lasers emit radiation in narrow line widths 10^{-4} cm^{-1} wave number. When operated at low pressure (10–100 mmHg) and temperature (<100°K), the spacing between spectral lines is larger than the widths of the features themselves (Fried & Sams, 1987). This, as well as the narrow line widths, ensures high selectivity. While there are variations on the designs of TDLs, basic principles are the same among most instruments. A sample is first introduced into a multipass White cell (path lengths are typically 10–150 m). The transmitted intensity is then measured using a liquid nitrogen cooled detector as the lased light is scanned through the absorption feature of interest.

Using known spectroscopic parameters, measured pressure, path length and transmitted radiation intensities, absolute CO concentrations can be calculated with Beers–Lambert's law (Fried & Sams, 1987). This is an effective method to verify the accuracy of gas calibration standards (Fried *et al.*, 1988).

For trace gas measurements in the atmosphere, the technique of second harmonic detection is often used to increase sensitivity. The laser is simultaneously modulated at kHz to mHz frequencies and the demodulated signal is detected. The procedure has several advantages over direct absorption. The most significant advantage is that it eliminates the need for detecting small differences between large signals, as is the case for direct absorption of trace atmospheric species. In a continuous sample, TDLs can have 1 s response times, detection limits of less than 1 ppbv CO and precisions of better than 0.5% (Sachse *et al.*, 1987, 1988). The TDL is a versatile and sensitive tool for measuring CO (as well as CH_4 and N_2O). However, they are very expensive (US$80 000–150 000), require considerable power to operate at liquid nitrogen temperatures and require a high degree of technical skill to maintain. TDLs are best suited for determining the fine vertical or horizontal structure of trace gases in the atmosphere (Sachse *et al.*, 1988; Fried *et al.*, 1991; Harriss *et al.*, 1992) and for micrometeorological flux measurements (e.g. Fan *et al.*, 1992).

NDIR gas filter correlation
This method also operates on the principle of absorption by the 4.7 μm band of CO in a multiple pass optical cell. Radiation from an IR source first passes through a beam chopper, then through a rotating gas filter wheel which alternates between two windows containing either CO or CO-free N_2, through a White cell containing the sample and finally to the IR detector (see Fig. 6.1). The CO gas filter produces a reference beam which is not affected by CO in the sample cell while the beam passing through the CO-free N_2 filter is absorbed by CO in the sample. Commercial instruments, which correct for pressure and temperature fluctuations in the sample cell, are available for continuous measurement of ppmv levels of CO in air. Sensitivity may be increased by a few simple modifications to the commercial instruments (which were developed to monitor workplaces for relatively high CO levels). The detection limit can be decreased to below 50 ppbv CO in the running air stream by using silver (Ag) or gold (Au) mirrors in the White cell (rather than polished aluminium), using a selected detector (which involves evaluating several detector units and finding one with the highest sensitivity) and by the

addition of a CaF_2 lens to focus the radiation leaving the sample cell on the detector (Dickerson & Delany, 1988). These sensitized instruments require frequent zeroing and water vapour, and to a lesser extent O_3, interferes with the measurement (Parrish et al., 1991). The instrument is linear to within 3% between 0.3 to 3 ppmv with a precision of about 10% for 1 min averages (Parrish et al., 1991; Poulida et al., 1991). At higher mixing ratios, or for longer averaging periods, precision of the order of 5% can be achieved (Poulida et al., 1991).

There are several methods based upon GC which can be used to measure CO in air. The chromatographic analysis typically uses a two-column configuration: a pre-column of silica gel or activated alumina is used to separate CH_4 and CO from other hydrocarbons and CO_2. The pre-column is followed by an analytical column, usually a molecular sieve (MS). Several detectors are available for quantification of CO after chromatographic separation. Argon ionization detectors permit detection of CO at levels of 80 ng to 9 µg. However, the lower detection limit requires cryogenic trapping of a large sample of the order of 250 ml (Popp & Oppermann, 1978). Helium ionization detectors are sensitive to CO, but N_2 interferes with the analysis and CO elutes as a broad peak on the N_2 tail (Andrawes & Gibson, 1980). Three GC detection methods are most commonly used for measurement of CO in environmental samples: (a) methanization with flame ionization detection; (b) mercuric oxide reduction; and (c) electron capture detection.

Methanization and GC-FID

As discussed above, the two major oxidized carbon species in the atmosphere (CO and CO_2) can be quantified using GC-FID after methanization. In this case, CO is reduced with H_2 at high temperatures in the methanizer to CH_4.

$$3H_2 + CO \rightarrow CH_4 + H_2O$$

Under ideal conditions this method has a detection limit (signal:noise ratio of 2) of about 20 pg CO (equivalent to a 1 ml sample of 18 ppbv CO). Precision at lower concentrations is only about 10% but improves at higher CO concentrations. This method has been used to monitor atmospheric CO levels (Khalil & Rasmussen, 1984; Fraser et al., 1986) and it has two noteworthy strengths. First, the GC-FID is linear over several orders of magnitude and thus is extremely useful when a wide range of concentrations will be encountered. Second, because CO, CH_4 and CO_2 can be measured on the same sample aliquot, this approach should prove useful in ecological studies.

GC and HgO reduction

Separation of atmospheric constituents using GC techniques as described in the previous section can be followed by the reaction of reduced gases with HgO at temperatures greater than 210°C (Robbins *et al.*, 1968). Reduced gases such as CO, H_2 and some hydrocarbons react with HgO at high temperatures to yield elemental Hg which is volatilized at the high temperatures. The HgO vapour is then swept into an optical cell by the carrier gas and detected by absorption. Run times are 2–3 min for H_2 and CO.

$$X + HgO \downarrow \rightarrow XO + Hg \uparrow$$

where X = reduced gas. An instrument based upon this technique is commercially available. Absolute response and response characteristics vary with individual HgO beds; non-linear responses of up to 5% have been observed at low amounts of CO (P.C. Novelli *et al.*, unpublished data). With rigorous attention to linearization and calibration, this technique is useful for measuring atmospheric background levels of CO because of the low detection limit (about 3 pg CO, the equivalent of 1 ml of 2 ppbv) and high precision of analysis (1–2% at 100 ppbv). Instrument response is often linear from about 400 pg to 25 ng but, as noted above, linearity varies widely among instruments. Molecular hydrogen (H_2) can also be measured by this method in conjunction with CO. Both H_2 and CO are useful tracers of fossil fuel combustion. GC followed by HgO reduction has been successfully used to measure tropospheric CO (Novelli *et al.*, 1992) and continuous measurements of tropospheric CO have been made by Seiler and co-workers with this method over the past 20 years (Seiler & Junge, 1970; Brunke *et al.*, 1990).

GC and ECD

The ECD can also be used for measuring trace levels of CO (Phillips *et al.*, 1979; Goldan *et al.*, 1982). Because CO does not rapidly absorb electrons, by itself it does not give a strong response in an ECD. But the addition of N_2O to N_2 carrier gas results in a sensitive response of the ECD to CO. The steady-state negative ion chemistry controlled by the β source and the N_2O in the carrier gas is disrupted by the presence of CO (Phillips *et al.*, 1979).

$$e^- + N_2O \rightarrow O^- + N_2$$

$$O^- + CO \rightarrow CO_2 + e^-$$

With N_2O doping the ECD detection limit is approximately 16 pg CO (1 ml of 13 ppbv CO) (Fehsenfeld *et al.*, 1981), of the order of the other chromatographic methods described above. ECDs have been reported linear to within 1% at up to 2.3 ng CO but some ECDs exhibit non-linear response which requires careful characterization. In addition to CO, CO_2, CH_4, H_2 and other hydrocarbons can be determined with N_2O-doped electron capture detection (e.g. Sievers *et al.*, 1979).

6.3.3 Reference gases and materials

With the exception of solar spectroscopic methods and the TDL absolute measurements, all of the CO measurement techniques described above require reference gases against which measurements of unknown samples are routinely compared. These reference gases should reflect the concentration range under investigation.

There are few commercial sources for certified CO standards, particularly at levels below 500 ppbv. The lowest certified CO in air standard reference material (SRM) available from NIST is about 10 ppmv. There are several commercial gas suppliers who provide certified CO in air mixtures to levels as low as 500 ppbv. Lower mixing ratios may be obtained from various government and university laboratories. The subject of CO standards has recently been discussed by Novelli *et al.* (1991).

The potential for contamination of samples is relatively high when measuring low CO levels, therefore all sampling routines must be evaluated for artefacts. There is little information available in the literature regarding the stability of CO in various materials. Novelli *et al.* (1992) evaluated several types of containers for use in atmospheric grab sampling. In general they found that the metal flasks tested grew CO over time (also see Mansbridge *et al.*, 1988). Other laboratories have used electropolished stainless steel containers for atmospheric sampling with apparent success (Kirchoff & Marinho, 1990; Greenberg *et al.*, 1990). Some rubber products and nylon also release CO (Novelli *et al.*, 1992; P.M. Crill, unpublished data). Glass containers can be used for CO sampling and flux measurements, but exposure to sunlight or even laboratory fluorescent lights causes CO levels to increase presumably due to the photo-oxidation of NMHC to CO.

6.3.4 Hydrocarbon analysis

The most common organic carbon compound in the atmosphere is CH_4 at a concentration of about 1.75 ppmv in the present day troposphere. Largely biogenically produced, it has a central role in the oxidation, water and radiation balances of the atmosphere. The C_2–C_5 NMHC are found at much lower concentrations, 0.1–50 ppbv, with higher concentrations near urban areas reflecting their anthropogenic sources. Some alkadienes, such as isoprene and terpenes, are exceptions since they are released from vegetation. NMHCs are important sources of tropospheric CO and, in the presence of NO_x, of tropospheric O_3.

Standard measurement techniques for CH_4 and NMHCs are mainly GC methods. Although a number of optical methods have been used for CH_4 because, like CO, it exhibits moderately strong IR absorption.

Optical methods

Solar spectroscopy total column mixing ratios (Dianov-Klokov et al., 1989) have been measured. TDL and He–Ne lasers have also been used for very fast and accurate measurements of CH_4 (e.g. McManus et al., 1989; Harriss et al., 1992). And a non-dispersive IR gas filter correlation (GFC) system was developed and used for field measurements of CH_4 (Sebacher & Harriss, 1982). None of these methods would be considered standard in ecological studies.

Chromatographic methods

GC-FID is perhaps the most common method for CH_4 analysis. CH_4 is often separated from air and CO_2 using porous polymer packings (Porapak Q or HayeSep Q) in usually 2–3 m of 3.2 mm o.d. stainless steel tubing (e.g. Crill, 1991). At a carrier gas flow rate of $30\,ml\,min^{-1}$ and column temperature of 45°, CH_4 will elute to the detector in less than a minute. Molecular sieve (MS5A) columns are sometimes used but these must be operated at a fairly high temperature (>150°) in order to achieve short retention times for CH_4. With a 1 ml standard loop injector, detection limits below 0.2 ppmv CH_4 are possible and precisions can be less than 0.2% for 1 ppmv CH_4 standards. The air sample is dried as it is loaded onto the loop cryogenically or by passing the sample through a short drying column filled with a solid moisture adsorbant (e.g. $CaSO_4$ or $Mg(ClO_4)_2$). GC-TCD may also be used at very high CH_4 (>0.1%) concentrations but unlike the FID's large linear dynamic range, the TCD has a more limited linear range. So care has to be taken to characterize the detector's response to a range of concentrations.

An FID without GC can be used to monitor directly total hydrocarbon concentration in a gas stream. This has the advantage of very rapid response to changes in concentration, fast enough for micrometeorological methods, but it ignores the details. Because of their low ambient concentrations, some sort of preconcentration step is required in the analysis of NMHCs. Most commonly, ambient samples are collected in electropolished stainless steel cans (e.g. Rasmussen & Lovelock, 1983). The samples are concentrated before injection onto the GC by cryotrapping the NMHCs onto loops that are immersed in a cryogen such as liquid Ar (l) or O_2 (l). N_2 (l) is used less frequently because of problems due to O_2 liquification in the trap. The loops may or may not be filled with an inert substance to increase surface area such as glass beads (e.g. Singh & Salas, 1982; Greenberg & Zimmerman, 1984). Chromatography packings have also been used in the cryogenic traps (e.g. Rudolph & Ehhalt, 1981). The sample is usually dried by passing through a cold trap at dry ice temperatures ($-58°$) or through a Nafion drier before the trapping loop. Sample sizes range from 400 to 1000 ml. The traps are warmed and the hydrocarbons are thermally desorbed and swept onto the analytical column. A variety of separation columns have been used. Porapak R or Q has worked for NMHCs up to C_5. For higher molecular weight hydrocarbons, liquid-coated Porasils or fused silica capillary columns have been preferred (e.g. Greenberg & Zimmerman, 1984). Coated wide-bore open tube columns are coming into use and they show a lot of promise. The column oven is temperature programmed and detection limits have been reported between 5 and 50 pptv (Rudolph & Ehhalt, 1981; Greenberg & Zimmerman, 1984).

6.4 Nitrogen gases

Trace gases of nitrogen are significant players in the chemistry and biochemistry of the atmosphere, aquatic systems and soils. N_2O, produced mainly by microbial nitrification and denitrification in soils and aquatic systems (McElroy & Wofsy, 1976; Khalil & Rasmussen, 1992), eventually migrates to the troposphere where it is chemically inert. N_2O can be consumed in anoxic environments but its only fate, once it reaches the atmosphere, is its conversion to odd nitrogen (NO and NO_2) by photolysis or reaction with singlet [O(1D)] in the stratosphere (McElroy & McConnell, 1971). Reactions with odd nitrogen in the stratosphere historically have regulated the levels of stratospheric O_3, although Cl plays an increasing part in this process every year (WMO, 1985). NO and NO_2, considered widely as noxious contaminants in the troposphere, are also important in regulating tropospheric O_3 and OH (e.g. Thompson *et al.*, 1990). In soils and aquatic environments, all of the trace nitrogenous gases are involved in microbial energy transfer, particularly in low oxic systems or microenvironments (Fenchel & Blackburn, 1979).

6.4.1 NO, NO_x, NO_y analysis

The reactive nitrogen species (NO, NO_x, NO_y) play a central part in the oxidation/reduction and the acid–base chemistry of the troposphere. Their concentration may vary over several orders of magnitude and they have a large anthropogenic source (e.g. Logan, 1983). NO_x is the sum of NO plus NO_2 whereas NO_y is the sum of NO_x plus all the other reactive N species including peroxyacetylnitrate (PAN) and HNO_3.

The most common analytical techniques for all three species or groups of species revolve around the chemiluminescent reaction of NO with O_3 (Fontijn *et al.*, 1970). The photomultiplier tube of the detector counts photons given off during the decay of excited NO_2, produced in the reaction of NO with O_3 (Fehsenfeld *et al.*, 1987). The intensity of the emitted light is proportional to the concentration of NO. Commercial detectors are available. The detection limits of the instrument are about 45 pptv for a 1 s integration time. This can be improved to 20 pptv for a 20 s integration period.

In order to measure the other reactive nitrogen species, converters are added to the basic instrument. For NO_2 conversion, ferrous sulphate or ultraviolet light from a Xe arc lamp converts NO_2 to NO before it enters the chemiluminescence detector (Kley & McFarland, 1980). The detection limit for the ultraviolet method is ~20 pptv for a 10 s integration time. A hot (300°) solid Au surface is the best means to reduce all the other reactive nitrogen species to NO. CO (0.3%) is added to the gas stream as a reducing agent (Fahy *et al.*, 1985). Conversion efficiencies for a number of tested species is near 100%. The detection limit for a 10 s integration approaches 10 pptv. (For more information on NO and NO_2 analyses, see Chapter 7, this volume.)

6.4.2 N_2O analysis

Although the presence of N_2O in the atmosphere has been known for some time from spectral studies of the sun (Adel, 1951), the first discrete measurements of N_2O in seawater and the marine atmosphere were conducted by Craig & Gordon (1963). Their approach, which required significant extraction and preconcentration of samples, involved isothermal GC with TCD. Each analysis required about 10 l of sample water or air. Later, after the introduction and development of the ECD (e.g. Lovelock & Lipsky, 1960; Wentworth & Freeman, 1973), which was 100–1000 times more sensitive to N_2O than the TCD, the measurement of N_2O by GC became a simpler matter. Continued refinement of the chromatography and detection of N_2O over the years has made this analysis highly precise (0.2% at 300 ppbv in air) and accurate (0.5%). It is now a routine measurement in laboratories throughout the world. Recent developments in optical techniques, using Fourier transform IR spectrometers (FTIR) and with TDLs, are continually improving and have significant potential as non-intrusive approaches for measuring background N_2O and N_2O fluxes. These techniques are discussed in Chapter 8 and will not be elaborated upon here because they cannot as yet be considered 'standard' methods.

The most common method for the analysis of N_2O is GC with ECD. The response for N_2O may or may not be linear depending upon the instrument and the range of concentrations being studied. N_2O produces the following sequence of reactions in an ECD cell:

$$N_2O + e^- \rightarrow N_2 + O^-$$

$$O^- + N_2O \rightarrow NO + NO^-$$

$$NO^- + N_2 \rightarrow NO + N_2 + e^-$$

The change in electron density in the cell depends mainly upon the rate constants for these reactions. If, for example, electrons were produced at the same rate as they were captured then there would be no response. As it turns out, if these were the only reactions involved, then N_2O would be a very difficult compound to measure. However, contaminants or dopants in the carrier gas can enhance considerably the response of the detector to N_2O. O_2, CO_2, CH_4 and H_2 will react with O^- produced in the first reaction above to form stable ions, thus shifting the equilibria and increasing the number of electrons captured (Grimsrud & Miller, 1978; Miller & Grimsrud, 1979; Phillips *et al.*, 1979; Sievers *et al.*, 1979), thus boosting detector response.

For this reason, it is not advisable to use N_2 as a carrier gas for the analysis of N_2O, unless it is doped with, say, O_2. A better alternative is the use of Ar blended with 5% CH_4 as a carrier gas. In addition to its reaction with N_2O products, CH_4 also significantly increases the electron drift velocity over that of N_2. Thus the response of an ECD to N_2O with Ar/CH_4 is enhanced 10–100 times over the response with ultra-high purity N_2 carrier gas. Furthermore, the presence of 5% CH_4 in the carrier swamps the effects of other contaminants or co-eluting peaks, such as CO_2, giving a more reliable and consistent response for N_2O.

Craig & Gordon (1963) originally used a 2.5 m column of silica gel to separate N_2O in their samples. Water and CO_2 were removed during pretreatment. Later, numerous investigators found molecular sieve 5A (MS5A) to be a suitable, long-lasting alternative to silica gel (e.g. Hahn, 1974; Cohen, 1977; Young & Cline, 1980; Weiss *et al.*, 1981). MS5A columns are typically short, about 1 m × 3.2 mm i.d., and are used at 250°C for N_2O. This represents a few thousand theoretical plates resulting in a retention time of 2–3 min and peak widths of 5–10 s. MS5A columns are excellent for light gases. One advantage of the MS5A columns is that CO_2 in the sample does not co-elute with N_2O, but rather comes after the N_2O peak (Weiss *et al.*, 1981). The disadvantage is that the CO_2 peak tails significantly and could contaminate ensuing samples. This effect is countered either by removing CO_2 from samples during pretreatment or by swamping the effect with CH_4.

For some time Porasil columns were employed routinely for the analysis of N_2O, particularly when certain halocarbons such as CCl_2F_2 and CCl_3F were also in the sample matrix. Column length ranges from 2 to 3 m × 3.2 mm i.d. and the column is maintained at 50–70°C. However, in these columns, CO_2 typically co-elutes with N_2O. When N_2 is used as carrier gas, the signal becomes more a measurement of CO_2 than N_2O, since over 95% of the response can be due to the reaction:

$$CO_2 + O^- \rightarrow CO_3^-$$

in the detector. Thus it is imperative that if one is to use any of the Porasil column packings (A–D) for separation of N_2O that either CO_2 should be scrubbed from the sample (e.g. with Ascarite) or Ar/CH_4 should be used as a carrier gas.

A third approach, and the one recommended here, is to use one of the polymeric bead column packings such as Porapak Q or HayeSep Q or N (e.g. Singh *et al.*, 1979; Elkins, 1980; Butler *et al.*, 1989). Although the columns must be longer (up to 6 m × 3.2 mm i.d.), they can be operated at much lower temperatures (45–70°C) than MS5A and still give the same number of theoretical plates. Compounds that elute from the column slower than N_2O can be backflushed by splitting the column into two lengths and combining flow reversal in the precolumn with sample injection. CO_2 precedes N_2O and, when Ar/CH_4 is used as carrier gas, contributes much less than 0.1% to the total signal (Butler & Elkins, 1991). It should be noted that Porapak Q, MS5A and Porasil D have all been used successfully for frequent, long-term measurements of N_2O with high precision (Weiss, 1981; Prinn *et al.*, 1983, 1990; Butler *et al.*, 1989) but proper consideration must be made for each of these columns.

N_2O can be measured precisely and accurately in air samples of 1–5 ml. With the packed columns described above, air samples can be swept from a sample loop of known volume directly onto the head of the column. Water samples, however, require some sample pretreatment to get N_2O into the gas phase. Samples as small as 5–15 ml can be extracted by stripping onto a cryogenically cooled trap packed with some adsorbant such as a molecular sieve (Cohen, 1977; Young & Cline, 1980). The trap is then heated either by immersion or conductive heating to drive the gases rapidly onto the column. Water must be removed during the stripping process to prevent clogging of the trap. CO_2 can also be removed during stripping with a tube of Ascarite (NaOH impregnated on an inert support) in line.

A second method for extraction of gases from water is an adaptation of the McAuliffe (1971) head space extraction called 'shake "n" bake'. Here a volume of sample water (typically about 25 ml) is equilibrated with an equal volume head space of an inert gas (N_2 or He) in a syringe (Elkins, 1980). A sample loop on the GC is then flushed and loaded with this head space and the sample injected onto the column. The sample must be dried during sample loop flushing and loading usually by passing the sample through a glass tube packed with dessicant (e.g. Aquasorb, Drierite or $Mg(ClO_4)_2$) before the sample loop. This method requires very little set-up and is ideal for exploratory work where few or infrequent measurements are required.

For frequent or continuous measurement of discrete water samples, an automated head space technique is preferred (Butler & Elkins, 1991). Although this approach requires considerably more expense, effort by the analyst is greatly reduced. Accuracy and precision are better or equal to those of the manual methods ($\pm 2\%$). Basically, this approach involves temperature-controlled equilibration of the sample in a capped vial with an inert head space. After a period to reach equilibration, a tray of vials is then processed by an automated head space sampler, such as the Hewlett Packard 19395A which can hold 24 vials at a time. Coupled with a GC configured for backflushing, such a system can run samples as frequently as every 6 min unattended.

An ECD does not always respond linearly to N_2O, so calibration of measurements is particularly important. A standard curve over a range of concentrations should be established and checked periodically on any instrument in operation. Routine analyses should always be run against at least one standard and, because ECDs are notorious for drifting during use, the standard should be run frequently. Reliable calibration gases can be obtained commercially, but if one is interested in absolute accuracy, the cylinders should be standardized with SRMs available from NIST. N_2O is stable in air and N_2 at ppbv levels in both steel and aluminium cylinders for many years.

6.5 Reduced sulphur gas analysis

The sulphur gases of interest in natural systems generally contain sulphur in the fully reduced sulphide form (S^{2-}). The most commonly discussed reduced sulphur gases are organic sulphides (e.g. $(CH_3)_2S$ or DMS), inorganic sulphides (H_2S, OCS), thiols (e.g. CH_3SH) and disulphides (e.g. CS_2, $(CH_3)_2S_2$ or DMDS). In addition, SO_2 is important in the environment, both as an oxidation product of the reduced biogenic sulphides and as an anthropogenic pollutant. Analysis of these gases poses special problems due to both their low concentrations in most environments and their chemical reactivity. The refinement of analytical systems for sulphur gases in natural systems continues to evolve at a fairly rapid pace, such that there are few methods that can be considered as standard. Choice of a suitable method for a given species involves consideration of:

1 sampling technique;
2 preconcentration method;
3 analytical technique;
4 detection system;
5 standardization; and
6 verification.

Of these steps, possibly the most important is the final one. Historically, realistic field testing of chosen sampling and analysis schemes has frequently been overlooked. Consequently, there are doubts concerning the accuracy of a significant portion of previously published data.

The reactivity of the reduced sulphur gases makes the potential for losses during sampling and analysis particularly acute. In addition to problems mentioned in the previous general discussion of sampling, in the case of SO_2, removal to an aqueous phase needs to be considered if moisture is present because SO_2, like CO_2, is highly soluble. Removal of interfering atmospheric oxidants and removal of sufficient atmospheric moisture are especially important. The latter two areas are complicated by the obvious requirement that the sulphur gases of interest are not also removed. On the positive side, not all of these problems will be evident for all the gases under all circumstances. Moisture may not pose a problem in the preconcentration of reduced gases by methods other than cryogenic preconcentration, oxidants may not be a problem for less reactive gases such as OCS and CS_2, and materials may not be a problem once oxidants and/or moisture have been removed from the gas stream.

Most sampling schemes for reduced sulphur gases involve pulling a sample through an oxidant scrubber and/or drying system prior to a preconcentration trap. This avoids the passage of interfering compounds through the pump before preconcentration. When this is not possible, e.g. in the case of bag samples or high altitude aircraft sampling, special inert pumps are needed to ensure sample integrity. These are generally constructed from stainless steel and FEP Teflon (e.g. Kuster & Goldan, 1987). The same inert materials, together with glass, are normally used for construction of all components of the sampling apparatus and flow system from chambers to sample loops.

Preconcentration systems for reactive sulphur gases include cryogenic methods, solid adsorbent methods and filter methods. The earliest systems used wet chemical and impregnated filter methods. These two systems are generally based on the same chemical principles. Filters that are still in routine use include the $AgNO_3$ method for the collection of H_2S (Natusch *et al.*, 1972) and basic (high pH) filter methods for the collection of SO_2 (e.g. Daum & Leahy, 1983). The former relies on trapping H_2S as Ag_2S, which is extracted in basic cyanide solution for analysis by the fluorescence quenching of fluorescein mercuric acetate. At low H_2S levels, interference from OCS occurs, which can be corrected as described by Cooper & Saltzman (1987). The latter method relies on the trapping of SO_2 as sulphite (SO_3^{2-}) on the moist, basic filter medium after passage through a prefilter to remove sulphate (SO_4^{2-}) aerosol. The carbonate filter is then extracted in hydrogen peroxide to convert all SO_3^{2-} to SO_4^{2-}, which is then analysed by ion chromatography.

Solid adsorbent methods are also generally specific to individual compounds or classes of compounds. Ammons (1980) developed the use of

Au to preconcentrate H_2S and DMS from air. These methods were used in the various field studies of Braman *et al.* (1978) and Andreae and co-workers (e.g. Andreae *et al.*, 1985; Ferek *et al.*, 1986). This procedure involves pulling air through a packed Au trap. The sulphur gases are subsequently desorbed at high temperature for analysis. This method is currently used exclusively for DMS analysis. A second class of solid adsorbent method is gas chromatographic packings such as molecular sieves and Tenax. These have the advantage of trapping a wide variety of compounds (Steudler & Kijowski, 1984) and have the capability of thermally desorbing the compounds at relatively low temperatures facilitating method automation (Cooper & Saltzman, 1992). The solid adsorbent methods have the advantage that samples can be taken and stored for later analysis, removing the need to have all analytical instrumentation close to field sites.

Cryogenic methods are more generally applicable than solid adsorbent methods, although they can be somewhat less desirable due to:
1 the need to have a constant supply of cryogen (not always easy outside industrialized nations);
2 the danger of liquid oxygen;
3 the high cost of liquid argon; and
4 the bulkiness of storage and delivery systems.
Liquid nitrogen cannot be used directly as a cryogen at ambient pressure due to its ability to condense liquid oxygen in sample traps. A solution to this problem is to lower the pressure across the trap. Cryogenic sample traps that have been used for collection of sulphur gases are constructed from surface-deactivated glass, Teflon or stainless steel, either in the form of an open tube or packed with deactivated glass beads, deactivated glass wool or Teflon wool.

Removal of moisture from sample streams prior to preconcentration involves using either subambient temperatures to freeze out water or Nafion tubing that allows water molecules to diffuse through its walls into a counterstream of dry air. Both methods appear successful although there have been suggestions that there may be conditioning effects for Nafion resulting in losses of sulphur compounds, particularly SO_2, under certain circumstances (Goldan *et al.*, 1987). Low temperature water traps have the disadvantage that periodic drying of the traps is necessary to avoid eventual blockage.

As mentioned above, SO_2 is particularly sensitive to moisture, due to its potential to dissolve in water. Cryogenic sampling methods require extremely efficient water removal to ensure sample integrity (Thornton *et al.*, 1987; Saltzman *et al.*, 1992). The applicability of current methods to low level SO_2 measurements in remote air remains in question due to poor agreement between methods in recent intercomparison studies

(Gregory *et al.*, 1993a). It is likely that these results reflect difficulties related to water removal and sampling losses. Conversely, the aqueous chemistry of SO_2 allows for alternate methods involving quantitative gas/liquid exchange systems such as mist chambers (Cofer *et al.*, 1985), porous membrane methods (Dasgupta *et al.*, 1986) and gas/liquid exchange coils (Saltzman *et al.*, 1992). The liquid chromatographic procedure described by Saltzman *et al.* (1992) appears particularly useful for the analysis of unpolluted air samples due to its extremely low detection limit of less than 10 pptv.

Removal of oxidants from gas sample streams prior to preconcentration of reduced sulphur compounds was first tested by Ammons (1980) and techniques are still evolving at the present time (Saltzman & Cooper, 1989; Goldan, 1990; Gregory *et al.*, 1993b). Among workers in this area the primary concern is in the quantification of atmospheric DMS. Consideration should also be given to the problem of oxidative sample loss as methods are developed for the quantification of other reactive gases such as H_2S, thiols and unsaturated hydrocarbons.

The first type of oxidant scrubber in widespread use was sodium carbonate coated on a solid support. Supports that have been used since include glass fibre filters and various chromatographic packings. These early systems were not completely successful, often resulting in unquantifiable losses of DMS (Andreae *et al.*, 1985; Saltzman & Cooper, 1989; Goldan, 1990). More recently, systems based on KOH impregnated supports have been used. The KOH systems appear to have a higher oxidant capacity than the carbonate-based systems (Goldan, 1990; Ferek & Hegg, 1992). An alternate system based on the removal of oxidants by neutrally buffered aqueous KI has also been described (Saltzman & Cooper, 1989; Cooper & Saltzman, 1993). This scrubber appears to have a much higher capacity than the base-treated solid support systems. Other scrubber systems that have received some attention include cotton (C. Leck, unpublished data), MnO_2 (Maroulis & Bandy, 1978; Ammons, 1980), solid KOH (Bandy & Thornton, 1993) and KOH coated cotton (A. Bandy, unpublished data).

A wide range of gas chromatographic systems have been developed for the analysis of sulphur compounds using both packed (e.g. Maroulis & Bandy 1978; Steudler & Kijowski, 1984; Johnson & Lovelock, 1988) and capillary columns (e.g. Cline & Bates, 1983; Turner & Liss, 1985; Goldan *et al.*, 1987). A number of specialized detection methods have been used for low level sulphur gas measurements including flame photometry, mass spectrometry, electron capture, photoionization and chemiluminescence. These detectors have generally been designed to achieve both selectivity and sensitivity to sulphur compounds.

The most commonly used sulphur detector is the sulphur-specific

FPD. As mentioned above, one of the major complicating factors in the use of a FPD is its squared response. One approach to linearizing FPDs and increasing their sensitivity and precision at low sulphur concentrations is to add sulphur to the flame gas mixture (doping). This generally involves using a sulphur compound blended in the hydrogen supply either directly in the cylinder, adding it via a make-up gas or by passing the fuel gas over a permeation tube containing a volatile sulphur compound. This approach is generally successful, with a few caveats. First, most reactive sulphur gases are at equilibrium with polymeric components of the flow system. Equipment used with high concentration mixtures has to be dedicated to that purpose to avoid subsequent contamination. Second, permeation tubes are temperature sensitive, requiring thermally controlled conditions, and can sometimes fail. Third, use of SF_6 as a doping agent leads to formation of HF in the flame, resulting in etching of the detector windows.

The other detector techniques mentioned above are more specialized than the FPD, which also has other applications such as phosphorus analysis. The only other commercially available instrument specifically engineered for sulphur detection is a sulphur chemiluminescence detector (Gaines *et al.*, 1990). This detector operates by drawing a gas stream from a normal FID flame (which is a much more O_2 rich than the flame used in a FPD) into a chamber where the gas stream is reacted with ozone (O_3) at low pressure. The theory is that SO molecules are formed as an intermediate combustion product in the FID flame. SO will react with O_3, emitting a photon.

$$SO + O_3 \rightarrow SO_2 + O_2 + h\nu$$

The sulphur concentration is then quantified by monitoring the photon emission with an ultraviolet band pass filter and a PMT. The detector appears to be specific to sulphur and has an extremely low detection limit of 0.5 pmol. The detector also has the advantage of being linear with respect to sulphur, unlike the FPD.

Of the other detectors listed above, the commercial versions are not specifically designed for sulphur detection, but they can be operated in ways such that either the response to non-sulphur compounds is minimized or relatively high sulphur levels are being analysed, e.g. in laboratory studies. The photoionization detector (PID) is commercially available and, though sensitive for sulphur with a detection limit of approximately 1 pmol, it is not sulphur specific. The PID uses a high energy ultraviolet lamp to ionize trace species in an inert carrier gas stream.

$$R + h\nu \rightarrow R^+ + e^-$$

The ions are then monitored through by measuring the current at a collector electrode. Some applications of this detector have been described (Cutter & Oatts, 1987; Dacey & Blough, 1987). However, environmental studies are likely to be limited due to interfering response of the instrument to gases such as NO, NO_2 and many organic compounds. Some selectivity can be achieved through the use of lower energy ultraviolet lamps (9.5–11.7 eV lamps are available).

Mass spectrometry has recently been applied to the analysis of sulphur compounds in environmental samples (Ridgeway *et al.*, 1991; Bandy & Thornton, 1992). These systems have the advantage of being both selective and sensitive. The use of an isotope dilution technique (Bandy & Thornton, 1992) for calibration of this system allows for quantification of sulphur gases even if sampling losses occur during preconcentration (provided that such losses are not severe). The high concentration isotope addition also provides conditioning of sample lines and analytical hardware, potentially reducing reactive losses. The drawbacks of mass spectrometry systems are their size, complexity and cost.

The ECD method described by Johnson & Lovelock (1988) is somewhat unique due to extremely high sensitivity. Although it already has been used successfully for field measurements (Johnson & Bates, 1993), this approach has the potential for significant future development. The method involves post-column reaction of sulphur compounds with fluorine (F_2) gas over a hot Ag catalyst to form fluorinated sulphur compounds (e.g. SF_6). Excess F_2 is removed with a palladium reduction catalyst. The ECD is extremely sensitive to the fluorinated compounds. A detection limit of the order of 10^{-14}–10^{-15} mol of sulphur may ultimately be possible if the reaction efficiency can be improved from the published rate. The major drawbacks of the system are its complexity, the hazard of using F_2 gas and the potential of HF to corrode the detector.

In general, while this discussion has focused largely on issues relevant to discrete samples, there are obviously additional considerations relating to automated sampling over longer time scales. Natural variability in trace gas sources is often extreme and full study of a system involves consideration of spatial, diurnal and seasonal components of this variability. Consideration of the factors discussed above will generally lead to the conclusion that there is rarely a single method that can be applied to all of the gases of interest, or even of two similar gases at widely different concentrations. The choice of a suitable analysis method will ultimately depend on a variety of considerations including which gases are to be quantified, how accessible the chosen sampling sites are, what sampling frequency is desired, what range in ambient concentrations is expected and what degree of automation is possible.

References

Adel A. (1951) Atmospheric nitrous oxide and the nitrogen cycle. *Science*, **113**, 624–625.

Ammons J.M. (1980) *Preconcentration methods for the determination of gaseous sulfur compounds in air.* PhD dissertation, University of South Florida.

Andrawes F.F. & Gibson E.K. (1980) Characteristic negative response of the helium ionization detector. *Analytical Chemistry*, **52**, 846–851.

Andreae M.O., Ferek R.O., Bermond F. *et al.* (1985) Dimethyl sulfide in the marine atmosphere. *Journal of Geophysical Research – Atmospheres*, **90**, 12891–12900.

Baldocchi D.D. & Meyers T.P. (1991) Trace gas exchange above the floor of a deciduous forest 1. Evaporation and CO_2 efflux. *Journal of Geophysical Research – Atmospheres*, **96**, 7271–7285.

Bandy A.R. & Thornton D.C. & Driedger III A.R. (1993) Airborne measurements of sulfur dioxide, dimethyl sulfide, carbon disulphide and carbonyl sulfide by Isotope Dilution Gas Chromatography/Mass Spectrometry. *Journal of Geophysical Research – Atmospheres* **98**, 23423–23434.

Bartlett D.S., Whiting G.J. & Hartman J.M. (1990) Use of spectral reflectance to estimate absorbed PAR and rates of photosynthesis in a grass canopy. *Remote Sensing of the Environment*, **30**, 115–128.

Braman R.S., Ammons J.M. & Bricker J.L. (1978) Preconcentration and determination of hydrogen sulfide in air by flame photometric detection. *Analytical Chemistry*, **50**, 992–996.

Brody S.S. & Chaney J.E. (1966) Flame photometric detector. The application of a specific detector for phosphorus and sulfur compounds sensitive to sub-nanogram quantities. *Journal of Gas Chromatography*, **4**, 42–46.

Brunke E.-G., Scheel H.G. & Seiler W. (1990) Trends of troposheric CO, N_2 and CH_4 as observed at Cape Point, South Africa. *Atmospheric Environment*, **24A**, 585–595.

Butler J.H. & Elkins J.W. (1991) An automated technique for the measurement of dissolved N_2O in natural waters. *Marine Chemistry*, **34**, 47–61.

Butler J.H., Elkins J.W., Thompson T.M. & Egan K.B. (1989) Tropospheric and dissolved N_2O of the West Pacific and East Indian oceans during the El Niño-Southern Oscillation event of 1987. *Journal of Geophysical Research – Atmospheres*, **94**, 14865–14877.

Cline J.D. (1969) Spectrophotometric determination of hydrogen sulfide in natural waters. *Limnology and Oceanography*, **14**, 454–458.

Cline J.D. & Bates T.S. (1983) Dimethyl sulfide in the equatorial Pacific Ocean: A natural source of sulfur to the atmosphere. *Geophysical Research Letters*, **10**, 949–952.

Cofer W.R., Collins V.G. & Talbot R.W. (1985) Improved aqueous scrubber for the collection of soluble atmospheric trace gases. *Environmental Science and Technology*, **19**, 557–560.

Cohen Y. (1977) Shipboard measurement of dissolved nitrous oxide in seawater by electron capture gas chromatography. *Analytical Chemistry*, **49**, 1238–1240.

Connor J. (1981) The electron-capture detector II. Design and performance. *Journal of Chromatography*, **210**, 193–210.

Cooper D.J. & Saltzman E.S. (1987) Uptake of carbonyl sulfide by silver nitrate impregnated filters: Implications for the measurement of low level atmospheric H_2S. *Geophysical Research Letters*, **14**, 206–209.

Cooper D.J. & Saltzman E.S. (1993) Measurements of atmospheric dimethylsulfide, hydrogen sulfide and carbon disulfide during GTE/CITE-3. *Journal of Geophysical Research*, **98**, 23397–23409.

Craig H. & Gordon L.I. (1963) Nitrous oxide in the ocean and the marine atmosphere. *Geochimica Cosmochimica Acta*, **27**, 949–955.

Crill P.M. (1991) Seasonal patterns of methane uptake and carbon dioxide release by a temperate woodland soil. *Global Biogeochemical Cycles*, **5**, 319–334.

Cutter G.A. & Oatts T.J. (1987) Determination of dissolved sulfide and sedimentary sulfur speciation using gas chromatography-photoionization detection. *Analytical Chemistry*, **59**, 717–721.

Dacey J.W.H. & Blough N.V. (1987) Hydroxide decomposition of dimethysulfoniopropionate to form dimethylsulfide. *Geophysical Research Letters*, **14**, 1246–1249.

Dampier W.C. (1971) *A History of Science*, 4th edn. Cambridge University Press, London.

Dasgupta P.K., McDowell W.L. & Rhee J.S. (1986) Porous membrane-based diffusion scrubber for the sampling of atmospheric gases. *Analyst (London)*, **111**, 87–90.

Daum P.H. & Leahy D.F. (1983) *The Brookhaven National Laboratory Filter Pack System for Collection and Determination of Air Pollutants*. Brookhaven National Laboratory Report Number BNL 31381R. Upton, New York.

Dianov-Klokov V.I., Yurganov L.N., Grechko E.I. & Dzola A.Z. (1989) Spectroscopic measurements of atmospheric carbon monoxide and methane. 1: Latitudinal distribution. *Journal of Atmospheric Chemistry*, **8**, 139–151.

Dickerson R.R. & Delany A.C. (1988) Modifications of a commercial gas filter correlation CO detector for enhanced sensitivity. *Journal of Atmospheric and Oceanic Technology*, **5**, 424–431.

Edwards N.T. & Ross-Todd B.M. (1983) Soil carbon dynamics in a mixed deciduous forest following clear-cutting with and without residue removal. *Soil Science Society of America Journal*, **47**, 1014–1021.

Elkins J.W. (1980) Determination of dissolved nitrous oxide in aquatic systems by gas chromatography using electron-capture detection and multiple phase equilibration. *Analytical Chemistry*, **52**, 263–266.

Ettre L.S. (1992) 1991: A year of anniversaries in chromatography, part 1, from Tswett to partition chromatography. *American Laboratory*, 48C–48J.

Fahy D.W., Eubank C.S., Hübler G. & Fehsenfeld F.C. (1985) Evaluation of a catalytic reduction technique for the measurement of total reactive odd-nitrogen NO_y in the atmosphere. *Journal of Atmospheric Chemistry*, **3**, 435–468.

Fan S.-M., Wofsy S.C., Bakwin P.S. *et al.* (1992) Micrometeorological measurements of CH_4 and CO_2 exchange between the atmosphere and subarctic tundra. *Journal of Geophysical Research – Atmospheres*, **97**, 16627–16644.

Farwell S.O. & Barinaga C.J. (1986) Sulfur-selective detection with FPD: Current enigmas, practical usage, and future directions. *Journal of Chromatographic Science*, **24**, 483–495.

Fehsenfeld F.C., Golden P.D., Phillips M.P. & Sievers R.E. (1981) Selective electron capture sensitization. *Journal of Chromatographic Science*, **20**, 69–90.

Fehsenfeld F.C., Dickerson R.R. & Hübler G. (1987) A ground-based intercomparison of NO, NO_x and NO_y measurement techniques. *Journal of Geophysical Research – Atmospheres*, **92**, 14710–14722.

Fenchel T. & Blackburn T.H. (1979) *Bacteria and Mineral Cycling*. Academic Press, San Francisco.

Ferek R.J., Chatfield R.B. & Andreae M.O. (1986) Vertical destribution of dimethyl sulphide in the marine atmosphere. *Nature*, **320**, 514–516.

Ferek R.J. & Hegg D.A. (1992) Measurement of DMS and SO_2 in GTE/CITE-3. *Journal of Geophysical Research – Atmospheres*, in review.

Ferm M. (1979) Method for determination of atmospheric ammonia. *Atmospheric Environment*, **13**, 1385–1393.

Fontijin A., Sabadell A.J. & Ronco R. (1970) Homogenous chemiluminescence measurement of nitric oxide with ozone. *Analytical Chemistry*, **42**, 575–579.

Fraser P.J., Hyson P., Rasmussen R.A., Crawford A.J. & Khalil M.A.K. (1986) Methane, carbon monoxide and methylchloroform in the Southern Hemisphere. *Journal of Atmospheric Chemistry*, **4**, 3–42.

Freijer J.I. & Bouten W. (1991) A comparison of field methods for measuring soil carbon dioxide evolution: Experiments and simulation. *Plant and Soil*, **135**, 133–142.

Fried A. & Sams R. (1987) *Tunable Diode Laser Absorption Spectrometry for Ultratrace Measurement and Calibration*, pp. 121–131. American Society for Testing and Materials

Special Technical Publication Number 957.

Fried A., Henry B., Parrish D.D., Carpenter J.R. & Buhr M.P. (1991) Intercomparison of tunable diode laser and gas filter correlation measurements of ambient carbon monoxide. *Atmospheric Environment*, **25A**, 2277–2284.

Fried A., Sams R., Dorko W., Elkins J.W. & Cai Z. (1988) Determination of nitrogen dioxide in air compressed gas mixtures by quantitative tunable diode laser spectroscopy and chemiluminescence detection. *Analytical Chemistry*, **60**, 394–403.

Gaines K.K., Chatham W.H. & Farwell S.O. (1990) Comparison of the SCD and FPD for HRGC determination of atmospheric sulfur gases. *Journal of High Resolution Chromatography*, **13**, 489–493.

Goldan P.D. (1990) Analysis of low concentration level gaseous sulfur compounds in the atmosphere. In: *Monitoring Methods for Toxics in the Atmosphere*. American Society for Testing Materials Standard Technical Publication Number 1052, Philadelphia.

Goldan P.D., Fehesenfeld F.C. & Phillips M.P. (1982) Detection of carbon monoxide at ambient levels with an N_2O-sensitized electron capture detector. *Journal of Chromatography*, **239**, 115–126.

Goldan P.D., Kuster W.C., Albritton D.L. & Fehsenfeld F.C. (1987) The measurement of natural sulfur emissions from soils and vegetation: Three sites in the eastern United States revisited. *Journal of Atmospheric Chemistry*, **5**, 429–467.

Golterman H.L. (ed) (1971) *Methods for Chemical Analysis of Fresh Waters*. IBP Handbook Number 8, Blackwell Scien, Oxford.

Greenberg J.P. & Zimmerman P.R. (1984) Nonmethane hydrocarbons in remote tropical, continental and marine atmospheres. *Journal of Geophysical Research – Atmospheres*, **89**, 4767–4778.

Greenberg J.P., Zimmerman P.R. & Haagensen P. (1990) Tropospheric hydrocarbon and CO profiles over the U.S. West Coast and Alaska. *Journal of Geophysical Research – Atmospheres*, **95**, 14015–14026.

Gregory G.L., Davis D.D., Beltz N., Bandy A.R., Ferek R.J. & Thornton D. (1993) An intercomparison of aircraft instrumentation for tropospheric measurements of sulfur dioxide. *Journal of Geophysical Research – Atmospheres*, **98**, 23325–23352.

Gregory G.L., Warren L., Davis D.D. *et al.* (1993) An intercomparison of aircraft instrumentation for tropospheric measurements of dimethylsulfide: results at the parts-per-trillion level. *Journal of Geophysical Research*, **98**, 23373–23388.

Grimsrud E.P. & Miller D.A. (1978) Oxygen doping of carrier gas in measurement of halogenated methanes by gas chromatography with electron capture detection. *Analytical Chemistry*, **50**, 1141–1145.

Hahn J. (1974) The North Atlantic Ocean as a source of atmospheric N_2O. *Tellus*, **26**, 160–168.

Hansen M.H., Ingvorsen K. & Jørgensen B.B. (1978) Mechanisms of hydrogen sulfide release from coastal marine sediments to the atmosphere. *Limnology and Oceanography*, **23**, 68–76.

Harley J., Nel W. & Pretorious V. (1958) Flame ionization detector for gas chromatography. *Nature*, **181**, 177–178.

Harris D.C. (1991) *Quantitative Chemical Analysis*. W.H. Freeman, New York.

Harriss R.C., Sachse G.W., Hill G.R. *et al.* (1992) Carbon monoxide and methane in the North American arctic and subarctic troposphere. *Journal of Geophysical Research – Atmospheres*, **97**, 16589–16599.

Jaeschke W. & Herrmann J. (1981) Measurement of H_2S in the atmosphere. *International Journal of Environmental Analytical Chemistry*, **10**, 107–110.

James A.T. & Martin A.J.P. (1952) Gas–liquid partition chromatography, A technique for the analysis of volatile materials. *Analyst (London)*, **77**, 915–932.

Johnson J.E. & Bates T.S. (1993) Atmospheric measurements of carbonyl sulfide, dimethyl sulfide and carbon disulfide using the electron capture detector. *Journal of Geophysical Research – Atmospheres*, in press.

Johnson J.E. & Lovelock J.E. (1988) Electron capture sulfur detector: Reduced sulfur species detection at the femtomole level. *Analytical Chemistry*, **60**, 812–816.

Keeling C.D., Bacastow R.B. & Whorf T.P. (1982) Measurements of the concentration of carbon dioxide at Mauna Loa Observatory, Hawaii. In: Clark W.C. (ed) *Carbon Dioxide Review: 1982*. Oxford University Press, New York.

Kelly T.J. & Kenny D.V. (1991) Continuous determination of dimethylsulfide at part-per-trillion concentrations by atmospheric pressure chemical ionization mass spectrometry. *Atmospheric Environment*, **25A**, 2155–2160.

Kerfoot H.B., Pierett S.L., Amick E.N., Bottrell D.W. & Petty J.D. (1990) Analytical performance of four portable gas chromatographs under field conditions. *Journal of the Air and Waste Management Association*, **40**, 1106–1113.

Khalil M.A.K. & Rasmussen R.A. (1984) Carbon monoxide in the earth's atmosphere: Increasing trend. *Science*, **224**, 5456.

Khalil M.A.K. & Rasmussen R.A. (1992) The global sources of nitrous oxide. *Journal of Geophysical Research – Atmospheres*, **97**, 14651–14660.

Kirchoff V.W.J.H. & Marinho E.V.A. (1990) Surface carbon monoxide measurements in Amazonia. *Journal of Geophysical Research – Atmospheres*, **95**, 16933–16943.

Klemm O. & Talbot R.W. (1991) A sensitive method for measuring atmospheric concentrations of sulfur dioxide. *Journal of Atmospheric Chemistry*, **13**, 325–342.

Kley D. & McFarland M. (1980) Chemiluminescence detector for NO and NO_2. *Atmospheric Environment*, **18**, 63–69.

Komhyr W.D., Gammon R.H., Harris T.B. *et al.* (1985) Global atmospheric CO_2 distribution and variations from 1968–1982 NOAA/GMCC CO_2 flask sample data. *Journal of Geophysical Research – Atmospheres*, **90**, 5567–5596.

Komhyr W.D., Waterman L.S. & Taylor W.R. (1983) Semiautomatic nondispersive infrared analyzer apparatus for CO_2 air sample analyses. *Journal of Geophysical Research – Atmospheres*, **88**, 1315–1322.

Ku R.T., Hinkley E.D. & Sample J.O. (1975) Long-path monitoring of atmospheric carbon monoxide with a tunable diode laser system *Applied Optics*, **14**, 854–861.

Kuster W.C. & Goldan P.D. (1987) Quantitation of the losses of gaseous sulfur compounds to enclosure walls. *Environmental Science and Technology*, **21**, 810–815.

LeBel P.J., Hoell J.M., Levine J.S. & Vay S.A. (1985) Aircraft measurements of ammonia and nitric acid in the lower troposphere. *Geophysical Research Letters*, **12**, 401–404.

Lee M.L., Yang F.J. & Bartle K.D. (1984) *Open Tubular Column Gas Chromatography: Theory and Practice*. Wiley, New York.

Logan J.A. (1983) Nitrogen oxides in the troposphere: Global and regional budgets. *Journal of Geophysical Research – Atmospheres*, **88**, 10785–10807.

Lovelock J.E. (1974) The electron capture detector, theory and practice. *Journal of Chromatography*, **99**, 3–12.

Lovelock J.E. & Lipsky S.R. (1960) Electron affinity spectroscopy – a new method for the identification of functional groups in chemical compounds separated by gas chromatography. *Journal of the American Chemical Society*, **82**, 431–433.

Lovelock J.E. & Watson A.J. (1978) The electron capture detector, theory and practice. II. *Journal of Chromatography*, **158**, 123–138.

Maroulis P.J. & Bandy A.R. (1978) Estimate of the contribution of biologically produced DMS to the global sulfur cycle. *Science*, **196**, 647–648.

McAuliffe C. (1971) GC determination of solutes by multiple phase equilibration. *Chemical Technology*, **1**, 46–51.

McElroy M.B. & McConnell J.C. (1971) Nitrous oxide: A natural source of stratospheric NO. *Journal of Atmospheric Chemistry*, **28**, 1095–1098.

McElroy M.B. & Wofsy S.W. (1976) Tropical forests: Interactions with the atmosphere. In: Prance G.T. (ed) *Tropical Rain Forests and World Atmospheres*, pp. 33–60. Westview Press, Boulder, Colorado.

McManus J.B., Kebabian P.L. & Kolb C.E. (1989) Atmospheric methane measurement instrument using a Zeeman-split He–Ne laser. *Applied Optics*, **28**, 5016–5023.

McWilliam I.G. & Dewar R.A. (1958) Flame ionization detector for gas chromatography. *Nature*, **181**, 760.

Migeotte M. & Neven L. (1952) Recents progres dans l'observation du spectra solaire a la station scientific du Jungfraujoch. *Societie de Sciences à Liege*, **12**, 165–169.

Miller D.A. & Grimsrud E.P. (1979) Correlation of electron capture response enhancements caused by oxygen with chemical structure for chlorinated hydrocarbons. *Analytical Chemistry*, **51**, 851–859.

Natusch D.F.S., Klonis H.B., Axelrod H.D., Teck R.J. & Lodge J.P. Jr (1972) Sensitive method for the determination of atmospheric hydrogen sulfide. *Analytical Chemistry*, **44**, 2067–2073.

Novelli P.C., Steele L.P. & Elkins J.E. (1991) The development and evaluation of a gravimetric reference scale for measurements of atmospheric carbon monoxide. *Journal of Geophysical Research – Atmospheres*, **96**, 13109–13121.

Novelli P.C., Steele L.P. & Tans P.P. (1992) Mixing ratios of carbon monoxide in the troposphere. *Journal of Geophysical Research – Atmospheres*, **97**, 20731–20750.

Parrish D.D., Trainer M., Buhr M.P., Watkins B.A. & Fehsenfeld F.C. (1991) CO concentrations and their relation to NO_x at two rural sites. *Journal of Geophysical Research – Atmospheres*, **96**, 9309–9320.

Phillips M.P., Sievers R.E., Goldan P.D., Kuster W.C. & Fehsenfeld F.C. (1979) Enhancement of electron capture detector sensitivity to nonelectron attaching compounds by addition of nitrous oxide to the carrier gas. *Analytical Chemistry*, **51**, 1819–1825.

Popp P. & Oppermann G. (1978) Determination of carbon monoxide concentrations in air by gas chromatography using an argon ionization detector. *Journal of Chromatography*, **148**, 265–268.

Poulida O., Dickerson R.R., Doddridge B.G., Holland J.Z., Wardell R.G. & Watkins J.E. (1991) Trace concentrations and meteorology in rural Virginia 1. Ozone and carbon monoxide. *Journal of Geophysical Research – Atmospheres*, **96**, 22461–22475.

Prinn R., Cunnold D. Rasmussen R. *et al.* (1990) Atmospheric emissions and trends of nitrous oxide deduced from 10 years of ALE–GAGE data. *Journal of Geophysical Research – Atmospheres*, **95**, 18369–18385.

Prinn R.G., Simmonds P.G., Rasmussen R.A. *et al.* (1983) The atmospheric lifetime experiment 1. Introduction, instrumentation and overview. *Journal of Geophysical Research – Atmospheres*, **88**, 8353–8367.

Raich J.W., Bowden R.D. & Steudler P.A. (1990) Comparison of two static chamber techniques for determining carbon dioxide efflux from forest soils. *Soil Science Society of America Journal*, **54**, 1754–1757.

Rasmussen R.E. & Khalil M.A.K. (1981) Tropospheric methane (CH_4): Trends and seasonal cycles. *Journal of Geophysical Research – Atmospheres*, **86**, 9826–9832.

Rasmussen R.E. & Lovelock J.E. (1983) The atmospheric lifetime experiment 2. Calibration. *Journal of Geophysical Research – Atmospheres*, **88**, 8369–8378.

Ridgeway R.G. Jr, Bandy A.R. & Thornton D.C. (1991) Determination of aqueous dimethyl sulfide using isotope dilution gas chromatography/mass spectrometry. *Marine Chemistry*, **33**, 321–334.

Robbins R.C., Borg K.M. & Robinson E. (1968) Carbon monoxide in the atmosphere. *Journal of the Air Pollution Control Association*, **18**, 106–110.

Rudolph J. & Ehhalt D.H. (1981) Measurements of C_2–C_5 hydrocarbons over the North Atlantic. *Journal of Geophysical Research – Atmospheres*, **86**, 11959–11964.

Sachse G.W., Harriss R.C., Fishman J., Hill G.F. & Cahoon D.R. (1988) Carbon monoxide over the Amazon Basin during the 1985 dry season. *Journal of Geophysical Research – Atmospheres*, **93**, 1422–1430.

Sachse G.W., Hill G.F., Wade L.O. & Perry M.G. (1987) Fast-response, high-precision carbon monoxide sensor using a tunable diode laser absorption technique. *Journal of Geophysical Research – Atmospheres*, **92**, 2071–2081.

Saltzman E.S. & Cooper D.J. (1989) Dimethylsulfide and hydrogen sulfide in marine air. In: Saltzman E.S. & Cooper W.J. (eds) *Biogenic Sulfur in the Environment*. American Chemical Society Symposium Series Number 393, American Chemical Society, Washington, DC.

Saltzman E.S., Yvon S.A. & Matrai P.A. (1993) Low-level atmospheric sulfur dioxide measurement using HPLC/fluorescence detection. *Journal of Atmospheric Chemistry* **17**, 73–90.

Sebacher D.I. (1978) Airborne nondispersive infrared monitor for atmospheric trace gases. *Reviews of Scientific Instrumentation*, **49**, 1520–1525.

Sebacher D.I. & Harriss R.C. (1982) A system for measuring methane fluxes from inland and coastal environments. *Journal of Environmental Quality*, **11**, 34–37.

Seiler W. (1974) The cycle of atmospheric CO. *Tellus*, **26**, 116–135.

Seiler W. & Junge C. (1970) Carbon monoxide in the atmosphere. *Journal of Geophysical Research*, **75**, 2217–2226.

Seiler W., Giehl W. & Roggendorf P. (1980) Detection of carbon monoxide and hydrogen by conversion of mercury oxide to mercury vapor. *Atmospheric Technology*, **12**, 40–45.

Siggia S. (1968) *Survey of Analytical Chemistry*. McGraw-Hill, New York.

Sievers R.E., Phillips M.P., Barkley R.M. *et al.* (1979) Selective electron-capture sensitization. *Journal of Gas Chromatography*, **186**, 3–14.

Singh H.B. & Salas L.J. (1982) Measurement of selected light hydrocarbons over the Pacific Ocean: Latitudinal and seasonal variations. *Geophysical Research Letters*, **9**, 842–845.

Singh H.B., Salas L.J. & Shigeishi H. (1979) The distribution of nitrous oxide (N_2O) in the global atmosphere and the Pacific Ocean. *Tellus*, **31**, 313–320.

Skoog D.A. (1985) *Principles of Instrumental Analysis*, 3rd edn. W.B. Saunders, Philadelphia.

Solorzano L. (1969) Determination of ammonia in natural waters by the phenolhypochlorite method. *Limnology and Oceanography*, **14**, 799–801.

Steele L.P., Fraser P.J., Rasmussen R.A. *et al.* (1987) The global distribution of methane in the troposphere. *Journal of Atmospheric Chemistry*, **5**, 125–171.

Steudler P.A. & Kijowski W. (1984) Determination of reduced sulfur gases in air by solid absorbent preconcentration and gas chromatography. *Analytical Chemistry*, **56**, 1432–1436.

Strobel H.A. & Heineman W.R. (1989) *Chemical Instrumentation: A Systematic Approach*. Wiley, New York.

Tacket J.L. (1968) Theory and application of gas chromatography in soil aeration reasearch. *Soil Science Society of America Proceedings*, **32**, 346–350.

Thompson A.M., Huntly M.A. & Stewart R.W. (1990) Perturbations to tropospheric oxidants, 1985–2035 1. Calculations of ozone and OH in chemically coherent regions. *Journal of Geophysical Research – Atmospheres*, **95**, 9829–9844.

Thornton D.C., Bandy A.R. & Driedger A.R. III (1987) Sulfur dioxide over the western Atlantic Ocean. *Global Biogeochemical Cycles*, **1**, 317–328.

Turner S.M. & Liss P.S. (1985) Measurements of various sulphur gases in the coastal environment. *Journal of Atmospheric Chemistry*, **2**, 223–232.

Weiss R.F. (1981) Determinations of carbon dioxide and methane by dual catalyst flame ionization chromatography and nitrous oxide by electron capture chromatography. *Journal of Chromatographic Science*, **19**, 611–615.

Weiss R.F., Keeling C.D. & Craig H. (1981) The determination of tropospheric nitrous oxide. *Journal of Geophysical Research – Atmospheres*, **86**, 7197–7202.

Wentworth W.E. & Freeman R.R. (1973) Measurement of atmospheric nitrous oxide using an electron capture detector in conjunction with gas chromatography. *Journal of Chromatography*, **79**, 322–324.

WMO (1985) Atmospheric Ozone 1985. World Meteorological Organization global ozone research and monitoring project. Report number 16.

Witkamp M. & van der Drift J. (1961) Breakdown of forest litter in relation to environmental factors. *Plant and Soil*, **15**, 295–311.

Wright N. & Herscher L.W. (1946) Recording infrared analyzers for butadiene and styrene plant streams. *Journal of the Optical Society of America*, **36**, 195–199.

Young A. & Cline J. (1980) *Precision Measurement of Atmospheric Nitrous Oxide Utilizing Electron Capture Gas Chromatography*, pp. 79–87. World Meteorological Organization Special Environmental Report Number 14.

Zander R., Demoulin P., Ehhalt D.H., Schmidt U. & Rinsland C.P. (1989) Secular increase of total vertical abundance of carbon monoxide above central Europe since 1950. *Journal of Geophysical Research – Atmospheres*, **94**, 11021–11028.

CHAPTER 7

Measurement of chemically reactive trace gases at ambient concentrations

F. FEHSENFELD

7.1 Introduction

A key element in the understanding of biosphere atmosphere exchange is the ability to make unequivocal measurements of the concentrations and fluxes of the compounds of interest. If these measurements are to be meaningful, it is necessary to have reliable instruments and techniques for which there are trustworthy estimates of the uncertainties in the observations, since they are the touchstones against which theoretical understanding is tested. With that estimate:

1 observations and theory can be compared meaningfully;

2 the results from separate field studies can be merged reliably;

3 spatial gradients from separate data sets can be characterized credibly; and

4 time series from different networks can be used to establish longer trend records.

The formal intercomparison and testing of individual instruments to measure air concentrations of chemically labile species has been an area of active investigation for the past 20 years. The goal of this instrument validation has been to develop a reliable suite of instruments to study the processes attendant to acid deposition, degradation of urban and regional air quality, global climate change and stratospheric ozone depletion. Such validations have rarely been done within the ecological community. However, the lessons learned in those studies provide valuable guidance in preparation for future intercomparison exercises.

Technique development follows a logical sequence. New techniques, as they are devised, are tested in the field. This testing demonstrates the field worthiness of the technique and establishes the necessary level of and frequency of housekeeping procedures (calibration, zeroing, routine maintenance) required for proper operation. Even at this early stage, the instrument may be called upon to make important measurements. However, these measurements should be regarded as preliminary until the instrument is fully validated.

The initial phase of instrument evaluation may also include informal instrument comparisons. In these comparisons, instruments are brought together and operated side by side. As differences are found they are noted and modifications are made to correct the causes of these differences. They are intended to alert the scientists who are developing the new techniques to problems that might be obscured by independent testing alone and to provide significant insights into instrument capability. Several such informal intercomparisons are drawn in later sections.

The final stage of instrument validation involves a formal instrument comparison for instruments that have demonstrated field worthiness. The instrument intercomparisons should include:

1 evaluation of calibration standards used by and available to the participants;

2 use of several different techniques for measuring the same species;

3 measurement at the same place and time and under typical operating conditions; and

4 comparison with accuracy and precision estimates set in advance.

Each investigator should prepare his/her results independently, but should jointly (or via an independent party) compile separate results and assess the state of agreement/disagreement for publication in a refereed journal.

There have been several field campaigns devoted specifically to the assessment of instrument reliability, as opposed to obtaining data solely to answer a scientific question. Since these intercomparisons provide the only objective assessment of instrument capability, our conclusions concerning the capability of present instrumentation depends heavily on these results.

The instruments and techniques available for measurement of the atmospheric concentration and fluxes of several chemically active trace compounds, the basic operating principle of these devices, and highlights of the tests done thus far to determine instrument reliability are summarized below. This discussion only deals with recent development of instruments that make rapid measurement of trace quantities of chemically active compounds, especially those of biogeochemical importance. (Information on standard measurement techniques for less reactive biogenic gases is presented in Chapter 6.) Informal intercomparisons are also briefly described in the sections devoted to some of those compounds. Finally, the issues of instrument calibration and the preparation of calibration standards for compounds of interest to atmospheric chemistry are briefly addressed.

7.2 The oxides of nitrogen and their oxidation products

7.2.1 Fate in the atmosphere

The reactive oxides of nitrogen enter the atmosphere as NO and NO_2. During daylight hours there is a rapid conversion of NO and NO_2 (together

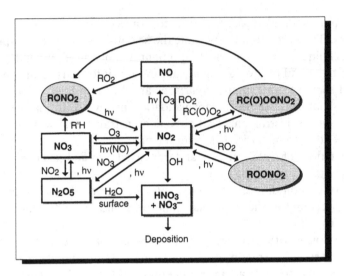

Fig. 7.1 Simplified schematic diagram outlining the chemical processing that occurs among the group of compounds that make up the nitrate family. (From Fehsenfeld & Liu, 1993.)

known as NO_x) to a variety of organic and inorganic nitrates. These are the compounds that make up the nitrate family, NO_y (= NO_x + organic nitrates + inorganic nitrates). The conversion of NO_x to the other members of the NO_y family takes place within a period of a few hours during photochemically active summer days to a few days during winter. One of the important byproducts of these conversions is ozone produced in the troposphere.

The oxidation of organic compounds in the presence of NO_x leads to the formation of organic nitrates, listed as $ROONO_2$, $RC(O)OONO_2$ and $RONO_2$ in Fig. 7.1. Although the oxidation of volatile organic compounds can form a large variety of organic nitrates, the best known and by far the most studied of these compounds is peroxy acetyl nitrate (PAN). PAN was first observed in the atmosphere as a component of urban photochemical smog. Since that time, surface and aircraft observations have confirmed its ubiquity in the atmosphere. These measurements have indicated that organic nitrates such as PAN act to preserve nitrogen oxides temporarily in the troposphere, especially at low temperatures, and release them at high temperatures or by photolysis by sunlight. The importance of organic nitrates, and in particular PAN, relative to NO_x increases rapidly at the lower temperatures characteristic of winter at high altitudes. Under these conditions PAN levels may substantially exceed NO_x levels. This implies a significant role for PAN and the other organic nitrates in the long range transport of nitrates through the atmosphere. Since the lifetimes of organic nitrates depend strongly on temperature and sunlight intensity, for a given organic nitrate the lifetime can range from less than 1 h to several weeks.

In addition, the nitrate radical, NO_3, is formed by the reaction of NO_2 with ozone. NO_3 is rapidly photolysed in sunlight so its concentration and reaction are significant only at night. However, during these hours, particularly in wintertime, NO_3 may have a significant role in the formation of HNO_3 (Calvert et al., 1985; Liu et al., 1987) and, under certain circumstances, in initiating the oxidation of volatile organic compounds (Atkinson et al., 1984, 1988; Dlugokensky & Howard, 1989). Because of its high reactivity, the concentration of NO_3 in the atmosphere is usually very small.

The principal inorganic nitrate present in the troposphere is HNO_3. HNO_3 is formed during the daytime by the association of NO_2 with the hydroxyl radical (OH). HNO_3 is rapidly incorporated into cloud water and aqueous aerosols to form nitrate ions (NO_3-). In addition, as noted above, at night NO_3 can react through homogenous and heterogeneous chemistry to form HNO_3 and nitrate ions. The wet and dry deposition of HNO_3 and nitrate ions represent the principal removal mechanisms for nitrate from the troposphere. HNO_3 is rapidly removed from the near-surface boundary layer by dry deposition. In addition, the removal of HNO_3 and nitrate ions occurs rapidly during precipitation. However, these species are relatively stable against photochemical destruction and in the absence of precipitation can persist much longer in the free troposphere.

7.2.2 Techniques for NO_y measurements

NO

The reliability of techniques to measure NO has been established rigorously. Recently, two fundamentally different methods have been compared: chemiluminescence and laser-induced fluorescence (LIF). The chemiluminescence technique (Fontijn et al., 1970; Ridley & Howlett, 1974) is based on the chemical reaction from NO in the sampled air with added ozone that generates emission of detectable radiation from the NO_2 product proportional to the concentration of the NO in the sampled air. The LIF technique (Bradshaw et al., 1985) detection of NO relies on the absorption of radiation by NO in the sampled air and then re-radiation at different wavelengths by the excited NO. During three separate intercomparison campaigns, chemiluminescence instruments and an LIF instrument were operated simultaneously to measure ambient concentrations at a rural ground-based site and in airborne studies (Hoell et al., 1985, 1987a; Gregory et al., 1990d). The data agreed within 30% in all of these chemically different environments, which spanned a concen-

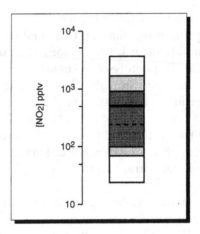

Fig. 7.2 Histogram showing the range of NO_2 measurements made during intercomparison of NO_2 measurement techniques (cf. Fehsenfeld *et al.*, 1990), 22 June to 21 July 1987. The bar contains all the NO_2 measurements made during the period. □ represents the central 90% of the measurements, and ■ the central 67% of the measurements; (——) average; (– – –) median.

ration range of 0.005–0.2 ppbv. These results strongly indicate that NO can be measured reliably by either of these techniques under most field conditions.

NO_2

Many techniques have been developed to measure NO_2, but few are capable of measuring NO_2 at the sub-ppbv level or have been demonstrated to be free of interference from other atmospheric constituents. The standard way to measure NO_2 in almost all air quality studies has been to use surface-conversion techniques to convert NO_2 to NO, and subsequently detect the NO by chemiluminescence. These conversion techniques include heated catalytic metal surfaces, and those coated with ferrous sulphate or other compounds. However, the development of the photolytic NO_2 to NO converter several years ago (Kley & McFarland, 1980) offered a potentially more specific conversion technique, albeit less simple. Fehsenfeld *et al.* (1987) made a detailed study of the performance of surface and photolytic methods. In this intercomparison, the ferrous sulphate and photolytic converters agreed well at NO_2 levels of 1 ppbv and greater. However, the ferrous sulphate converter systematically reported higher values at lower NO_2 levels, reaching a factor of 2 higher at 0.1 ppbv. Spiking tests showed that the ferrous sulphate converter was also converting PAN to NO, thus resulting in an overestimate of NO_2. A heated molybdenum oxide surface converter was found to convert NO_2,

HNO$_3$ and PAN to NO, thereby indicating that heated-surface converters also could not be considered specific for NO$_2$ (and hence the nitrogen oxides).

Newer technology is emerging to measure NO$_2$. Three of the newer techniques which show considerable promise are photofragmentation/2-photon LIF, tunable diode laser absorption spectrometry (TDLAS) and luminol fluorescence. The LIF technique detection of NO$_2$ relies on the absorption of ultraviolet radiation by NO$_2$ in the sampled air. This NO$_2$ photodissociates and yields an excited NO molecule that re-radiates at different wavelengths from the initial exciting radiation's wavelength. The TDLAS (Walega et al., 1984; Schiff et al., 1987) measures the absorption of laser-generated infrared radiation by the NO$_2$ molecules in an air sample within the White cell of the laser spectrometer. The luminol instrument (Schiff et al., 1986) detects NO$_2$ by measuring the chemiluminescence produced by the oxidation of the luminol by NO$_2$. The LIF and TDLAS techniques provide specific spectroscopic methods to measure the NO$_2$ level, while the luminol technique provides a sensitive, portable method with low power requirements. Two recent intercomparisons have tested these techniques against the photolysis/chemiluminescence technique.

A ground-based intercomparison (Fig. 7.1) (Fehsenfeld et al., 1990) tested the photolysis/chemiluminescence technique against the TDLAS and the luminol techniques. Typical data from this intercomparison are shown in Fig. 7.2. Clearly, the NO$_2$ mixing ratios measured using the luminol technique with ozone removed from the sampled air (Fig. 7.3c) agree much better with the ozone mixing ratios measured using the photolysis/chemiluminescence technique (Fig. 7.3a). The results from this intercomparison indicated that interferences from PAN as well as ozone influenced the NO$_2$ measurements made using the luminol technique. Strategies are currently being developed to remove or separate these interferences from the ambient air prior to sampling into the luminol detector. For discussion of other limitations identified by the intercomparison see Fehsenfeld et al. (1990).

An airborne intercomparison of the TDLAS technique with the LIF and photolytic converter/chemiluminescence techniques was conducted (Gregory et al., 1990a). The intercomparison of these three instruments in ambient air for [NO$_2$] > 0.1 ppbv indicated a general level of agreement among the instruments of the order of 30–40%. Intercomparison of the TDLAS, LIF and photolysis/chemiluminescence technique for [NO$_2$] < 0.05 ppbv indicated the TDLAS overestimated the NO$_2$ mixing ratio, presumably due to the reliance on correlation coefficients as the data selection criteria. At these low levels, the agreement between LIF and photolysis/chemiluminescence measurements was within 0.02 ppbv with

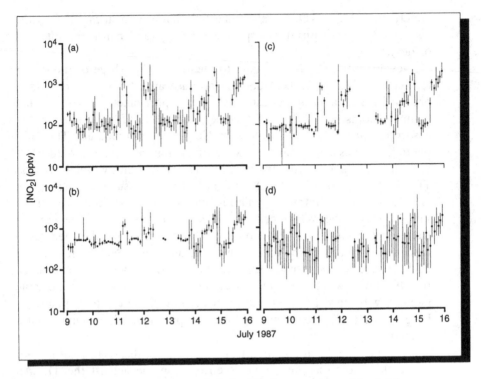

Fig. 7.3 Simultaneous measurements made during NO_2 intercomparison: (a) using the photolysis/chemiluminescence technique; (b) using the Luminox technique; (c) using the Luminox technique preceded by an ozone destruction trap; (d) using the tunable diode laser absorption spectroscopy technique. Each of the symbols gives the reported measurements averaged over 2 h intervals. The vertical bars represent the range of variability in the NO_2 levels seen during the averaging period. (From Fehsenfeld *et al.*, 1990.)

an equal tendency for one to be high or low compared to the other. This 0.02 ppbv agreement is typically within the stated uncertainties of the two techniques at NO_2 mixing ratios < 0.05 ppbv.

It is currently believed that properly used, the LIF, TDLAS and photolysis/chemiluminescence techniques are capable of the measurement of NO_2 to levels well below 0.1 ppbv free of significant artefact or interference. These techniques should therefore be capable of measuring the NO_2 levels in most locations.

PAN

Two instruments, both employing cryogenically enriched sampling with electron capture gas chromatography detection (Singh & Salas, 1983; Ridley *et al.*, 1990), have been intercompared in the remote maritime free troposphere (Gregory *et al.*, 1990c).

The intercomparison showed that at mixing ratios <0.1 ppbv, the two instruments differed on average by 0.017 ppbv with a 95% confidence interval of ± 0.009 ppbv. At mixing ratios of PAN of 0.1–0.3 ppbv, the difference between the instruments was of the order of 25% ± 6%. A linear regression fit comparing all data below 0.3 ppbv from the two techniques gave a slope of 1.34 ± 0.12 with an intercept of 0.0004 ± 0.012 ppbv. Although one instrument was consistently high relative to the other for ambient measurements, these levels of agreement were usually within the stated accuracy and precision of the two instruments. These results are reassuring. Nevertheless, their significance is reduced because of the similarity in the two instruments.

NO_3

Two methods have been used to measure the concentration of NO_3 in the atmosphere: (a) long path differential optical absorption spectroscopy (DOAS); and (b) matrix isolation with electron spin resonance (MI-ESR) detection. The DOAS detection is accomplished by the measured adsorption occurring in the NO_3 absorption bands that lie between 600 and 700 nm. Total and height resolved NO_3 abundances have been measured using the moon and sky light near dawn as light sources (Noxon *et al.*, 1978; Solomon *et al.*, 1987, 1989a,b). Measurements of NO_3 have also been made in the troposphere with the DOAS technique using lamps and a horizontal long-path (Platt *et al.*, 1979, 1980, 1984; Noxon *et al.*, 1980). The MI-ESR technique (Mihelcic *et al.*, 1978) relies on the cryogenic trapping of NO_3 and peroxy free radicals (HO_2 and RO_2) in a water matrix followed by the detection of the free radical using ESR. Problems with interferences in the ESR spectra have been overcome by using D_2O, instead of H_2O, as the isolation matrix. This substitution has improved the signal:noise ratio and spectral resolution, allowing the identification of different free radical species during field measurements (Mihelcic *et al.*, 1985, 1989, 1990). No attempt has yet been made to compare the measurements of NO_3 made using these techniques.

HNO_3

A comparison was made focusing on the status of capabilities of measuring HNO_3 (Hering *et al.*, 1988). Over an 8-day period at an urban–suburban site, six methods made simultaneous measurements: (a) filter pack; (b) denuder difference; (c) annular denuder; (d) transition flow reactor; (e) TDL; and (f) Fourier transform infrared (FTIR) spectrometer. The reported HNO_3 concentrations varied among methods by more than a factor of 2. These differences were substantially larger than the estimated precision of these instruments. The tests indicated that artefacts or interferences existed for some of the sampling methods that were associated with either the field sampling components (e.g. inlet/lines), operating procedures, detector specificity or alteration during sampling in the physical or chemical make-up of the ambient air, such as shifts in the gas- and solid–aqueous-phase equilibrium of HNO_3, NH_3 and ammonium nitrates (see Hering *et al.*, 1988).

Several conclusions could be drawn from this data set. The larger percentage differences in the techniques that were observed at higher HNO_3 concentrations and the dependence of the differences on day or night sampling suggest uncontrollable shifts of the equilibrium (i.e. ammonium nitrate evaporation) in samples obtained by some instruments. The annular denuder exhibited poor intramethod precision for HNO_3, and its average value was substantially below the means of the spectroscopic methods and that of all methods. The results from tungstic acid adsorption tubes and filter packs (>8-h sample) deviated substantially from those two means. The filter packs exhibited a positive bias that increased as the sampling time average increased, indicating an artefact due to ammonium nitrate particle evaporation to release HNO_3 (and NH_3). The denuder difference, transition-flow reactor, filter pack (<8-h sample), and spectroscopic methods were in good agreement. This comparison provides a valuable start in assessing the problems of reliable measurement of this challenging species.

A more recent intercomparison involved three different measurement techniques: nylon filter collection, tungstic oxide denuder and TDLAS (Gregory *et al.*, 1990b). In general, filter measurements were high relative to those reported by the denuder with correlation observed between the filter and denuder techniques for $HNO_3 < 0.15$ ppbv. Comparing the denuder and TDLAS techniques, TDLAS measurements were consistently higher; for $HNO_3 > 0.3$ ppbv, by a factor of approximately 2. There was only one instance of overlap among all three techniques at levels of HNO_3 well above detection limits. In that case, the measurements from filter and TDLAS were in agreement, while those from the denuder (with only a 35% overlap) were about a factor of 2 lower. It was clear that there was substantial disagreement among the three techniques, even at mixing ratios well above their respective detection limits.

Finally, an informal intercomparison of the nylon filter technique with the mist chamber technique was reported by Talbot *et al.* (1990). Both the mist chamber and standard nylon filter techniques were found to exhibit high collection efficiency and excellent agreement when measuring HNO_3 vapours from a permeation source. However, during simultaneous sampling of ambient air, the nylon filter measured on average 70% higher HNO_3 concentration than the mist chamber technique. Talbot *et al.* (1990) ascribed a major portion of the difference to an unidentified nitrogen species, perhaps an organic nitrate, in the atmosphere which presumably reacted on the filter to form NO_3-. Talbot *et al.* (1990) suggested that under certain conditions this interference could produce significant errors in the quantification nylon filter measurements of HNO_3 vapour.

These intercomparisons clearly indicate that present techniques do not allow the unequivocal determination of HNO_3 in the range of concentrations expected for that compound in the non-urban atmosphere.

NO_y

Our understanding of the reaction pathways can be aided by the measurement of the total abundance of atmospheric nitrogen compounds, NO_y, as well as by the measurement of the individual NO_y species. For example, it is $[NO_y]$, rather than components like NO_2, that is of primary interest in establishing the inflow/outflow regional budgets in the tropospheric acid transport/deposition problem.

Several NO_y measurement techniques have been proposed. In general, all these techniques rely on the reduction of NO_y species to NO followed by the detection of NO. A ground-based intercomparison of two of these techniques, the Au-catalysed conversion of NO_y to NO in the presence of CO and the reduction of NO_y to NO on a heated molybdenum oxide surface, has been done (Fehsenfeld et al., 1987). Figure 7.4 presents the results from this NO_y intercomparison. Here $[NO_y]$ measured using the molybdenum oxide converter, $[NO_y]_{MoO}$, is plotted versus that obtained using the gold converter, $[NO_y]_{Au}$. It is clear from the data in Fig. 7.4 that, except for a few cases, these instruments were found to give similar results for the measurement of $[NO_y]$ in ambient air under conditions that varied from typical urban air to clean continental background air, with NO_y ranging between 0.4 and 100 ppbv. The points indicating a larger difference in the measurements were attributed to a calibration error in the molybdenum oxide system that was discovered after the final data from that measurement were submitted to the referee.

However, it was found that when the molybdenum oxide converter was operated for extended periods (i.e. several hours) with NO_y levels greater than 100 ppbv the conversion efficiency dropped significantly. For this reason the Au-catalysed converter was judged to have greater reliability when used in a polluted environment.

7.2.3 Measurement of NO_y emission rates and fluxes

The emission of nitric oxides from soils is one of the largest global sources of reactive nitrogen oxides to the atmosphere (cf. Williams et al., 1992a). For this reason, the development of techniques and approaches to measure these emissions has been the subject of considerable research. This has generated an effort to assess critically the capability of the various approaches used to measure the soil emissions of NO_y.

Enclosure, gradient techniques and eddy correlation techniques have

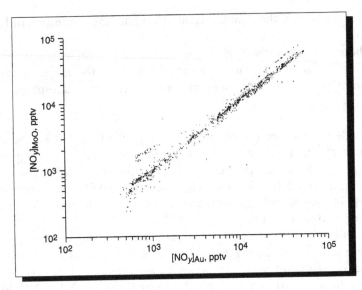

Fig. 7.4 Results from comparison of NO_y measurement techniques. This is a plot of mixing ratios of NO_y measured using a molybdenum oxide NO_y converter, $[No_y]_{MoO}$, vs. those made using a gold catalysed converted, $[NO_y]_{Au}$. Each point represents a 3 min average of the simultaneous measurements of the NO mixing ratio in ambient air that were made by the two instruments during the comparison. (From Fehsenfeld *et al.*, 1987.)

been utilized to measure NO_y (principally, NO) fluxes from soils. The enclosure technique measures the NO_y flux from a relatively small area of soil ($\approx 1\,m^2$), whereas the gradient techniques determine the average compound flux from soils over a large area (typically $10^2\,m^2$ or more). In the enclosure technique (cf. Slemr & Seiler, 1984; Williams *et al.*, 1987), the mass accumulation of NO_y is determined from the difference in the concentrations in the enclosure over a measured period of time. This method is illustrated in a simplified schematic diagram shown in Fig. 7.5. The gradient technique was used by Parrish *et al.* (1987), Kaplan *et al.* (1988) and Johansson *et al.* (1988) to measure the night-time fluxes of NO from soil. The technique relies on the measured NO concentration gradient with height above the ground to define the vertical mixing and, thus, avoids the usual requirement for simultaneous measurement of meteorological parameters. This technique is illustrated in a simplified diagram in Fig. 7.6. The method, which is described in detail by Parrish *et al.* (1987), is based on the premise that during the night when photo-chemical production of NO from NO_2 is inoperative, net soil emission of

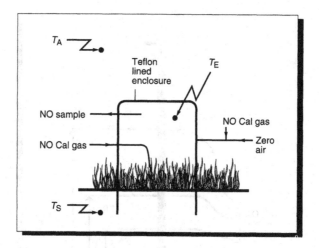

Fig. 7.5 Simplified schematic drawing illustrating the use of the enclosure method to measure NO fluxes from the soil. T_A, T_S and T_E indicate the air, soil and enclosure temperatures, respectively. These temperatures influence the flux of NO from the soil or the interpretation of those fluxes (cf. Parrish *et al.*, 1987; Williams *et al.*, 1987).

NO is equal to its integrated loss in the lower atmosphere due to reaction with O_3. The eddy correlation technique has also been used to measure the flux of NO_y (Delany *et al.*, 1986; Hicks *et al.*, 1986; Wesely *et al.*, 1989; Zeller *et al.*, 1989). The technique relies on the fast (1–10 Hz) measurement of vertical wind, temperature, humidity and trace gas concentrations. The flux is calculated by averaging the deviations of the product of the vertical air motion and gas concentration. To date, however, the results of NO_x fluxes from soils obtained using this technique have not been compared with the other methods.

Although no formal intercomparison of these NO measurement techniques has been attempted, an informal comparison of the enclosure and gradient methods for the measurement of NO fluxes at night has been done (Parrish *et al.*, 1987; Williams *et al.*, 1988). The results of this intercomparison are shown in Table 7.1. The first column indicates the date of the night of the measurement, while the second and third columns indicate the results obtained using the enclosure and the gradient techniques. The agreement between the average flux inferred from enclosure measurements and the simultaneous co-located measurement by the gradient method was good.

7.3 NH₃

7.3.1 Fate in the atmosphere
The atmospheric NH_3 concentration is determined by a combination

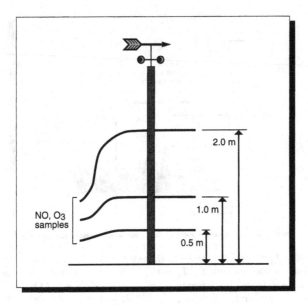

Fig. 7.6 Simplified schematic drawing illustrating the use of the gradient method to measure NO fluxes from the soil (cf. Parrish *et al.*, 1987). In these experiments air was simultaneously sampled 0.5, 1.0 and 2.0 m above the soil, and the NO flux was deduced from the gradient in the NO mixing ratio.

atmosphere–biosphere exchange processes near the surface and gas-to-particle transformations in the atmosphere. Gaseous NH_3 does not absorb in the actinic ultraviolet and reacts only slowly with OH (DeMore *et al.*, 1990); the lifetime with respect to this process is ≈ 3 months. This, coupled with the low concentrations of gaseous NH_3 above the boundary layer, suggests that photochemical destruction is a relatively minor sink.

The most likely fate for gaseous NH_3 escaping the surface layer is incorporation into acidic aerosols and cloud-water droplets in the boundary layer and lower troposphere (Seinfeld, 1986). The incorporation of NH_3 into these liquid and/or solid phases displaces H^+ as the primary cation and lowers the acidity thereby facilitating the uptake of weaker acids, like HNO_3, HCl or organic acids. The sequestering of NH_3 by aerosols and cloud-water occurs primarily through association with H_2SO_4 to form NH_4HSO_4 and $(NH_4)_2SO_4$. The equilibrium vapour pressure of NH_3 above sulphuric acid is extremely small (Scott & Cattell, 1979; Tang, 1980) so that this process can be considered irreversible. NH_3 can also react with gas phase HNO_3 to form NH_4NO_3. This association is reversible under atmospheric conditions (Stelson & Seinfeld, 1982) and ammonium nitrate can decompose releasing NH_3. Because of the greater affinity of sulphate for ammonium, little nitrate will be incorporated into

Table 7.1 Results for the intercomparison of enclosure and gradient measurements of NO flux. (From Parrish *et al.*, 1987.)

Date	Enclosure	Gradient
Aug. 24	0.35	1.0 ± 0.1
Aug. 25	0.28	0.2 ± 0.1
Aug. 26	0.83	0.5 ± 0.1
Aug. 27	0.85	1.0 ± 0.1
Aug. 28	—	0.6 ± 0.3
Aug. 29	0.28	0.08 ± 0.05
Aug. 30	0.37 ± 0.05	0.13 ± 0.02
Average Aug. 24–Aug. 30	0.49 ± 0.28	0.50 ± 0.36
Aug. 31 p.m.	2.8	2.1 ± 0.5
Aug. 31 a.m.	—	21 ± 4
Sept. 1	20 ± 4	16 ± 9
Sept. 2	—	3.2 ± 1.6
Sept. 3	2.0 ± 0.4	2.2 ± 1.5
Oct. 4	0.19	0.5

Units of flux are $ng\,N\,m^{-2}\,s^{-1}$.

the particulate phase until the sulphate is almost completely neutralized (Tang, 1980). In coastal areas, a similar reaction with HCl forms ammonium chloride (NH_4Cl) (Harrison *et al.*, 1989).

The above processes lead to atmospheric lifetimes for gaseous NH_3 of the order of several hours to a few days (Georgii & Gravenhorst, 1977; Asman & Janssen, 1987). In general, gaseous NH_3 is thought to decrease with altitude above land surfaces and is larger over the continents than the oceans (Georgii & Müller, 1974; Georgii & Gravenhorst, 1977; Hoell *et al.*, 1980; Lenhard & Gravenhorst, 1980; Levine *et al.*, 1980; Tanner, 1982; LeBel *et al.*, 1985; Ziereis & Arnold, 1986).

7.3.2 Techniques to measure NH₃

Much of the uncertainty associated with atmospheric NH_3 is due to the scarcity of reliable measurements. This in turn is due to a lack of sensitive measurement techniques which are easy to use in the field and can operate automatically and continuously to establish large databases for statistical analysis. A major difficulty associated with NH_3 measurements is the presence of ammonia in gaseous (NH_3), particulate (e.g. $(NH_4)_2SO_4$ and NH_4NO_3), and liquid (i.e. NH_4OH in cloud and fog droplets) forms in the atmosphere. The partitioning between these three phases is highly variable and a complete picture generally requires simultaneous, but separate measurements of at least the gaseous and particulate phases. A second major difficulty results from the tendency of NH_3 to form strong hydrogen bonds with water and thus adsorb on the surfaces of most materials exposed to air. This can lead to high backgrounds and memory effects due to retention on sampling tubing and other plumbing.

Most gaseous NH_3 measurements reported in the literature trap ammonia with an acidic medium: acid bubblers (Junge, 1957; Breeding *et al.*, 1973), acid-coated filters (Shendriker & Lodge, 1975) or acid-coated denuder tubes (Ferm, 1979). The solution from the bubblers or rinsed filters or tubes is then analysed for ammonium via colorimetry, ion-specific electrode or ion chromatography. The diffusion denuders developed by Ferm (1979) separate gaseous NH_3 from particulate and droplet NH_{4^+} by exploiting the large difference in diffusion coefficient; gaseous NH_3 in an airstream passing through the tube diffuses to and adheres to the acid-coated wall while the heavier particles pass straight through. Variants of this technique based on both cylindrical and annular designs have become the *de facto* standards for gaseous NH_3 measurements. The detection limit depends on the sample time and denuder preparation, but levels below $\approx 0.01\,ppbv$ appear to be possible for a 2-h sample. Similar limits may be obtained with acid-coated filters which are also widely used, typically with a Teflon prefilter to collect the particulate phase. Filter packs are generally considered the standard technique for measuring particulate NH_{4^+} and related species although more sophisticated techniques are under development and evaluation (Sickles *et al.*, 1990).

Several intercomparisons of the diffusion denuder and filter pack techniques for measuring gaseous NH_3 have been conducted. A field comparison of oxalic acid-coated paper filters and diffusion denuders described by Gras (1984) indicated that volatilization of particulate ammonium collected on the Teflon prefilter led to high apparent gaseous NH_3 in the filter pack technique. This result was also indicated in an intercomparison of filter pack, diffusion denuder, FTIR and tungsten oxide denuder detectors in Claremont, California, during 1986 (Appel *et al.*, 1988; Wiebe *et al.*, 1990). In this case, NH_3 concentrations measured with the filter packs were approximately 1.5 times those measured with the oxalic acid denuders. Harrison & Kiitto (1990) also found evidence for particulate volatilization in the filter packs, but concluded that the differences will typically be small compared to the natural variability of ambient NH_3. These workers also stressed that the differences depend on the particular site, operator and apparatus and thus do not reflect intrinsic limitations of the filter pack technique.

Both filter packs and denuders offer the advantages of simplicity and relatively low cost, but the preparation and analysis of the sampler is labour intensive. Furthermore, great care must be taken to prevent contamination during storage and handling, since NH_3 is present in human breath and perspiration as well as a variety of cleaning agents found in most laboratories. Thus, the reproducibility and detection limit (and hence the necessary exposure time) obtained by these techniques largely depends on the skill and consistency of the operator. These restrictions make the acquisition of large data sets with sufficient temporal resolution to probe the parameters which influence atmospheric concentrations of NH_3 difficult.

Several groups (Bos, 1980; Keuken *et al.*, 1988) have developed automated diffusion denuder systems in which the acid solution coating is continuously replenished. The system developed at The Netherlands Energy Research Foundation (ECN) by Keuken *et al.* (1988) can operate for up to 8 h without user intervention and detect other species as well as NH_3. Unfortunately, the detection limit for a 30-min sample time is relatively high (≥ 0.3 ppbv). An alternate approach to automated detection uses 'permanent' acid coatings and thermal desorption. McClenny and co-workers (McClenny & Bennet, 1980; Harward *et al.*, 1982) collected NH_3 on Teflon beads or tungsten oxide tubes and detected the thermally desorbed product by photoacoustic spectroscopy with a nominal detection limit of ≈ 0.2 ppbv. Braman *et al.* (1982) developed a denuder tube coated with tungsten oxide/tungstic acid to collect NH_3. The collected NH_3 is thermally desorbed and subsequently converted to NO on a hot Au surface. The NO is measured with a standard chemiluminescence detector. Field measurements based on this system are described by McClenny *et al.* (1982). LeBel *et al.* (1985) used a similar instrument to measure vertical profiles of NH_3 above coastal Virginia and Maryland from an airplane, and Roberts *et al.* (1988) used this technique at Niwot Ridge, Colorado and Pt. Arena, California. A similar instrument, developed by Appel *et al.* (1984) and tested in the Claremont intercomparison described above, agreed with the diffusion denuder measurements to within about a factor of 2 but exhibited a temperature-dependent sensitivity.

Thermal denuder systems based on other metal oxides have also been developed. Keuken *et al.* (1989) developed a system based on vanadium pentoxide while Langford *et al.* (1989) used molybdenum oxide. The former instrument employs an oxide-coated annular quartz denuder while the latter uses the oxidized surface of a molybdenum rod suspended in a quartz tube. Both vanadium and molybdenum oxide directly oxidize the NH_3 to NO or NO_x upon desorption, thus eliminating the need for a secondary converter. In ambient sampling tests, both of these instruments compared well with conventional denuders at mixing ratios above 1–2 ppbv. In general, metal oxide systems are reasonably simple to automate with good reproducibility and detection limits below 0.1 ppbv for 0.3–0.5 h sample times. The major drawbacks are related to the expense and long-term stability of the denuders in the field. These systems may also respond to aliphatic amines, although these species are generally present at very low levels relative to NH_3 (Hutchinson *et al.*, 1982).

Several other promising sampling techniques have been developed to measure gaseous NH_3. Abbas & Tanner (1981) described a fluorescence derivatization technique that collects NH_3 in a continuously flowing acidic solution and monitors fluorescence from 1-alkylthiosoindole formed in the subsequent reaction with *o*-phthalaldehyde. The detection limit is rather high (≈ 0.3 ppbv for a 5-min response time), however, and the specificity toward gaseous NH_3 has not been established. Rapsomanikis *et al.* (1988) used this analytical technique to quantify NH_3 collected on filters or denuders. Farmer & Dawson (1982) used a cold plate to condense NH_3 and other soluble gases in water, analysing the condensate by colorimetry or ion-selective electrode. The extraction of atmospheric concentrations by this method is rather involved, however, and the estimated detection limit of 0.15 ppbv requires sampling times of several hours. Genfa *et al.* (1989) described an instrument in which NH_3 diffuses through a porous membrane into water. The ammonium is then reacted with sodium sulphite and *o*-phthaldialdehyde and the resulting fluorescence from 1-sulfonatoisoindole is monitored. The technique offers great potential for general use since it is relatively inexpensive, appears to be highly specific towards NH_3, and has a detection limit of $\leqslant 0.05$ ppbv for a 5-min sample time. At present, this technique has not been compared to other standard methods. Other instruments based on mist chamber techniques (Cofer *et al.*, 1985) are under development in several laboratories and offer similar advantages.

It is clear, however, that direct spectroscopic detection of gaseous NH_3 without the intervention of a collecting medium would be most desirable as this could provide both continuous measurements and unequivocal identification of the compound. Long-path absorption techniques using CO_2 differential absorption lidar (DIAL) (Force *et al.*, 1985) and FTIR spectroscopy (Biermann *et al.*, 1988) have been used, but have very high detection limits (5 and 1.5 ppbv, respectively). TDLAS, an *in situ* spectroscopic technique (MacKay & Schiff 1984), has excellent sensitivity and specificity but the application at low concentrations is limited by NH_3 losses on absorption cell walls. Recently, a photofragmentation (PF)/LIF (Schendel *et al.*, 1990) has been used which overcomes this problem by using very high flow rates ($\geqslant 1000 \, l \, min^{-1}$) through the excitation cell. This technique appears to offer detection limits approaching 0.001 ppbv for a 1 min integration time with no known interferences. However, in its present form the system is large, expensive and complicated. Nevertheless, the technique offers great potential as a transfer standard against which other measurements can be evaluated.

The PF/LIF technique was recently compared with citric acid-coated diffusion denuder/ion chromatograph system (CAD/IC), a filter pack/ colorimetry system (FP/COL), a tungsten oxide denuder (DARE), and a molybdenum oxide denuder system (MOADS) in Boulder, Colorado, during 1989 (Williams *et al.*, 1992b). Figure 7.7 shows a sample of the results of a set of simultaneous measurements of NH_3 made during the intercomparison with these instruments. In general, during this period agreement to better than a factor of 2 was seen among all the ammonia instruments at all times, even when large changes in the ammonia mixing ratio occurred over very short time intervals.

The conclusion from this study was that the PF/LIF and CAD/IC systems agreed within the stated accuracies for all tests with no evidence of interferences. The FP/COL system was systematically lower due to temperature related losses on the Teflon prefilter. The DARE system agreed for separate inlet ambient sampling, but not for common inlet sampling due to retention of NH_3 on the (Teflon) transfer line from the main sampling manifold. The MOADS system showed a very strong correlation with the PF/LIF data for both common and independent sampling configurations under clear conditions, but had a systematic offset due to a calibration error. This system also exhibited a background level of ≈ 0.07 ppbv attributed to retention of NH_3 on the Teflon inlets and responded to both gaseous NH_3 and free NH_{4+} dissolved in cloud or fog droplets when the latter were present.

A main conclusion of the above study is that even short inlet lines retain NH_3 and cause memory effects. In particular, Teflon appears to be unsuitable for ammonia sampling inlets. Thus, sampling through any type of inlet should be done using high flow rates (i.e. $\geq 100 \, \text{l s}^{-1}$) if low mixing ratios of NH_3 ($[NH_3] \leq 1$ ppbv) are to be measured.

7.3.3 NH₃ flux measurement techniques

The techniques currently available for the measurement of NH_3 fluxes from terrestrial ecosystems are based on enclosure methods, or gradient diffusion. More sophisticated micrometeorological methods such as eddy correlation await the development of a sufficiently sensitive and fast (≈ 0.1 s integration time) NH_3 detector. The techniques used to measure NH_3 fluxes from agricultural systems have been reviewed by Harper (1990) and by Denmead (1983).

Enclosure methods are widely used to measure trace gas fluxes from soils and plants because of their relative simplicity. Fluxes are calculated from changes in the gaseous concentration within a small enclosure placed over the ground and purged with air containing a known concentration of the compound of interest. Since the enclosure technique measures fluxes from only a small area, replicate measurements spread over a larger plot are required to address the question of spatial variability in the

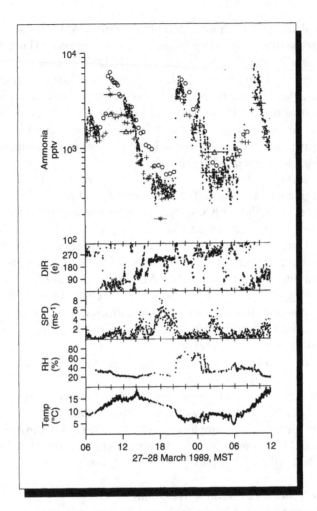

Fig. 7.7 Ammonia mixing ratios deduced measurements made during the ambient air sampling period of 0600 MST on 27 March to 1200 MST on 28 March 1989 by the five techniques involved in the intercomparison: photofragmentation/laser-induced fluorescence (PF/LIF), citric acid-coated diffusion denuder (CAD/IC), filter pack/colorimetry (FP/COL), tungsten oxide denuder (DARE) and molybdenum oxide denuder (MOADS). The horizontal bars on the FP/COL and CAD/IC data show the sampling times for each point. The figure also indicates the wind direction, wind speed per cent relative humidity, and temperatures that were recorded during the sampling period. (From Williams *et al.*, 1992b.)

field. The chief criticism of the enclosure technique arises from the fact that the microenvironment within the box is not representative of the open field. For NH_3, this leads to serious consequences since soil fluxes are known to be sensitive not only to the atmospheric concentrations, but also environmental quantities such as wind speed, temperature, evaporation rates and dew formation (Freney *et al.*, 1983). Since NH_3 is readily adsorbed on the materials used to make enclosure walls and sampling tubing (especially when condensation is present) the reliability of this technique at the low emission rates believed to be typical of unmodified field soils is highly questionable.

Gradient diffusion techniques have been widely used to measure NH_3 fluxes from agricultural systems (Denmead, 1983). In this approach NH_3 fluxes are determined from simultaneous measurements of an NH_3 concentration gradient $\Delta[NH_3]/\Delta z$ and the heat or momentum flux. This approach does not perturb the plant–soil system or the influence of environmental factors on the exchange process. Gradient diffusion also horizontally averages the large spatial variability of NH_3 fluxes over several hundred square metres. This characteristic is a disadvantage in that it necessitates a large fetch and relatively simple terrain. As a result, most measurements have been made in grassland ecosystems. In some locations, the difficulty in the interpretation of gradient measurements is compounded by gas-to-particle conversion of NH_3 in the atmosphere (Harrison *et al.*, 1989). None the less, this technique appears to be the most reliable method available.

7.4 CO

7.4.1 Fate in the atmosphere

The oxidized carbon species in the atmosphere are CO and CO_2. All the reduced organic compounds can eventually be oxidized to CO and CO_2. However, CO, unlike CO_2, is chemically active in the atmosphere. On a global basis, the oxidation of CO has a key role in the photochemical formation of ozone and, thereby, the oxidizing capacity of the atmosphere. For this reason the sources of CO and its distribution in the atmosphere are of concern to both the atmospheric chemistry and biogeochemical cycling communities. Consequently, the methods for its measurement are discussed in this section.

7.4.2 Techniques for measurements of CO

CO is an ubiquitous gas in the atmosphere with a variety of natural (oxidation of CH_4 and other natural hydrocarbons and biomass burning) and anthropogenic (combustion processes) sources. Its lifetime is long enough (on the order of a few months) that it is distributed globally and its concentration ranges from 50 to 150 ppbv in the remote troposphere. Levels above this background can indicate air masses that have had recent anthropogenic pollution input. Given the relatively unreactive nature of the gas and its high concentration levels, it is one of the more easily measured trace atmospheric species.

Four techniques currently used for measurement of background CO levels have been subject to intercomparisons. These are: (a) TDLAS; (b) collection of grab samples followed by gas chromatographic (GC) analysis (Rassmussen & Khalil, 1982); (c) differential absorption CO measurement (DACOM); and (d) gas filter correlation, non-dispersive infrared absorption spectroscopy (NDIR). The NDIR (Dickerson & Delaney, 1988) technique relies on the absorption of infrared light from a continuum light source, while the TDLAS (Fried *et al.*, 1991) measures the absorption of laser-generated infrared radiation by the CO molecules in an air sample within the White cell of the laser spectrometer. The DACOM method (Sachse *et al.*, 1987) is a refinement of diode laser spectroscopy that relies on the difference in absorption by CO in sampled air of infrared light from a TDL compared to the absorption by a known amount of CO contained in a reference cell.

Two of these methods, GC and DACOM, have been compared in two rigorous intercomparisons (cf. Hoell *et al.*, 1985, 1987a, and the references therein). The later (Hoell *et al.*, 1987a) airborne intercomparison involved the DACOM system and two GC systems; one of these systems trapped a sample at ambient atmospheric pressure, GC(amb), while the other cryogenically compressed the sample, GC(cryo). By the time this intercomparison was made, the techniques had been refined to the point that for mixing ratios of CO between 60 and 140 ppbv the level of agreement observed for the ensemble of measurements was well within the overall accuracy stated for each instrument. A sample of the data obtained during this campaign is shown in Fig. 7.8. Table 7.2 gives the parameters the regression line for each of the paired set during this portion of the intercomparison. For the entire intercomparison, the correlation observed between the measurements from respective pairs of instruments ranged from 0.85 to 0.98, with neither an interference nor a constant or proportional bias among any of these instruments. Thus, the reliability of the measurement of CO has also been rigorously established.

In addition, an informal intercomparison (Fried *et al.*, 1991) has been done held to compare the TDLAS and the NDIR techniques. The intercomparison was carried out at a suburban location at the western edge of the Denver metropolitan area. Thus winds from the east would bring pollution from the urban area to the east with attendant high levels of CO, while winds from the west would generally bring relatively clean continental air. For CO measurements in ambient air at CO mixing ratios

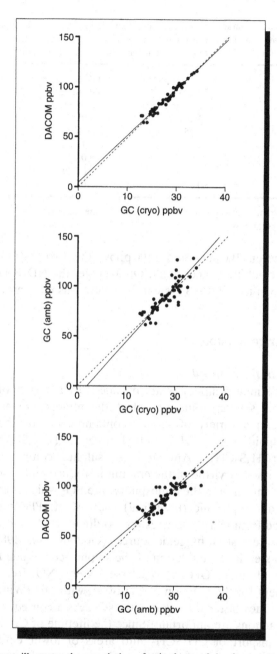

Fig. 7.8 This figure illustrates the correlation of paired sets of simultaneous measurements made by these three instruments from flights made during the latter portion of the study (cf. Hoell *et al.*, 1987a). (——) On each scattergram is the linear least squares fit to the measurements. (– – –) Shown for reference, is the 1:1 correlation line.

Table 7.2 Regression parameters for the paired correlations of simultaneous CO measurements made by three instruments during CITE 2 (fall flights). In the table σ is the standard deviation, r is the correlation coefficient, and n is the number of measurements

Parameter	GC (cryo) vs. DACOM	GC (cryo) vs. GC (amb)	GC (amb) vs. DACOM
Intercept, ppbv	3.46	−10.47	11.74
σ (intercept, ppbv)	±3.45	±12.01	±8.79
Slope	0.96	1.13	0.86
σ (slope)	±0.04	±0.14	±0.10
r	0.978	0.849	0.861
n	60	60	60

DACOM, differential absorption of carbon monoxide measurement;
GC, gas chromatography.

ranging between 100 pptv and 1500 ppbv, the two instruments agreed within their combined uncertainly. On average, the NDIR technique was 6% higher than the TDLAS system, but there was no constant systematic offset.

7.5 Sulphur compounds

7.5.1 Fate in the atmosphere

Sulphur-containing compounds are produced by a variety of natural and anthropogenic sources. Sulphur enters the atmosphere as SO_2, COS, H_2S, CS_2, and a variety of sulphur-containing hydrocarbons including dimethyl sulphide (CH_3SCH_3), methyl mecaptan (CH_3SH) and dimethyl disulphide ($CH_3S_2CH_3$). Among these sulphur compounds the most important, either by virtue of the amount introduced into the atmosphere or their concentration in the atmosphere, are SO_2 which is emitted principally from anthropogenic (industrial) sources (cf. Placet *et al.*, 1990), DMS emitted from marine sources (cf. Andreae, 1986) and COS which is emitted from terrestrial biogenic sources (Goldan *et al.*, 1988).

The first step in the oxidation of the sulphur-containing hydrocarbons is reaction with either OH or, to a lesser extent, NO_3 free radicals. The rate coefficient for the reaction of CH_3SCH_3 with OH (Wine *et al.*, 1981) is only five times faster than that with NO_3. As a consequence both OH and NO_3 reactions are important. Since the lifetime of CH_3SCH_3 is a few days the transport of CH_3SCH_3 into areas of lower emission rates is possible (Winer *et al.*, 1984). The products formed by the reactions of OH and NO_3 with DMS is uncertain (DeMore *et al.*, 1987; Lovejoy *et al.*, 1990; Murrells *et al.*, 1990; Tyndall & Ravishankara, 1991). Indirect evidence suggests that SO_2 and methane sulphonic acid (MSA) are the end products in this oxidation (Hatakeyama & Akimoto, 1983). In this connection, since MSA can be removed from the atmosphere without forming sulphuric acid, it is very important to know the relative yields of MSA and SO_2 from the oxidation of DMS.

COS presents an interesting exception to the generalizations drawn above for the oxidation of sulphur-containing hydrocarbons. Among the major reduced sulphur species present in air, COS has the highest abundance. Although its sources are reasonably small its abundance in the atmosphere is attributed to its slow reaction with the common oxidants in the troposphere. For example, its reaction with OH leads to a chemical lifetime of approximately 25 years (Cheng & Lee, 1986; Wahner & Ravishankara, 1987). The major loss mechanism for carbonyl sulphide in the troposphere appears to be its uptake by vegetation (Goldan *et al.*, 1988).

The gas phase oxidation of SO_2 is initiated by the association reaction with OH (Baulch *et al.*, 1984; DeMore *et al.*, 1987). This reaction is the rate-limiting step in the gas phase oxidation process. The subsequent conversion of $HOSO_2$ to H_2SO_4 occurs through the sequential abstraction of hydrogen to form SO_3 followed by the association of SO_3 with H_2O to form sulphuric acid as proposed by Stockwell & Calvert (1983). The conversion of SO_3 to H_2SO_4 also occurs very rapidly on any surface exposed to the atmosphere. Sulphuric acid is a strong nucleating agent and rapidly forms atmospheric aerosols. It provides one of the most important source of aerosols in the atmosphere.

7.5.2 Techniques for the measurement of sulphur-containing compounds

SO_2 can be measured (cf. Parrish *et al.*, 1991) using ultraviolet absorpion, pulsed fluorescence techniques. In this method SO_2 absorbs ultraviolet light at one wavelength while the emission from the excited SO_2 is detected at a different (longer) wavelength. The sensitivity of the technique for the detection of SO_2 depends on the competitive destruction of the excited SO_2 by collisional quenching especially by nitrogen, oxygen and water vapour in the absorption/detection cell. In addition, interference with the detection of SO_2 is produced by the fluorescence from other molecules, such as aromatic hydrocarbons, that also absorb in the ultraviolet and emit in the same wavelength region as the excited SO_2. Traps are required to remove the interfering hydrocarbons. Currently available commercial instruments using this approach have detection limits for SO_2 of approximately 0.2 ppbv with 5 min integration. In sampling SO_2 from the atmosphere care must also be taken to reduce the length of the sample line to prevent the loss of SO_2 by surface reactions (e.g. with H_2O_2) in the inlet.

Several techniques are also available to measure reduced sulphur compounds. Considerable effort was extended during the 1980s as part of the US National Acid Precipitation Assessment Program to develop techniques to measure the flux of sulphur compounds from natural sources. Instruments were designed, tested and used to measure a variety of reduced sulphur species, including H_2S, COS, SO_2, CS_2, CH_3SH, C_2H_5SH, CH_3CSH_3 (DMS) and CH_3SSCH_3 (DMDS). These techniques include: GC with flame photometric detection (GC-FID) (Goldan et al., 1987; Lamb et al., 1987), metal foil collection followed by flash desorption and FPD (MFC-FD-FPD) (MacTaggart et al., 1987) and filter collection followed by extraction and spectrographic analysis (Natusch et al., 1972; Lamb et al., 1987).

The techniques mentioned above are adequate to measure sulphur compounds at the higher concentrations observed for SO_2 in areas impacted by human-made pollution or for reduced sulphur compounds in the effluent from flux chambers. However, much lower detection limits are required to measure these compounds in the remote troposphere. Recently, an airborne intercomparison of sulphur-containing compounds has been carried out (Hoell et al., 1993). The primary objective of this study was to evaluate the capability of instrumentation for the airborne measurement of ambient concentrations of SO_2, H_2S, CS_2, DMS and COS in the remote troposphere. The study involved an intercomparison of sulphur calibration standards followed by airborne measurements of the ambient concentrations of the sulphur compounds of interest over the North Atlantic and the tropical Atlantic.

Six instruments designed to measure DMS participated in the intercomparison. The instruments utilized four different measurement approaches: (a) GC followed by mass spectrometric (MS) detection (Thornton et al., 1990); (b) GC followed by FPD; (c) GC followed by fluorination followed by electron capture detection (ECD) (Johnson & Bates, 1993); and (d) Au wool absorption followed by collection followed by FPD (Ferek & Negg, 1993). The six instruments involved three fundamentally different detection principals (FPD, MS and ECD), three collection/preconcentration methods (cryogenic, Au wool absorption and polymer absorbent), and three types of oxidant scrubbers (solid phase alkaline, aqueous reactor and cotton). The results (Gregory et al., 1993c) indicated that the instruments agreed to within a few pptv for DMS mixing ratios less than 50 pptv, and to within 15% for DMS mixing ratios above 50 pptv. The intercomparison indicated that all the instruments compared were capable of measuring DMS mixing ratios from a few pptv to 100 pptv (the upper range of DMS mixing ratios encountered during the intercomparison).

Four instruments designed to measure COS, H_2S or CS_2 participated in the intercomparison (Gregory et al., 1993b): GC-FPD for COS, H_2S

and CS_2, GC/MS for COS and CS_2 (Thornton *et al.*, 1990), GC-F-ECD for COS and CS_2, and the Natusch technique for H_2S. During the intercomparison, the COS mixing ratios generally varied between 400 and 600 pptv. The COS measurement techniques agreed on average to within about 5%, and individual measurements generally agreed to within 10%. The mixing ratios for H_2S reached a maximum of 100 pptv but were generally less than 25 pptv. For H_2S at mixing ratios greater than 25 pptv, the instruments agreed on average to about 15%. At mixing ratios less than 25 pptv, the agreement was about 5 pptv. In the case of CS_2, mixing ratios encountered during the intercomparison were typically less than 100 pptv with a maximum mixing ratio of 50 pptv. For CS_2 at mixing ratios less than 50 pptv, the GC-FPD and GC-MS agreed on average to about 4 pptv with the GC-FPD measurements consistently higher by 4 pptv. For the same mixing ratio range, the GC-F-ECD technique measured consistently higher relative to the other two techniques by 3–7 pptv. At the lowest mixing ratios encountered (a few pptv or less), the CS_2 levels were below the detection limits of the GC-F-ECD.

Five instruments designed to measure SO_2 participated in the intercomparison (Gregory *et al.*, 1993a): GC-FPD, GC-MS (Thornton *et al.*, 1986; Driedger *et al.*, 1987), chemiluminescence (Meixner & Jaeschke, 1981; Maser *et al.*, 1991), filter collection with ion chromatographic analysis and filter collection with the above-mentioned chemiluminescence analysis. For SO_2 mixing ratios in the range of a few pptv to 200 pptv, there was no general agreement among techniques and the results from the techniques were uncorrelated. The intercomparison did not unambiguously validate any of the measurement techniques involved as capable of providing valid SO_2 measurements, but identified the range of 'potential' uncertainty in SO_2 measurements reported by currently available instrumentation and as measured under realistic aircraft environments.

7.5.3 Sulphur flux measurement techniques

The technique presently available for the measurement of sulphur fluxes from terrestrial ecosystems is based on enclosure methods. More sophisticated micrometeorological methods such as eddy correlation await the development of sufficiently sensitive and fast (≈ 0.1 s integration time) sulphur detectors. The techniques used to measure sulphur fluxes from a variety of ecosystems were described by Adams *et al.* (1981), Goldan *et al.* (1987), Lamb *et al.* (1987) and MacTaggart *et al.* (1987). The technique involves the placing of an open bottom enclosure over an area of soil, mud or water, which may include or exclude vegetation as the experimenter desires. A purge or sweep gas is passed through the chamber and the increase in concentration of the sulphur compound of interest in the sampled effluent from the chamber is monitored. This increased sulphur concentration is related to the flux of sulphur compounds from the soil covered by the enclosure or the biomass enclosed by the chamber.

The chief criticism of the enclosure technique arises from the fact that the microenvironment within the box is not representative of the open field. For sulphur emissions from soils this could lead to misinterpretation of results since soil fluxes are known to be sensitive to environmental quantities such as temperature, evaporation rates and dew formation (Guenther *et al.*, 1989). Moreover, since the sulphur species of interest can be adsorbed on the materials used to make enclosure walls and sampling tubing, care must be taken in the choice of material for the enclosure walls and allowances made from the possible losses of the sulphur compounds to the walls of the chamber.

Kuster & Goldan (1987) measured the wall loss rates for H_2S, COS, SO_2, CS_2, CH_3SH, C_2H_5SH, DMS and DMDS from chambers with various wall-surface composition, including FEP Teflon, TFE Teflon, Pyrex and polycarbonate. They found that COS and CS_2 exhibited negligible losses in all these enclosures. However, other species were lost more or less rapidly depending on the surface and the relative humidity in the chamber. At relative humidities between 40 and 70%, the FEP Teflon enclosure exhibited the smallest losses for all species tested with only SO_2 and CH_3SH showing large losses on this material.

7.6 Ozone

7.6.1 Fate in the atmosphere
The most abundant atmospheric oxidant is ozone. Ozone plays a key part in controlling tropospheric photochemistry since the photolysis of ozone by sunlight initiates almost all tropospheric photochemistry (Fehsenfeld & Liu, 1993). The stratosphere contains a preponderance of the Earth's ozone, approximately 90%. The downward migration of this stratospheric ozone into the troposphere is estimated to contribute on average a flux of about $5 \times 10^{10} \, cm^{-2} \, s^{-1}$, or a total flux of 13×10^{28} molecules s^{-1} to the trophosphere in the Northern Hemisphere (NH) and about half of that in the Southern Hemisphere (SH). Ozone is also produced chemically within the troposphere by the oxidation of CO and other volatile organic compounds in the presence of the oxides of nitrogen. By contrast, at low levels of NO_x, ozone is destroyed by reaction with oxidizing radicals (cf. Fehsenfeld & Liu, 1992). In addition, ozone is efficiently removed from the atmosphere by deposition to soils and vegetation.

7.6.2 Techniques for the measurement of ozone
Over the years, a variety of techniques has been developed to measure

ozone. These include absorption of ultraviolet light, chemiluminescence and chemical titration methods, particularly electrochemical techniques. Each of these techniques has special advantages for certain types of tropospheric ozone measurements.

The absorption of ultraviolet light by the ozone molecule provides a reasonably straightforward and accurate means to measure ozone. Most instruments rely on the 254 nm emission line of mercury emitted from a mercury discharge lamp as the ultraviolet light source. The techniques which have been incorporated into several high quality commercially available instruments are reliable and interferences that occur because of the absorption of the ultraviolet light by molecules other than ozone can generally be corrected. Most high quality, routine *in situ* measurements of ozone have been made using this technique.

The chemiluminescence produced by the reaction of ozone with NO forms the basis for a very sensitive specific ozone detection method (the reactions of ozone with unsaturated non-methane hydrocarbons such as ethylene have also been used but are somewhat less sensitive). Although the technique tends to be more involved than the ultraviolet absorption method, because of the sensitivity, the chemiluminescence approach has the capability of fast response ozone measurements. For this reason the technique has been used to measure ozone fluxes which can be deduced from the correlation of ozone variation with meteorological variations.

Electrochemical sondes rely on the conversion of chemicals in a solution by ozone contained in sample air to alter the electric conductivity of the solution. A typical instrument, such as the electrochemical concentration cell (Komhyr, 1969), is composed of platinum electrodes immersed in neutral buffered potassium iodide solutions of different concentrations in anode and cathode chambers. When ozone containing air is pumped into the cathode region of the cells a current is generated which is proportional to the ozone flux through the cell. These sondes can be made very light and can therefore be lifted by small balloons. They are generally used to measure ozone profiles in the atmosphere. However, the measurements made by these instruments may suffer interferences from compounds other than ozone and tend to increase the instrument uncertainty (Barnes *et al.*, 1985).

Over the years several formal and informal intercomparisons have been made of these techniques (cf. Attmannspacher & Dütsch, 1981; Aimedieu, 1983; Robbins, 1983; Aimedieu *et al.*, 1983; Hilsenrath *et al.*, 1986). Although the most recent and comprehensive of these intercomparisons was aimed at evaluation of instruments used to measure stratospheric ozone, many of the findings were obtained in or can be applied to tropospheric ozone measurements. The consensus to be drawn from these intercomparisons indicate that the best ultraviolet absorption instruments are probably reliable for measurement of tropospheric ozone with uncertainties less than 3%. Chemiluminescence instruments should be equally good. The electrochemical sondes have susceptibilities to interferences that reduce their intrinsic accuracy somewhat. However, it must be emphasized that all these techniques when used in routine measurement will be subject to much larger uncertainties.

7.7 Calibration

The demonstration of instrument capability through rigorous intercomparison is needed to provide measurement tools of assured reliability. However, in order to use these tools successfully in the field, reliable and accurate methods of *in situ* instrument calibration are required. Experience has demonstrated that such calibration must be done with a known quantity of the molecule of interest, added by standard addition to ambient air at the inlet of the system. It is desirable that the standard addition be done at levels representative of the ambient levels of the analyte compound in the atmosphere to be sampled. The small mixing ratios of these labile gases found in the atmosphere can be achieved by dilution of the calibration standard using a dynamic dilution system (Goldan, 1990) or by simply adding a small flow of the calibration gas mixture to a very large ambient gas flow (Williams *et al.*, 1992b). A sampling strategy using large carrier gas flows (i.e. $>10^3 \, l \, min^{-1}$) has proven to be relatively effective at conditioning inlets for sampling compounds such as HNO_3 and NH_3 at low mixing ratios (i.e. ppbv or pptv levels) through relatively long sampling lines (cf. Williams *et al.*, 1992b).

7.7.1 Standard reference gas mixture (SRM)

The preferred method of supplying standards is in the form of a dilute mixture (ppmv or ppbv range) of the compound of interest in a suitable carrier gas (usually diluted by a synthetic mixture of N_2, O_2, i.e. 'zero' air). The number of mixing ratios of the calibration standards is limited to those compounds that can be stored and remain stable at trace concentration (ppmv or less) in a high pressure gas mixture. The development of specially treated, chemically passive, high pressure cylinders, careful gas handling procedures and make-up gases free of contamination are required to produce low concentration standards for most of the compounds of interest to the atmospheric chemist.

A variety of reference standards are available from national standards laboratories, for example the National Institute Standard Technology (NIST) in the USA or the National Standards Laboratory in the UK. Generally, these standards, when available, are prepared by microgravimetric techniques and supplied in specially treated cylinders at high pressure (cf. Fried & Hodgeson, 1982). The stability of the SRM standards in these cylinders is assured by the supplier as a result of monitoring over a period of years. The development and maintenance of such SRM standards is time-consuming and costly. Usually the potential customers for these standards are limited. As a consequence, the number of SRM standards available from the national standards laboratory is limited.

In addition, a variety of other standard mixtures are supplied by compressed gas manufacturers. These secondary standards are sometimes traceable to the primary SRM standards supplied by the national standards laboratories. However, there is no set procedure or protocol provided by the national standards laboratories to assure or even to gauge the quality of these manufactured secondary standards.

7.7.2 Permeation tube

Another method to provide a small but known addition of a compound of interest is through permeation tubes. This source for a known flow of an analyte of interest has been particularly useful for compounds that are difficult to store at trace concentration in high pressure cylinders. A variety of compounds in permeation tubes are available commercially. These 'low loss' permeation tubes are usually gravimetrically calibrated. The permeable membrane is made of a suitable inert material such as Teflon. The output of the tube is determined by the permeation rate of the material at a given temperature and area of the permeating surface. The rate of permeation can be altered by varying the temperature over a limited range of temperatures. The output of the tube is diluted by a suitable carrier gas, often a synthetic mixture of N_2, O_2 ('zero' air) and CO_2 at normal atmospheric mixing ratios. Water vapour may be admixed in the final stage to simulate relative humidities from near zero to approximately 90%.

The versatility of the permeation tube calibration sources is somewhat offset by some inherent stability problems. Since the permeation rate is a strong function of temperature the tubes must be held at a constant, known temperature. In addition, the calibrated rate of analyte permeation from these tubes at a given temperature is subject to irreversible change. Commonly this is a result of overheating. Moreover, the mixture inside the tube can change due to permeation of atmospheric gases into the tube and chemical reactions that may occur within the tube. In this case part of the emission from the tube and hence the weight loss may be associated with the permeation of the volatile adulterants. This problem has been documented with absorption of water vapour into permeation tubes containing HNO_3. In this case concentrated sulphuric acid acting as a non-volatile, drying agent is sometimes added along with the HNO_3 to the permeation tube.

7.7.3 Diffusion/effusion sources

There are a variety of labile compounds that are sufficiently unstable that they cannot be prepared, stored and maintained in high pressure cylinders or permeation tubes. In certain cases these compounds have been synthesized, stored at low temperature and then supplied in known amounts from diffusion or effusion sources. Documented examples in the literature are calibration sources that have been developed for PAN and N_2O_5. The accepted calibration source for PAN is a capillary diffusion source containing PAN, chemically synthesized (Gaffney *et al.*, 1984), and dissolved in tridecane solvent ($C_{13}H_{28}$). The PAN emission from this source can be held constant by submersion of the source in an ice-water bath and control of the system pressure. For N_2O_5, a flow of an inert carrier gas can be used to elute N_2O_5 vapour from an N_2O_5 storage trap (Fahey *et al.*, 1985). The carrier gas and the gas handling system must be carefully purged of condensible vapors, primarily H_2O. N_2O_5 can be synthesized in the form of crystals produced by oxidizing NO with O_3 and cryogenically trapped at 192°K (Davidson *et al.*, 1978; Viggiano, 1981) but will sublime at higher temperatures.

7.7.4 Analytical interferences

Unfortunately, obtaining a sample through the inlet with no significant losses of the species of interest is no guarantee of interference-free analysis. For example, the detection of sulphur compounds is often done by GC followed by FPD. This detection scheme depends upon the detection of photons emitted by excited sulphur molecules, S_2^*, produced in the combustion of sulphur compounds reaching the detector. Consequently, any species that can collisionally de-excite the S_2^* and that co-eludes with a sulphur species of interest can quench the detector and interfere with quantitation. Obviously, being aware of the presence of such quenching species is important in ensuring the meaningfulness of sulphur species quantitation. Since CO_2 is an efficient quencher present in all atmospheric samples, calibration gas streams for species eluting on a particular column close to and following CO_2 must be provided with ambient levels of CO_2 to ensure calibration accuracy.

Another form of interference takes place as a result of the sample enrichment. Since the enrichment step, be it cryogenic enrichment or adsorption on a solid substrate at ambient temperature, immobilizes the chemical species of interest, chemical reactions with oxidants such as ozone may occur much more rapidly than in the dilute gas phase of the original sample.

7.8 Conclusions

The preceding sections have outlined the intercomparisons that have been carried out to evaluate the measurement techniques that are currently being used to determine fluxes of chemically active trace compounds. No attempt has been made to describe the operating principles nor configurations of particular techniques in detail. The tables contained in this section summarize these techniques and their capabilities. The list of techniques is not comprehensive. Many older techniques, which are not being used currently by the atmospheric sciences community, and many promising new techniques, which have not yet been field tested, are not included.

The tables note when commercial versions of the instruments are available. However, the commercial versions of these instruments may not achieve the performance characteristics that are indicated in the tables and were achieved with specially constructed research instruments specifically constructed to match measurement location and platform. In addition, the performance of any instrument will depend heavily on the skill and experience of the operator.

The information in these tables make it clear that reliable techniques for the measurement of nitrogen oxides, even at the low concentrations, have been developed during the past decade. Intercomparisons of methods to measure NO, NO_2, PAN, and the total concentration of NO_y have been completed. On the other hand, further development of measurement techniques for HNO_3 are required before the measurements of that compound can be considered reliable.

NO emissions from soils have been measured using enclosure as well as meteorological techniques. Informal intercomparisons of gradient and enclosure methods to measure soil fluxes of NO have been made. With the wide range of sensitive, fast response, reliable NO detectors currently available, scientists are now in a position to attempt more formal tests of the technology currently available to measure these NO emissions.

There are a variety of measurement techniques for ammonia that have been intercompared. Of the instruments intercompared, PF followed by LIF and ammonia collection on citric acid-coated denuder tubes followed by ion chromatography agreed within the stated accuracies for all tests with no evidence of interferences.

Several techniques appear to be capable of CO measurement under most circumstances. The intercomparisons performed to date indicate that the dual absorption CO monitor and GC analysis of grab samples agree well under all conditions tested. Reliable techniques to measure ozone also are available. However, the uncertainty of routine measurements using these techniques is probably of the order of 10–20%.

Finally, techniques are available to measure SO_2 and reduced sulphur compounds at higher levels encountered in the atmosphere near pollution sources in environments or where the sulphur compounds are more concentrated such as the effluent from chamber studies. At reduced levels found in remote regions a recent airborne intercomparison found favourable agreement among several techniques for the measurement of reduced sulphur compounds. However, this intercomparison was unable to identify a technique capable of measuring SO_2 mixing ratios in the remote atmosphere.

In order to use the techniques described above successfully in the field, reliable and accurate methods of *in situ* instrument calibration are required. Calibration should be done with a known quantity of the molecule of interest, added by standard addition to ambient air at the inlet of the system. It is desirable that the standard addition be done at levels representative of the ambient levels of the analyte compound in the atmosphere to be sampled. The preferred method of supplying standards is in the form of a dilute mixture (ppmv or ppbv range) of the compound of interest in a suitable carrier gas. For more stable compounds, a variety of such mixtures are currently available from various manufacturers, but no uniform code or protocol exists to ensure the integrity of these standards. For more chemical labile compound gases other methods, such as permeation tubes, diffusion tubes or on-line chemical generation may be required.

Clearly, there remains much to be done and learned about instrument reliability. Since individual instrument development and tracking down will-o-the-wisp discrepancies are arduous, time-consuming and costly tasks, it should continue to be recognized that (a) multiple techniques are essential (and are not 'wasteful' duplication); and that (b) intercomparisons are vital (and are indeed as much a part of doing atmospheric science as is gathering data to test a geophysical hypothesis). The need is as simple as being able to demonstrate unequivocally that what is measured is indeed correct.

Acknowledgements
The author wishes to acknowledge the support received by NOAA's Global Change Program and the Southern Oxidant Study. The author would also like to acknowledge those colleagues who have contributed to the development of the instruments described in this chapter, including Martin Buhr, David Fahey, Gerhard Hübler, Paul Goldan, William Kuster, Richard Norton, David Parrish, James Roberts and Eric Williams.

Table 7.3 Measurement techniques for reactive nitrogen oxides

Species/technique	Sample method	Limit (time)	Comments	References
NO				
NO/O_3 chemiluminescence	Continuous	10 pptv (10 s)	Lower detection limit can be obtained by integration, or shorter response time with increased detection limit. Has been operated successfully at 5 Hz. Commercial version available	Carroll et al., 1985; Fehsenfeld et al., 1987; Hoell et al., 1987a; Ridley et al., 1987
Two-photon laser/induced fluorescence	Continuous	12 pptv (60 s)	Lower detection limit can be obtained by integration, or shorter response time with increased detection limit	Bradshaw et al., 1985; Davis et al., 1987; Hoell et al., 1987a
NO_2				
TDLAS	Continuous	125 pptv (30 min)	Lower detection limit can be obtained by integration, or shorter response time with increased detection limit	Schiff et al., 1987, 1990; Fehsenfeld et al., 1990
NO/O_3 chemiluminescence and photolysis	Continuous	20 pptv (10 s)	Requires subtraction of ambient NO. Lower detection limit can be obtained by integration. Commercial version available	Kley & McFarland, 1980; Fehsenfeld et al., 1990
Photofragmentation/two-photon laser-induced fluorescence	Continuous	10 pptv (6 min)	Lower detection limit can be obtained by integration, or shorter response time with increased detection limit	Bradshaw et al., 1985; Sandholm et al., 1990
Luminol	Continuous	5 pptv (3 s)	Lower detection limit can be obtained by integration, or shorter response time with increased detection limit. PAN and O_3 interferences. Commercial version available	Schiff et al., 1986; Fehsenfeld et al., 1990

continued on p.240

Table 7.3 *Continued*

Species/technique	Sample method	Limit (time)	Comments	References
Long-path absorption	Direct	50 pptv (3 min)	Sensitivity and stability dependent on atmospheric conditions. Lower detection limit can be obtained by integration. No field intercomparison	Platt *et al.*, 1984; Harder & Mount, 1991
NO/O_3 chemiluminescence and surface (e.g. MoO_3 and $FeSO_4$) conversion	Continuous	45 pptv as NO (1 s)	Requires subtraction of ambient NO. Interferences from PAN and other nitrates. Commercial version available	Dickerson *et al.*, 1984; Fehsenfeld *et al.*, 1987
PAN				
GC with luminol detection	Batch	5 pptv (60 STP cm^3)	No known interferences in GC method. No compressed gases required. Non-linear calibration necessary at low end. No field intercomparison. Sensitivity depends on sample size. Commercial version available	Blanchard *et al.*, 1990; Drummond *et al.*, 1991
GC-ECD	Batch	4 pptv (50 STP cm^3)	No known interferences in GC method. Sensitivity depends on sample size	Singh & Salas, 1983; Penkett & Brice, 1986; Ridley *et al.*, 1990; Gregory *et al.*, 1990c

HNO₃

TDLAS	Continuous	75 pptv (3 min)	Lower detection limit can be obtained by integration	Schiff et al., 1990; Gregory et al., 1990b
Tungstic acid	Batch	20 pptv (10 min)	Conversion of HNO_3 to NO after collection. Potential interference from organic nitrates	Braman et al., 1982; Gregory et al., 1990b; LaBel et al., 1990
HNO_3 by nylon or	Batch	8 pptv (90 min)	Direct trade between speed and sensitivity. Ion chromatographic analysis. Nylon filters are available commercially	Huebert & Lazrus, 1980; Goldan et al., 1983; Gregory et al., 1990b
NO_y				
Molybdenum converter followed by NO/O₃ chemiluminescence	Continuous	45 pptv (1 s)	Conversion efficiency reduced when operated in polluted enviroments	Dickerson et al., 1984; Fehsenfeld et al., 1987
Au/CO converter followed by NO/O₃ chemiluminescence	Continuous	10 pptv (1 s)	Lower detection limit can be obtained by integration, or shorter response time with increased detection limit. Has been operated successfully at 2 Hz	Fahey et al., 1985; Fehsenfeld et al., 1987
Au/CO converter followed by two-photon laser-induced fluoescence	Continuous	1 pptv (90 s)	Lower detection limit can be obtained by integration, or shorter response time with increased detection limit	Bradshaw et al., 1985; Sandholm et al., 1993

ECD, electron capture detector; GC, gas chromatography; TDLAS, tunable diode laser absorption spectroscopy.

Table 7.4 Measurement techniques for ammonia

Species/technique	Sample method	Time/limit	Comments	References
Tungstic acid denuder	Batch	40 pptv; 10 l (STP)	Ammonia can be lost on inlet lines	Braman et al., 1982; LeBel et al., 1985; Williams et al., 1992
MO$_3$ denuder	Batch	100 pptv; 15 l (STP)	Ammonia can be lost on inlet lines. Also, collects ammonia in aerosols and fog droplets	Langford et al., 1988; Williams, et al., 1992
Wet trapping	Batch	1–2 h/0.5 ppbv	—	Denmead et al., 1976, 1977
Oxcilic acid-coated filter	Batch	0.2 µg N m^{-3}	—	Ferm et al., 1988; Ferm 1979
Photofragmentation/two-photon laser-induced fluorescence	Continuous	5 pptv (5 min)	Lower detection limit can be obtained by integration, or shorter response time with increased detection limit. No interferences found in intercomparison	Schendel et al., 1990; Williams et al., 1992
Citric acid-coated denuder/ion chromatography.	Batch	25 pptv; 240 l (STP)	No interferences found in intercomparison	Ferm et al., 1988; Williams et al., 1992
Oxcilic acid coated filter/ colorimetric analysis	Batch	36 pptv; 6000 l (STP)	Ammonia may be lost on aerosol prefilter	Quinn & Bates 1989; Williams et al., 1992

Table 7.5 Measurement technique for carbon monoxide

Species	Sampling method	Time/limit	Comments	References
Closed-path laser differential absorption	Continuous	1.4 ppbv, 1 s	Lower detection limit can be obtained by integration, or shorter response time with increased detection limit. No interferences found in intercomparison	Sachse et al., 1987, 1988; Hoell et al., 1985, 1987b
GC with flame ionization detection	Batch	20 ppbv (1000 STP cm^3)	—	Rassmussen & Khalil, 1982; Hoell et al., 1985, 1987a; Zimmerman et al., 1988; Greenberg et al., 1992; Heidt et al., 1980
TDLAS	Continuous		Oxidants such as O$_3$ can generate small CO interference in White cell through reactions with hydrocarbons absorbed on wall. Lower detection limit can be obtained by integration, or shorter response time with increased detection limit. Commercially available	Fried et al., 1991
Non-dispersive infrared absorption	Continuous	50 ppbv, 5 min	Commercial version available	Fried et al., 1991; Dickerson & Delaney, 1988
Mercuric oxide detector after GC separation	Batch	(2 STP cm^3) 1 ppbv, 1 s	CO must be separated from other gases that will reduce HgO (e.g. H$_2$ and NMHC). Commercially available	Seiler et al., 1976, 1984; Novelli et al., 1991
Resonance-fluorescence	Continuous	1 ppbv, 1 s	Small interference (~0.1 ppbv equivalent CO) due to photolysis of CO$_2$	Voltz & Kley, 1985

GC, gas chromatography; TDLAS, tunable diode laser absorption spectroscopy.

Table 7.6 Measurement techniques for sulphur-containing compounds

Species	Sample method	Time/limit	Comments	References
Total sulphur				
Flame photometric detection	Continuous sampled or batch	~ 0.5 s/0.5 ppb @ 0.02 l/min^{-1} (~ 2 pg S s^{-1}; ~ 10 min 3 pg^{-5} S/ sample 2 pptv 1 l^{-1} sample	SF$_6$ doped FPD. Careful flow control required on all flows. Sensitivity is species dependent. Hydrocarbon interferences	D'Ottavio et al., 1981
Conversion to SF$_6$/electron capture detection	Continuous sampled or batch	~ 0.5 s/~ 50 pptv @20 cm^3 min^{-1}; ~ 10 min ~ 0.2 pg^{-1} S s^{-1} ~ 0.03 pg S sample^{-1}	Estimates from GC application below. Conversion to SF$_6$ followed by F$_2$ removal by catalysis	Johnson & Bates, 1993; Johnson & Lovelock, 1988
Metal foil collection/flash vaporization/flame photometric detection	Batch	~ 10 min ~ 20 pg^{-1} S sample^{-1}	Collects predominantly reduced sulphur. Non-speciated flash vapourization from arbitrarily large sample. Some H$_2$O interference	Kagel & Farwell 1986
Vacuum ultraviolet flash photolysis/laser-induced fluorescence	Direct, no sampling	0.1–2000 s 150 pptv^{-1}	Underdevelopment	Rodgers et al., 1980
CH$_3$SH				
Gas chromatography/flame photometric detection	—	—		
Methane sulphonic acid				
Ion chromatography	—	—		

Dimethyl sulphoxide and dimethyl sulphonic acid

Method	Type	Detection	Comments	References
GC/hall conductivity detector	Batch	2–30 min⁻¹	Requires preconcentration	Harvey & Lang 1986; Lang et al., 1990; Pszenny et al., 1990
Filter collection FMA fluorescence quenching	Batch	10–60 min ~3 ng⁻¹ S sample⁻¹	Large filter-to-filter background variability	Natusch et al., 1972
Gold collection/flame photometric detection	Batch	10–100 min⁻¹ 10 pg S sample⁻¹	Large chance for variability of H₂S collection efficiency. Requires SO₂ removal. Interference from other reduced sulphur compounds	Braman et al., 1978
OCS				
GC/mass spectrometer	Batch	3 min, 0.2 pptv	Isotopically labelled/requires GC separation	Bandy et al., 1992, 1993
GC flame photometric detection	Batch	—	No O₃ interference. Requires GC separation	Goldan et al., 1987
Conversion to SF₆ electron capture detection	Batch	—	Requires GC separation	Johnson & Bates, 1993; Johnson & Lovelock, 1985
CS₂				
GC/mass spectrometer	Batch	3 min, 0.2 pptv	Isotopically labelled/requires GC separation	Bandy et al., 1992, 1993
GC flame photometric detection	Batch	—	Requires GC separation. Small O₃ interference	Johnson & Bates, 1993; Jonhson & Lovelock, 1985
Conversion to SF₆ electron capture detection	Batch	—	Requires GC separation	

continued on p.246

Table 7.6 Continued

Species	Sample method	Time/limit	Comments	References
Dimethyl sulphide				
GC/mass spectrometer	Batch	3 min, 1 pptv	Isotopically labelled/requires GC separation	Bandy et al., 1992, 1993
GC flame photometric detection	Batch	~10 min 3 pg^{-1} S sample^{-1} ~pptv l^{-1} sample	Requires GC separation. SF$_6$ doped FPD. Very important O$_3$ interference	Goldan et al., 1987; Andreae et al., 1985
Chemiluminescent detection	Continuous samples or batch	~1 s/~200 pptv; 10 min ~pptv^{-1} l^{-1} sample	Interferences from other reduced sulphur compounds or requires GC separation. New technique. Little experience	Ryerson & Saevers, 1994; Ryerson et al., 1994
Gold wool collection/ thermal desorption/flame photometric detection	Batch	~10 min 20 pg^{-1} S sample^{-1}	Requires GC separation. Large O$_3$ interference	Bernard et al., 1982
Conversion to SF$_6$ electron capture detection	Batch	~10 min ~0.03 pg^{-1} S sample^{-1}	Requires GC separation. Conversion to SF$_6$ with F$_2$ catalytic removal of excess F$_2$. New technique	Johnson & Lovelock 1988
H$_2$S				
GC flame photometric detection	Batch		No significant O$_3$ interference	Goldan et al., 1987
Conversion to SF$_6$ electron capture detection	Batch		Requires GC separation	Johnson & Bates, 1993; Johnson & Lovelock, 1988

SO$_2$

Method	Sampling	Performance	Comments	Reference
GC/mass spectrometer	Batch	3 min, 0.2 pptv	Isotopically labelled/requires GC separation	Bandy et al., 1992, 1993
Chemiluminescence	Batch	>10 min ~4 ng^{-1} S batch^{-1} ~3 ppb sample^{-1}	Impregnated filters, wet chemical handling, capable of large samples	Meixner & Jaeschke, 1981
Laser-induced fluorescence	Direct, no sampling	3–5 s 200 pptv^{-1}	Near maximum achievable sensitivity	Bradshas et al., 1985
Vacuum ultraviolet flash photolysis/laser-induced fluorescence	Direct, no sampling	0.1–2000 s 150 pptv^{-1}	Underdevelopment	Rodgers et al., 1980
Differential ultraviolet absorption	Continuous	>1 s ~20 ppbv^{-1}	Has been used for eddy correlation measurements in very polluted regions (H. Nestler, Heidelberg University). Sensitivity of 0.01 ppbv for sampling sensitivity times of ~20 min	Platt & Perner 1980
Tunable diode laser	Direct, no sampling or continuous	0.1–100 s ppbv^{-1} (closed cell)	Underdevelopment; may be improved by factor of 20 with astimatic off-axis resonator	Reid et al., 1978

Table 7.7 Measurement technique for ozone

Species	Sample method	Time/limit	Comments	References
Gas/liquid chemiluminescence (Eosin and relatives)	Batch	0.1 s/0.3 ppbv	Still under development, U. of Denver. A 'conventional' luminol detector with a different reagent solution insensitive to NO_2 but sensitive to ozone. Lightweight, low power. No known interferences – also may be under development at Scintrex	Ray et al., 1986
Ultraviolet absorption	Continuous	1–10 s/1–3 ppbv	Commercial versions available	—
NO chemiluminescence	Continuous	0.05 s/0.5 ppbv	Slight interference from H_2O vapour, correctable from simultaneous measurements of moisture time series	Pearson & Stedman 1980; Lenschow et al., 1981
Ethylene chemiluminescence	Continuous	0.5 s/0.5 ppbv	Commercial version available	—

References

Abbas R. & Tanner R.L. (1981) Continuous determination of gaseous ammonia in the ambient atmosphere using fluorescence derivization. *Atmospheric Environment*, **15**, 277–281.

Adams D.F., Farwell S.O., Robinson E., Pack M.R. & Bamesberger W.L. (1981) Biogenic sulfur source strengths. *Environmental Science and Technology*, **15**, 1493–1498.

Aimedieu P. (1983) Ozone profile intercomparison based on simultaneous observations between 20 and 40 km. *Planetary and Space Science*, **31**, 801–807.

Andreae M.O. (1986) The ocean as a source of atmospheric sulfur compounds. In: Buat-Menard P. (ed) *The Role of Air–Sea Exchange in Geochemical Cycling*, pp. 331–362. D. Reidel, Hingham, Massachusetts.

Appel B.R., Tokiwa V., Kothny E.L. *et al.* (1984) *Determination of Acidity in Ambient Air*, Report, Research Division, California Air Resource Board, Sacramento.

Appel B.R., Tokiwa V., Kothny E.L. *et al.* (1988) Evaluation of procedures for measuring atmospheric nitric acid and ammonia. *Atmospheric Environment*, **22**, 1565–1573.

Asman W.A.H. & Janssen A.J. (1987) A long-range transport model for ammonia and ammonium in Europe. *Atmospheric Environment*, **21**, 2099–2119.

Atkinson R., Aschmann S.M. & Pitts J.N. Jr. (1988) Rate constants for the gas-phase reactions of the NO_3 radical with a series of organic compounds at $296 \pm 2 K$. *Journal of Physical Chemistry*, **92**, 3454–3457.

Atkinson R., Carter W.P.L., Plum C.N., Winer A.M. & Pitts J.N. Jr (1984) Kinetics of the gas-phase reactions of NO_3 radicals with a series of aromatics at 296°K. *International Journal of Chemical Kinetics*, **16**, 887–898.

Attmannspacher W. & Dütsch H.U. (1981) Second International ozone sonde intercomparison at the Observatory Hohenpeissenberg, 5–20 April 1978. *Berichte – Deutschen Wetterdienstes*, **157**, 1.

Bandy A.R., Thornton D.C. & Driedger A.R. III. (1993) Airbrone Measurements of Sulfur Dioxide, Dimethyl Sulfide, Carbon Disulfide, and Carbonyl Sulfide by Isotope Dilution Gas Chromatography/Mass Spectrometry. *Journal Geophysical Research*, **98(D12)**, 23423–23433.

Barnes R.A., Bandy A.R. & Torres A.L. (1985) ECC ozonesonde accuracy and precision. *Journal of Geophysical Research*, **90**, 7881.

Baulch D.L., Cox R.A., Hampson R.F. Jr, Kerr J.A., Troe J. & Watson R.T. (1984) Evaluated kinetic and photochemical data for atmospheric chemistry: Supplement II CODATA Task Group on gas phase chemical kinetics. *Journal of Physical Chemistry Reference Data*, **13**, 1259–1380.

Bernard W.R., Andreae M.O. & Watkins W.E. (1982) The flux of dimethyl sulfide from oceans to the atmosphere. *Journal Geophysical Research*, **87**, 8787–8793.

Biermann H.W., Tuazon E.C. & Wilner A.M. (1988) Simultaneous absolute measurements of gaseous nitrogen species in urban ambient air by long pathlength infrared and ultraviolet-visible specroscopy. *Atmospheric Environment*, **22**, 1545–1554.

Blanchard P., Shepson P.B., So K.W. *et al.* (1990) A comparison of calibration and measurement techniques for gas chromatographic determination of atmospheric peroxyacetyl nitrate (PAN). *Atmospheric Environment*, **24A**, 2839–2846.

Bos R. (1980) Automatic measurement of atmospheric ammonia. *Journal of Air Pollution Control Association*, **30**, 1222–1224.

Bradshaw J.D., Rodgers M.O. & Davis D.D. (1982) Single photon/laser-induced fluorescence detection of NO and SO_2 for atmospheric conditions of composition and pressure. *Applied Optics*, **21**, 2493–2500.

Bradshaw J.D., Rodgers M.O., Sandholm S.T., KeSheng S. & Davis D.D. (1985) A two-photon laser-induced fluorescence field instrument for ground-based and airborne measurements of atmospheric NO. *Journal of Geophysical Research*, **90**, 12861–12874.

Braman R.S., Ammons J.M. & Bricker J.L. (1978) Preconcentration and determination of hydrogen sulfide in air by flame photometric detection. *Analytical Chemistry*, **50**, 992–996.

Braman T.S., Shelly T.J. & McClenny W.A. (1982) Tungstic acid for preconcentration and determination of gaseous and particulate ammonia and nitric acid in ambient air. *Analytical Chemistry*, **52**, 358–364.

Breeding R.J. *et al.* (1973) Background trace gas concentrations in the central United States. *Journal of Geophysical Research*, **78**, 7057–7064.

Carroll M.A., McFarland M., Ridley B.A. & Albritton D.L. (1985) Ground-based nitric oxide measurements at Wallops Island, Virginia. *Journal of Geophysical Research*, **90**, 12853–12860.

Calvert J.G., Lazrus A., Kok G.L. *et al.* (1985) Chemical mechanisms of acid generation in the troposphere. *Nature*, **317**, 27–38.

Cheng B.M. & Lee Y.P. (1986) Rate constant of OH + OCS reaction over the temperature range 255–483 K. *International Journal of Chemical Kinetics*, **18**, 1303–1304.

Cofer W.R., Collins V.G. & Talbot R.W. (1985) Improved aqueous scrubber for collection of soluble atmospheric trace gases. *Environmental Science and Technology*, **19**, 557–560.

Davidson J.A., Viggiano A.A., Howard C.J. *et al.* (1978) Rate constants for the reactions of O_2-, NO_2+, H_3O+, CO_3-, NO_2-, and halide ions with N_2O_5 at 300 K. *Journal of Chemical Physics*, **68**, 2085–2087.

Davis D.D., Bradshaw J.D., Rodgers M.O., Sandholm S.T. & KeSheng S. (1987) Free tropospheric and boundary layer measurements of NO over the central and eastern North Pacific Ocean. *Journal of Geophysical Research*, **92**, 2049–2070.

Delany A.C., Fitzjarrald D.R., Lenschow D.H., Pearson R. Jr, Wendel G.J. & Woodruff B. (1986) Direct measurements of nitrogen oxides and ozone fluxes over grassland. *Journal of Atmospheric Chemistry*, **4**, 429–444.

DeMore W.B., Molina M.J., Sander S.P. *et al.* (1990) *Chemical kinetics and photochemical data for use in stratospheric modeling*. Report, NASA Jet Propulsion Laboratory, Pasadena, California.

DeMore W.B., Molina M.J., Sander S.P. *et al.* (1987) *Chemical kinetics and photochemical data for use in stratospheric modeling, evaluation 8. NASA panel for data evaluation*. Jet Propulsion Laboratory, Publication 87-41, Pasadena, California.

Denmead O.T. (1983) Micrometeorological methods for measuring gaseous losses of nitrogen in the field. In: Freney J.R. & Simpson J.R. (eds) *Gaseous Losses of Nitrogen from Plant-Soil Systems*, pp. 133–157. Martinus Nijhoff/ W. Junk, The Hague.

Dickerson R.R., Delany A.C. & Wartburg A.F. (1984) Further modification of a commercial NO_x detector for high sensitivity. *Review of Scientific Instruments*, **55**, 1995–1998.

Dickerson R.R. & Delany A.C. (1988) Modification of a commercial gas filter correlation CO detector for enhanced sensitivity. *Journal of Atmospheric and Oceanic Technology*, **5**, 424–431.

Dlugokencky E.J. & Howard C.J. (1988) Laboratory studies of NO_3 radical reactions with some atmospheric sulfur compounds. *Journal of Physical Chemistry*, **92**, 1188–1193.

Dlugokencky E.J. & Howard C.J. (1989) Studies of NO_3 radical reactions with some atmospheric organic compounds at low pressures. *Journal of Physical Chemistry*, **93**, 1091–1096.

D'Ottavio T., Garber T., Tanner R.L. & Newman L. (1981) Determination of ambient aerosol sulfur using a continuous flame photometric detection system – II. The measurement of low-level sulfur concentrations under varying atmospheric conditions. *Atmospheric Environment*, **15**, 197–203.

Driedger A.R. III, Thornton D.C., Laleviv M. & Bandy A.R. (1987) Determination of part-per-trillion levels of atmospheric sulfur dioxide by isotope dilution gas chromatography/mass spectrometry. *Analytical Chemistry*, **59**, 1196–1200.

Drummond J.W., Mackay G.I. & Schiff H.I. (1991) Measurement of peroxyacetyl nitrate,

NO$_2$, and NO$_x$ by using a gas chromatograph with a luminol based detector. In: Schiff H.J. (ed) *Measurement of Atmospheric Gases*. Proceedings of the International Society for Optical Engineers, **1433**, 242–252.

Fahey D.W., Eubank C.S., Hübler G. & Fehsenfeld F.C. (1985) A calibrated source of N$_2$O$_5$. *Atmospheric Environment*, **19**, 1883–1890.

Farmer J.C. & Dawson G.A. (1982) Condensation sampling of atmospheric trace gases. *Journal of Geophysical Research*, **87**, 8931–8942.

Fehsenfeld F.C. & Liu S.C. (1993) Tropospheric ozone: Distribution and sources. In: Hewitt C.N. & Sturges W.T. (eds) *Global Atmospheric Chemical Change*, pp. 169–231. Elsevier Applied Science, New York.

Fehsenfeld F.C., Dickerson R.R., Hübler G. *et al.* (1987) A ground-based intercomparison of NO, NO$_x$, and NO$_y$ measurement techniques. *Journal of Geophysical Research*, **92**, 14710–14722.

Fehsenfeld F.C., Drummond J.W., Roychowdhury U.K. *et al.* (1990) Intercomparison of NO$_2$ measurement techniques. *Journal of Geophysical Research*, **95**, 3579–3597.

Ferek R.J. & Negg D.A. (1993) Measurements of DMS by gold absorption and SO$_2$ by carbonate-impregnated filters during GTE/CITE-3. *Journal of Geophysical Research*, in press.

Ferm M. (1979) Method for determination of atmospheric ammonia. *Atmospheric Environment*, **13**, 1385–1393.

Fontijn A., Sabadell A.J. & Ronco R. (1970) Homogeneous chemiluminescence measurement of nitric oxide with ozone. *Analytical Chemistry*, **42**, 575–579.

Force A.P., Killinger D.K., Defed W.E. *et al.* (1985) Laser remote sensing of atmospheric ammonia using a CO$_2$ lidar system. *Applied Optics*, **24**, 2837–2841.

Freney J.R. & Simpson J.R. (1983) *Volatilization of Ammonia. Gaseous Losses of Nitrogen from Plant–Soil Systems* Freny J.R. & Simpson R.J., pp. 1–32. Martinus Nijhoff/W. Junk, The Hague.

Fried A., Drummond J.R., Henry B. & Fox J. (1991) A Versatile Integrated Tunable Diode Laser System for High Precision Application for Ambient Measurement of OCS. *Applied Optics*, **30(15)**, 1916–1933.

Fried A. & Hodgeson J. (1982) Laser photoacoustic detection of nitrogen dioxide in the gas-phase titration of nitric oxide with ozone. *Analytical Chemistry*, **54**, 278–282.

Fried A., Henry B., Parrish D.D., Carpenter J.R. & Buhr M.P. (1991) Intercomparison of tunable diode laser and gas filter correlation measurements of ambient carbon monoxide. *Atmospheric Environment*, **25A(10)**, 2277–2284.

Gaffney J.S., Fajer R. & Senum G.I. (1984) An improved procedure for high purity gaseous peroxyacetyl nitrate production: Use of heavy lipid solvents. *Atmospheric Environment*, **18**, 215–218.

Genfa Z., Dasgupta P.K. & Dong S. (1989) Measurement of atmospheric ammonia. *Environmental Science Technology*, **23**, 1467–1474.

Georgii H.-W. & Gravenhorst G. (1977) The ocean as a source or sink of reactive trace-gases, *Pure Applied Geophysics*, **115**, 503–511.

Georgii H.-W. & Müller W.J. (1974) On the distribution of ammonia in the middle and lower troposphere. *Tellus*, **26**, 180–184.

Goldan P.D. (1990) Analysis of low-concentration-level gaseous sulfur compounds in the atmosphere. *American Society for Testing and Materials, Standard Technical Publication*, **1052**, 114–132.

Goldan P.D., Kuster W.C., Albritton D.L. *et al.* (1983) Calibration and Tests of the Filter-Collection Method for Measuring Clean-Air, Ambient Levels of Nitric Acid. *Atmospheric Environment*, **17(7)**, 1355–1364.

Goldan P.D., Fall R., Kuster W.C. & Fehsenfeld F.C. (1988) Uptake of COS by growing vegetation: A major tropospheric sink. *Journal of Geophysical Research*, **93**, 14186–14192.

Goldan P.D., Kuster W.C., Albritton D.L. & Fehsenfeld F.C. (1987) The measurement of natural sulfur emissions from soils and vegetation: Three sites in the eastern United States revisited. *Journal of Atmospheric Chemistry*, **5**, 439–467.

Gras J.L. (1984) A field comparison of two atmospheric ammonia sampling techniques. *Tellus, Series B*, **36**, 38–43.

Greenberg J.P., Zimmerman P.R., Pollock W.F., Lueb R.A. & Heidt L.E. (1992) Diurnal variability of atmospheric methane, nonmethane hydrocarbons, and carbon monoxide at Mauna Loa. *Journal of Geophysical Research*, **97**, 10395–10413.

Gregory G.L., Hoell J.M. Jr, Carroll M.A. *et al.* (1990a) An intercomparison of airborne nitrogen dioxide instruments. *Journal of Geophysical Research*, **95**, 10103–10128.

Gregory G.L, Hoell J.M. Jr, Huebert B.J. *et al.* (1990b) An intercomparison of airborne nitric acid measurements. *Journal of Geophysical Research*, **95**, 10089–10102.

Gregory G.L., Hoell J.M. Jr, Ridley B.A. *et al.* (1990c) An intercomparison of airborne PAN measurements. *Journal of Geophysical Research*, **95**, 10077–10087.

Gregory G.L., Hoell J.M. Jr, Torres A.L. *et al.* (1990d) An intercomparison of airborne nitric oxide measurements: A second opportunity. *Journal of Geophysical Research*, **95**, 10129–10138.

Gregory G.L., Davis D.D., Beltz N., Bandy A.R., Ferek R.J. & Thornton D.C. (1993a) An intercomparison of aircraft instrumentation for tropospheric measurements of sulfur dioxide. *Journal of Geophysical Research*, **98(D12)**, 23325–23352.

Gregory G.L., Davis D.D., Thornton D.C. *et al.* (1993b) An intercomparison of aircraft instrumentation for tropospheric measurements of carbonyl sulfide (COS), hydrogen sulfide (H_2S), and carbon disulfide (CS_2). *Journal of Geophysical Research*, **98(D12)**, 23353–23372.

Gregory G.L., Warren L.S., Davis D.D. *et al.* (1993c) An intercomparison of instrumentation for tropospheric measurements of dimethyl sulfide: Aircraft results for concentrations at the parts-per-trillion level. *Journal of Geophysical Research*, **98(D12)**, 23373–23388.

Grosjean D. (1984) Photooxidation of methylsulfide, ethylsulfide and methanethiol. *Environmental Science and Technology*, **18**, 460–468.

Guenther A., Lamb B. & Westberg H. (1989) US National Biogenic Sulfur Emissions Inventory. In: Salzman E. & Cooper W. (eds) *Biogenic Sulfur in the Environment*. American Chemical Society, Washington, DC.

Harder J. & Mount G. (1991) Long path differential absorption measurements of tropospheric molecules. *Proceedings, SPIE Conference, Orlando, Florida, April 1991. Remote Sensing of Atmospheric Chemistry*, **1491**, 33–42.

Harper L.A. (1990) *Comparisons of Methods to Measure Ammonia Volatilization in the Field. Ammonia Volatilization from Urea Fertilizers*. Muscle Shoals, Alabama, National Fertilizer Development Center, Tennessee Valley Authority, 93–109.

Harrison R.M. & Kiitto A.-M.N. (1990) Field intercomparison of filter pack and denuder sampling methods for reactive gaseous and particulate pollutants. *Atmospheric Environment*, **24A**, 2633–2640.

Harrison R.M., Rapomanikis S. & Turnbull A. (1989) Land-surface exchange in a chemically-reactive system; surface fluxes of HNO_3, HCl, and NH_3. *Atmospheric Environment*, **23**, 1795–1800.

Harvey G.R. & Lang R.F. (1986) Dimethylsulfoxide and dimethylsulfone in the marine atmosphere. *Geophysical Research Letters*, **12**, 49–51.

Harward C.N., McClenny W.A., Hoell J.M. *et al.* (1982) Ambient ammonia measurements in coastal southeastern Virginia. *Atmospheric Environment*, **16**, 2497–2500.

Hatakeyama S. & Akimoto H. (1983) Reactions of OH radicals with methanethiol, dimethyl sulfide, and dimethyl disulfide in air. *Journal of Physical Chemistry*, **87**, 2387–2395.

Heidt L.E. (1978) Whole air collection and analysis. *Atmospheric Technology*, **9**, 3–7.

Hering S.V., Lawson D.R., Allegrini I. *et al.* (1988) The nitric acid shootout: Field comparison of measurement methods. *Atmospheric Environment*, **22**, 1519–1539.

Hicks B.B., Wesely M.L., Coulter R.L. *et al.* (1986) An experimental study of sulfur and NO$_x$ fluxes over grassland. *Boundary Layer Meteorology*, **34**, 103–121.

Hilsenrath E., Attmannspacher W., Bass A. *et al.* (1986) Results from the balloon ozone intercomparison campaign (BOIC). *Journal of Geophysical Research*, **90**, 13137–13152.

Hoell J.M. Jr, Gregory G.L., McDougal D.S. *et al.* (1985) An intercomparison of nitric oxide measurement techniques. *Journal of Geophysical Research*, **90**, 12843–12852.

Hoell J.M. Jr, Gregory G.L., McDougal D.S. *et al.* (1987a) Airborne intercomparison of nitric oxide measurement techniques. *Journal of Geophysical Research*, **92**, 1995–2008.

Hoell J.M. Jr, Gregory G.L., McDougal D.S. *et al.* (1987b) Airborne intercomparison of carbon monoxide measurement techniques. *Journal of Geophysical Research*, **92**, 2009–2019.

Hoell J.M., Harward C.N. & Williams B.S. (1980) Remote infrared heterodyne radiometer measurements of atmospheric ammonia profiles. *Geophysical Research Letters*, **7**, 313–316.

Hoell J.M. Jr, Davis D.D., Gregory G.L. *et al.* (1993) Operational overview of the NASA GTE/CITE-3 airborne instrument intercomparisons for sulfur dioxide, hydrogen sulfide, carbonyl sulfide, dimethyl sulfide, and carbon disulfide. *Journal of Geophysical Research*, in press.

Huebert B.J. & Lazrus A.L. (1980) Tropospheric gas-phase and particulate nitrate measurements. *Journal Geophysical Research*, **85**, 7322–7328.

Hutchinson G.L., Mosier A.R. & Andre C.E. (1982) Ammonia and amine emissions from a large cattle feedlot. *Journal of Environmental Quality*, **11**, 288–293.

Johansson C., Rodhe H. & Sanhueza E. (1988) Emission of NO in a tropical savanna and a cloud forest during the dry season. *Journal of Geophysical Research*, **93**, 7180–7192.

Johnson J.E. & Bates T.S. (1993) Atmospheric measurements of dimethyl sulfide and carbon disulfide using the electron capture sulfur detector. *Journal of Geophysical Research*, **98(D12)**, 23411–23421.

Johnson J.E. & Lovelock J.E. (1988) The electron capture sulfur detector: Reduced sulfur species detection at the femtomole level. *Analytical Chemistry*, **60**, 812–816.

Junge C.E. (1957) Chemical analysis of aerosol particles and of gas traces on the island of Hawaii. *Tellus*, **9**, 529–537.

Kagel R.A. & Farwell S.O. (1986) Evaluation of metallic foils for the preconcentration of sulfur-containing gases with subsequent flash desorption/flame photometric detection. *Analytical Chemistry*, **58**, 1197–1202.

Kaplan W.A., Wofsy S.C., Keller M. & DaCosta J.M. (1988) Emission of NO and deposition of O$_3$ in a tropical forest system. **93**, 1389–1395.

Keuken M.P., Schoonebeek C.A.M., Van Wensveen-Louter A. *et al.* (1988) Simultaneous sampling of NH$_3$, HCl, SO$_2$, and H$_2$O$_2$ in ambient air by a wet annular denuder system. *Atmospheric Environment*, **22**, 2541–2548.

Keuken M.P., Wayers-Ijpelahn A., Mols J.J. *et al.* (1989) The determination of ammonia in ambient air by an automated thermodenuder system. *Atmospheric Environment*, **23**, 2177–2185.

Kley D. & McFarland M. (1980) Chemiluminescence detector for NO and NO. *Atmospheric Technology*, **12**, 63–69.

Komhyr W.D. (1969) Electrochemical concentration cells for gas analysis. *Annals of Geophysics*, **25**, 203.

Kuster W.C. & Goldan P.D. (1987) Quantitation of the losses of gaseous sulfur compounds to enclosure walls. *Environmental Science and Technology*, **21**, 810–815.

Lang R.F., Brown C. & Harvey G.R. (1991) Determination of dimethyl sulfoxide and dimethyl sulfoxide and dimethyl sulfone in air. *Analytical Chemistry*, **63(2)**, 186–189.

Lamb B., Westberg H., Allwine G., Bamesberger L. & Guenther A. (1987) Measurement of biogenic sulfur emissions from soils and vegetation: Application of dynamic enclosure methods with Natusch filter and GC/FPD analysis. *Journal of Atmospheric Chemistry*, **5**, 469–491.

Langford A.O., Goldan P.D. & Fehsenfeld F.C. (1989) A molybdenum oxide annular denuder system for gas phase ambient ammonia measurements. *Journal of Atmospheric Chemistry*, **8**, 359–376.

LeBel P.J., Hoell J.M., Levine J.S. *et al.* (1985) Aircraft measurements of ammonia and nitric acid in the lower troposphere. *Geophysical Research Letters*, **12**, 401–404.

Lenhard U. & Gravenhorst G. (1980) Evaluation of ammonia fluxes into the free atmosphere over West Germany. *Tellus*, **32**, 48–55.

Lenschow D.H., Pearson R. Jr. & Stankov B.B. (1981) Estimating the ozone budget in the boundary layer by use of aircraft measurements of ozone eddy flux and mean concentration. *Journal Geophysical Research*, **86**, 7291–7297.

Levine J.S., Augustsson T.R. & Hoell J.M. (1980) The vertical distribution of tropospheric ammonia. *Geophysical Research Letters*, **7**, 317–320.

Liu S.C., Trainer M., Fehsenfeld F.C. *et al.* (1987) Ozone production in the rural troposphere and the implications for regional and global ozone distributions. *Journal of Geophysical Research*, **92**, 4191–4207.

Lovejoy E.R., Murrells T.P. & Ravishankara A.R. (1990) Oxidation of CS_2 by reaction with OH: II. Yields of HO_2 and SO_2 in oxygen. *Journal of Physical Chemistry*, **94**, 2386–2393.

MacKay G.I. & Schiff H.I. (1984) *Field Measurements with the TAMS-2B and Development of Methods for Measuring Formaldehyde, Ammonia, and Nitrous Acid in Real Air.* Report, Unisearch Association, Concord, Ontario, Canada.

MacTaggart D.L., Adams D.F. & Farwell S.O. (1987) Measurement of biogenic sulfur emissions from soil and vegetation using dynamic enclosure methods: Total sulfur gas emissions via MFC/FD/FPD determination. *Journal of Atmospheric Chemistry*, **5**, 417–437.

Maser R., Obenland H., Jaeschke W., Beltz N. & Herrmann J. (1991) A new method for continuous measurements of SO_2 in the pptv-range. *Proceedings EUROTRAC Symposium 1990*, pp. 557–558. Garmisch-Partenkirchen, Germany.

McClenny W.A. & Bennet C.A.J. (1980) Integrative technique for detection of atmospheric ammonia. *Atmospheric Environment*, **14**, 641–645.

McClenny W.A., Galley P.C., Braman R.S. *et al.* (1982) Tungstic acid technique for monitoring nitric acid and ammonia in ambient air. *Analytical Chemistry*, **54**, 365–369.

Meixner F.X. & Jaeschke W. (1981) The determination of low atmospheric SO_2 concentrations with a chemiluminescence technique. *International Journal of Environmental and Analytical Chemistry*, **10**, 51–67.

Mihelcic D., Ehhalt D.H., Kulessa G., Klomfass J., Trainer M. & Schmidt U. (1978) Measurements of free radicals in the atmosphere by matrix isolation and electron paramagnetic resonance. *Pageoph*, **116**, 530–536.

Mihelcic D., Musgen P. & Ehhalt D.H. (1985) An improved method of measuring tropospheric NO_2 and RO_2 by matrix isolation and electron spin resonance. *Journal of Atmospheric Chemistry*, **3**, 341–361.

Mihelcic D., Pätz H.W., Kley D. & Volz-Thomas A. (1989) Improved analysis of ESR Spectra from atmospheric samples. *EOS Transactions*, **70**, 1007.

Mihelcic D., Volz-Thomas A., Pätz H.W., Kley D. & Mihelcic M. (1990) Numerical analysis of ESR spectra from atmospheric samples. *Journal of Atmospheric Chemistry*, **11**, 271.

Murrells T.P., Lovejoy E.R. & Ravishankara A.R. (1990) Oxidation of CS_2 by reaction with OH: I. Equilibrium constant for the reaction OH + CS_2 | CS_2OH and the kinetics of the CS_2OH + O_2 reaction. *Journal of Physical Chemistry*, **94**, 2381–2386.

Natusch D., Konis H., Axelrod H., Teck R. & Lodge J. (1972) Sensitive method for measurement of atmospheric hydrogen sulfide. *Analytical Chemistry*, **44**, 2067–2070.

Novelli P.C., Elkins J.W. & Steele L.P. (1991) The development and evaluation of a gravimetric reference scale for measurements of atmospheric carbon monoxide. *Journal of Geophysical Research*, **96**, 13109–13121.

Noxon J.F., Norton R.B. & Henderson W.R. (1978) Observation of atmospheric NO_3. *Geophysical Research Letters*, **5**, 675–678.

Noxon J.F., Norton R.B. & Marovich E. (1980) NO_3 in the troposphere. *Geophysical Research Letters*, **7**, 125–128.

Parrish D.D., Trainer M., Buhr M.P., Watkins B.A. & Fehsenfeld F.C. (1991) Carbon monoxide concentrations and their relation to concentrations of total reactive oxidized nitrogen at two rural US sites. *Journal of Geophysical Research*, **96**, 9309–9320.

Parrish D.D., Williams E.J., Fahey D.W., Liu S.C. & Fehsenfeld F.C. (1987) Measurement of nitrogen oxide fluxes from soils: Intercomparison of enclosure and gradient measurement techniques. *Journal of Geophysical Research*, **92**, 2165–2171.

Pearson R. Jr. & Stedman D.H. (1980) Instrumentation for fast response ozone measurements from aircraft. *Atmospheric Technology*, **12**, 51–55.

Penkett S.A. & Brice K.A. (1986) The spring maximum in photoxidants in the northern hemisphere troposphere. *Nature*, **319**, 655–657.

Placet M. (1990) *Emissions involved in acidic deposition processes*. NAPAP SOS/T Report 1, NASA, Washington, DC.

Planetary and Space Science (1983) Le campaign d'intercomparison d'ozonemeters. Gap France, 1981. *Planetary and Space Science*, **31**.

Platt U. & Perner D. (1980) Direct Measurements of Atmospheric CH_2O, HNO_2, O_3, NO_2, and SO_2 by Differential Optical Absorption in the Near UV. *Journal Geophysical Research*, **85(C12)**, 7453–7458.

Platt U., Perner D. & Pätz H.W. (1979) Simultaneous measurement of atmospheric CH_2O, O_3, and NO_2 by differential optical absorption. *Journal of Geophysical Research*, **84**, 6329–6335.

Platt U., Perner D., Winer A.M., Harris G.W. & Pitts J.N. Jr (1980) Detection of NO_3 in the polluted troposphere by differential optical absorption. *Geophysical Research Letters*, **7**, 89–92.

Platt U., Winer A.M., Biermann H.M., Atkinson R. & Pitts J.N. Jr (1984) Measurements of nitrate radical concentrations in continental air. *Environmental Science and Technology*, **18**, 365–369.

Pszenny A.A.P., Harvey G.R., Brown C.J. *et al.* (1990) Measurements of Dimethyl Sulfide Oxidation Products in the Summertime North Atlantic Marine Boundary Layer. *Global Biogeochemical Cycles*, **4(4)**, 367–379.

Rapsomanikis S., Wake M., Kitto A.-M.N. *et al.* (1988) Analysis of atmospheric ammonia and particulate ammonium by a sensitive fluorescence method. *Environmental Science and Technology*, **22**, 948–952.

Rasmussen R.A. & Khalil M.A.K. (1982) Atmospheric trace gases at Point Barrow and Arctic haze. In: Bodhaine B.A. & Harris J. (eds) *Geophysical Monitoring for Climatic Change*. US Department of Transport, Washington, DC.

Ray J.R., Stedman D.H. & Wendel G.J. (1986) Fast Chemiluminescent Method for Measurement for Ambient Ozone. *Analytical Chemistry*, **58**, 598–600.

Reid J., Shewchun J., Garside B.K. & Ballik E.A. (1978) Point Monitoring of Ambient Concentration of Atmospheric Gas Using Tunable Lasers. *Optical Engineering*, **17**, 56–62.

Ridley B.A. & Howlett L.C. (1974) An instrument for nitric oxide measurements in the stratosphere. *Review of Scientific Instruments*, **45**, 742–746.

Ridley B.A., Carroll M.A. & Gregory G.L. (1987) Measurements of nitric oxide in the boundary layer and free troposphere over the Pacific Ocean. *Journal of Geophysical Research*, **92**, 2025–2047.

Ridley B.A., Shetter J.D., Gandrud B.W. *et al.* (1990) Ratios of peroxyacetyl nitrate to active nitrogen observed during aircraft flights over the eastern Pacific Ocean and continental United States. *Journal of Geophysical Research*, **95**, 10179–10192.

Roberts J.M., Norton R.B., Goldan P.D. *et al.* (1988) Ammonia measurements at Niwot Ridge, Colorado and Point Arena, California using the tungsten oxide denuder tube technique. *Journal of Atmospheric Chemistry*, **7**, 137–152.

Robbins D.E. (1983) NASA-Johnson Space Center (JSC) measurements during 'La campagne d'intercomparison d'ozonemetres', Gap France, 1981. *Planetary and Space Science*, **31**, 761–765.

Rodgers M.O., Asai K. & Davis D.D. (1980) Photofragmentation-laser induced fluroescence: A new method for detecting atmospheric trace gases. *Applied Optics*, **19**, 3597–3605.

Ryerson T.B., Dunham A.J., Barkley R,M. & Sievers R.E. (1994) Sulfur-selective Detector for Liquid Chromatography Based on sulfur Monoxide-Ozone Chemilumi-nescence. *Analytical Chemistry*, **66(18)**, 2841–2851.

Sachse G.W., Hill G.F., Wade L.O. & Perry M.G. (1987) Fast-response, high-precision carbon monoxide sensor using a tunable diode laser absorption technique. *Journal of Geophysical Research*, **92**, 2071–2082.

Sachse G.W., Harriss R.C., Fishman J., Hill G.F. & Cahoon D.R. (1988) Carbon monoxide over the Amazon Basin during the 1985 dry season. *Journal of Geophysical Research*, **93**, 1422–1430.

Sandholm S.T., Bradshaw J.D., Dorris K.S., Rodgers M.O. & Davis D.D. (1990) An airborne compatible photofragmentation two-photon laser-induced fluorescence instrument for measuring background tropopsheric levels of NO, NO_x, and NO_2. *Journal of Geophysical Research*, **95**, 10155–10161.

Sandholm S., Olson J., Bradshaw J. *et al.* (1994) Summertime partitioning and budget of NO_y compounds in the troposphere over Alaska and Canada: ABLE 3B. *Journal of Geophysical Research*, **99(D1)**, 1837–1861.

Schendel J.S., Stickel R.E., Van Dijk C.A. *et al.* (1990) Atmospheric ammonia measurement using a VUV/photofragmentation laser-induced fluorescence technique. *Applied Optics*, **29**, 4924–4937.

Schiff H.I., Harris G.W. & Mackay G.I. (1987) Measurement of atmospheric gases by laser absorption spectrometry. In: Johnson R.W. & Gordon G.E. (eds) *The Chemistry of Acid Rain – Sources and Atmospheric Processes*, pp. 274–288. ACS Symposium Series 349.

Schiff H.I., Karecki D.R., Harris G.W., Hastie D.R. & Mackay G.I. (1990) A tunable diode laser system for aircraft measurements of trace gases. *Journal of Geophysical Research*, **95**, 10147–10153.

Schiff H.I., Mackay G.I., Castledine C., Harris G.W. & Tran Q. (1986) Atmospheric measurements of nitrogen dioxide with a sensitive luminol instrument. *Water Air Soil Pollution*, **30**, 105–114.

Scott W.D. & Cattell F.C.R. (1979) Vapor pressure of ammonium sulfates. *Atmospheric Environment*, **13**, 307–317.

Seiler W., Giehl H. & Ellis H. (1976) A method for monitoring of background CO and first results of continuous CO registrations on Mauna Loa Observatory. *WMO Special Environmental Report*, **10**, 31–39.

Seiler W., Giehl H., Brunke E.-G. & Halliday E. (1984) The seasonality of CO abundance in the Southern Hemisphere. *Tellus*, **36B**, 219–231.

Seinfeld J.H. (1986) *Atmospheric Chemistry and Physics of Air Pollution*. John Wiley, New York.

Shendrikar A.D. & Lodge J.P. (1975) Microdetermination of ammonia by the ring oven technique and its application to air pollution studies. *Atmospheric Environment*, **9**, 431–435.

Sickles J.E. II., Hodson L.L., McClenny W.A. *et al.* (1990) Field comparison of methods for the measurement of gaseous and particulate contributors to acidic dry deposition. *Atmospheric Environment*, **24**, 155–165.

Singh H.B. & Salas L.J. (1983) Methodology for the analyses of peroxyacetyl nitrate (PAN) in the unpolluted atmosphere. *Atmospheric Environment*, **17**, 1507–1516.

Slemr F. & Seiler W. (1984) Field measurements of NO and NO_2 emissions from fertilized and unfertilized soils. *Journal of Atmospheric Chemistry*, **2**, 1–24.

Solomon S., Schmeltekopf A.L., & Sanders R.W. (1987) On the interpretation of zenith sky absorption measurements. *Journal of Geophysical Research*, **92**, 8311.

Solomon S., Miller H.L., Smith J.P., Sanders R.W., Mount G.H. & Schmeltekopf A.L. (1989a) Atmospheric NO_3, 1. Measurement technique and the annual cycle at 40° north. *Journal of Geophysical Research*, **94**, 11041–11048.

Solomon S., Sanders R.W., Mount G.H., Carroll M.A., Jakoubek R.O. & Schmeltekopf A.L. (1989b) Atmospheric NO_3, 2. Observations in polar regions. *Journal of Geophysical Research*, **94**, 16423–16427.

Stelson A.W. & Seinfeld J.H. (1982) Relative humidity and temperature dependence of the ammonium nitrate equilibrium constant. *Atmospheric Environment*, **13**, 983–992.

Stockwell W.R. & Calvert J.G. (1983) The mechanism of the HO-SO_2 reaction. *Atmospheric Environment*, **17**, 2231–2235.

Talbot R.W., Vijgen A.S. & Harriss R.C. (1990) Measuring tropospheric HNO_3: Problems and prospects for nylon filter and mist chamber techniques. *Journal of Geophysical Research*, **95**, 7553–7561.

Tang I.N. (1980) On the equilibrium partial pressures of nitric acid and ammonia in the atmosphere. *Atmospheric Environment*, **14**, 819–828.

Tanner R.L. (1982) An ambient experimental study of phase equilibrium in the atmospheric system: Aerosol H^+, NH_{4^+}, SO_4^{2-}, NO_3-, $-NH_{3(g)}$, $HNO_{3(g)}$. *Atmospheric Environment*, **16**, 2935–2942.

Thornton D.C., Driedger A.R. III & Bandy A.R. (1986) Determination of parts-per-trillion levels of sulfur dioxide in humid air. *Analytical Chemistry*, **58**, 2688–2691.

Thornton D.C., Ridgeway R.E., Bandy A.R., Driedger A.R. III & Lalevic M. (1990) Determination of part-per-trillion levels of atmospheric dimethyl sulfide by isotope dilution gas chromatography/mass spectrometry. *Journal of Atmospheric Chemistry*, **11**, 299–308.

Tyndall G.S. & Ravishankara A.R. (1991) Atmospheric oxidation of reduced sulfur species. *International Journal of Chemical Kinetics*, **23**, 483–527.

Viggiano A.A. (1981) *The ion chemistry of N_2O_5 and its application for measuring the thermal decomposition rate of N_2O_5*. PhD thesis, University of Colorado, Boulder.

Volz A. & Kley D. (1985) A resonance-fluorescence instrument for the *in situ* measurement of atmospheric carbon monoxide. *Journal of Atmospheric Chemistry*, **2**, 345–357.

Wahner A. & Ravishankara A.R. (1987) The kinetics of the reaction of OH with COS. *Journal of Geophysical Research*, **92**, 2189–2194.

Walega J.G., Stedman D.H., Shetter R.E., Mackay G.I., Iguchi T. & Schiff H.I. (1984) Comparison of a chemiluminescent and a tunable diode laser absorption technique for the measurement of nitrogen oxide, nitrogen dioxide, and nitric acid. *Environmental Science and Technology*, **18**, 823–826.

Wallington T.J., Atkinson R., Winer A.M. & Pitts J.N. Jr (1986) Absolute rate constants for the gas-phase reactions of the NO_3 radical with CH_3SH, CH_3SCH_3, CH_3SSCH_3, H_2S, SO_2, and CH_3OCH_3 over the temperature range 280–350 K. *Journal of Physical Chemistry*, **90**, 5393–5396.

Wesely M.L., Sisterson D.L., Hart R.L., Drapcho D.L. & Lee I.Y. (1989) Observations of nitric oxide fluxes over grass. *Journal of Atmospheric Chemistry*, **9**, 447–463.

Wiebe H.A., Anlauf K.G., Tuazon E.C. *et al.* (1990) A comparison of measurements of atmospheric ammonia by filter packs, transition-flow reactors, simple and annular denuders and fourier transform infrared spectroscopy. *Atmospheric Environment*, **24A**, 1019–1028.

Williams E.J., Hutchinson G.L. & Fehsenfeld F.C. (1992a) NO$_x$ and N$_2$O emissions from soil. *Global Biogeochemical Cycles*, **6**, 351–388.

Williams E.J., Parrish D.D. & Fehsenfeld F.C. (1987) Determination of nitrogen oxide emissions from soils: Results from a grassland site in Colorado, USA. *Journal of Geophysical Research*, **92**, 2173–2179.

Williams E.J., Parrish D.D., Buhr M.P. & Fehsenfeld F.C. (1988) Measurement of soil NO$_x$ emissions in central Pennsylvania. *Journal of Geophysical Research*, **93**, 9539–9546.

Williams E.J., Sandholm S.T., Bradshaw J.D. *et al.* (1992b) An intercomparison of five ammonia measurement techniques. *Journal of Geophysical Research*, **97**, 11591–11611.

Wine P.H., Kreutter N.M., Gump C.A. & Ravishankara A.R. (1981) Kinetics of OH reactions with the atmospheric sulfur compounds H$_2$S, CH$_3$SH, CH$_3$SCH$_3$, and CH$_3$SSCH$_3$. *Journal of Physical Chemistry*, **85**, 2660–2665.

Winer A.M., Atkinson R. & Pitts J.N. Jr (1984) Gaseous nitrate radical: Possible nighttime atmospheric sink for biogenic organic compounds. *Science*, **224**, 156–159.

Zeller K.F., Massman W.J., Stocker D., Fox D.G., Stedman D.H. & Hazlett D. (1989) *Initial Results from the Pawnee Eddy Correlation System for Dry Acid Deposition Research*. Research Paper RM-282, 30 pp., US Department of Agriculture, Forest Service, Rocky Mountain Forest and Range Experimental Station, Ft. Collins, Colorado.

Ziereis H. & Arnold F. (1986) Gaseous ammonia and ammonium ions in the free troposphere. *Nature*, **321**, 503–505.

Zimmerman P.R., Greenberg J.P. & Westberg C. (1988) Measurements of atmospheric hydrocarbons and biogenic emissions in the Amazon boundary layer. *Journal of Geophysical Research*, **93**, 1407–1416.

CHAPTER 8

Recent advances in spectroscopic instrumentation for measuring stable gases in the natural environment

C.E. KOLB, J.C. WORMHOUDT & M.S. ZAHNISER

8.1 Introduction

Traditionally, most trace chemical species measurements performed in ecological and environmental studies have utilized grab sampling techniques, whereby discrete time series or spatially dispersed samples are collected and transported in canisters, flasks, syringes or other containers to an analytical instrument for quantification at some later time. While some of the standard analytical methods for trace gas measurement (see Chapter 6, this volume), can be configured for nearly continuous, real-time monitoring, most were developed for and are normally used in off-line, discrete sample analysis.

Recent advances in optics, electro-optics and control electronics, including microprocessors and control computers, have revolutionized our ability to detect and monitor trace chemical species using a variety of real-time spectroscopic techniques. The capability to perform precise real-time measurements, often with very high temporal resolution, has revolutionized the field of atmospheric chemistry by greatly expanding the types of field measurements it is possible to perform. Recent reviews which emphasize modern spectroscopic-based measurement techniques for atmospheric trace species and catalogue their utility in various types of field investigations have been published (Albritton *et al.*, 1990; Kolb, 1991; Grant *et al.*, 1992; Schiff, 1992).

8.1.1 Gas species coverage

In this chapter we will review the application of advanced spectroscopic instrumentation for measuring the stable biogenically active trace gases generally of interest in ecological studies of trace gas processing and atmosphere–biosphere exchange. Trace species explicitly considered include CO, CO_2, N_2O, NH_3, CH_4, simple non-methane hydrocarbons (NMHC) and reduced sulphur species (OCS, DMS, CS_2, H_2S and related compounds). While we will discuss, only in passing, oxidative atmospheric trace species such as ozone, hydrogen peroxide, NO, NO_2, SO_2, etc. or inorganic atmospheric acid gases such as HNO_3, HCl, H_2SO_4, etc., many of these species can also be monitored using techniques discussed in this chapter. Furthermore, the same techniques used to monitor CH_4 and NMHC are often suitable for low molecular weight halogenated (e.g. CH_3Cl, CH_3Br) and partially oxygenated (e.g. CH_2O, CH_3OH, CH_3COOH, etc.) hydrocarbons.

One additional advantage of many of the spectroscopic measurement techniques discussed below is their capability to distinguish between isotopic variants of many trace species, e.g. $^{12}CH_4$-$^{13}CH_4$-$^{12}CH_3D$. Thus, they can, in principle, be utilized in the isotopic source characterization and tracer studies discussed in Chapter 9. While most isotopic ratio measurements are currently performed using a dedicated fixed site mass spectrometer after extensive sample concentration/preparation, several of the mobile techniques presented below could be used for reasonably precise isotopic measurements if suitable trace species concentration techniques can be adapted for field use.

8.1.2 Types of measurements

We will focus our review on instrumentation suitable for measuring trace gas concentrations in soils, water and air and the flux of trace gases between the atmosphere and the Earth's surface. While spectroscopic trace species measurement techniques can often be adapted to both remote sensing and local, *in situ* sampling measurements (Kolb, 1991), we will concentrate on the latter in this review.

Using the techniques discussed below, trace species concentration measurements can be performed continuously and directly on atmospheric samples and can also be used to quantify trace gas levels in gases drawn or purged from soil and water samples as long as a discrete sample cell or loop is employed. Those spectroscopic instruments which monitor atmospheric trace gases without explicit sampling, e.g. with an 'open' optical path, cannot, in general, also be used to measure trace gases drawn from soil or water samples.

While most of the specific applications presented in this chapter utilize point source or small gas volume measurements, some techniques can also be adapted to long-path measurements which yield the average concentration measurement over a line-of-sight path which can be as long as a few kilometres. Such techniques are most applicable where low concentrations make very long path sampling necessary or when a concentration average over a significant path length is the ecological parameter of interest.

Since trace gas fluxes are of special interest to ecologists, we will highlight instruments capable of performing the very fast response or high precision trace species concentration measurements necessary to utilize the micrometeorological flux measurement techniques discussed in Chapter 5. (Some additional discussion of trace gas flux techniques and instrumentation can be found in Lenschow & Hicks, 1989.)

8.1.3 Instrumentation types

We will cover four specific types of measurements in this review:
1 infrared absorption using tunable lasers;
2 infrared absorption using Fourier transform infrared (FTIR) spectrometers;
3 ultraviolet/visible laser photofragmentation/fluorescence spectroscopy; and
4 advanced, field portable mass spectrometer systems.

In our judgement, most of the advanced spectroscopic gaseous trace species concentration measurement techniques currently being utilized or developed for routine field use fall into one of these four categories. Of the four, we will present the most detail on the first and second techniques, both of which rely on infrared absorption. There are several (interrelated) reasons for this emphasis, the most important is that infrared absorption is the most convenient spectroscopic 'handle' for most of the stable biogenic trace gases of greatest current interest to ecologists and climatologists. Second, there are more research and development groups actively pursuing atmospheric trace species instrumentation based on infrared absorption than other techniques and this activity is driving their rapid improvement, and finally, during the past decade our own personal research has emphasized infrared trace gas detection techniques.

8.2 Tunable infrared laser differential absorption spectroscopy (TILDAS)

8.2.1 Method overview

The first type of advanced spectroscopic trace gas measurement system discussed is based on the differential absorption achieved when an infrared laser is tuned on and off a spectral feature, usually a single vibrational/rotational line, of the target species. For trace gas sampling at atmospheric pressure, the fractional absorption at line centre, $I_0 - I$, divided by the intensity of the unabsorbed probe laser beam I_0, is given by Beer's law. In the limit of small absorptions this becomes:

$$\frac{I_0 - I}{I_0} = \frac{\Delta I}{I_0} = \frac{nlS}{\pi b P} \tag{8.1}$$

where n is the number density of the target trace species, S is the absorption line strength, b is the pressure-broadening coefficient of the line width (half width at half maximum), P is the sampling pressure, and l is the path length through the sample.

As evident from Eqn. 8.1, the change in transmitted laser probe intensity, and therefore the sensitivity of this technique, is proportional to the absorption line strength and the path length of the measurement and

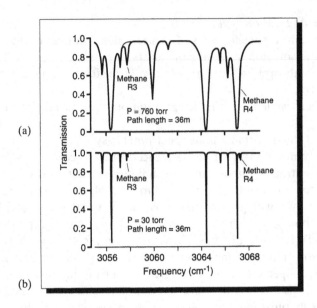

Fig. 8.1 Atmospheric transmission through a 36 m path in region of strong methane lines at sampling pressures of (a) 760 mmHg and (b) 30 mmHg. The upper traces are calculated without CH_4 to show background absorption due to water vapour (1.3%) and CO_2 (320 ppmv).

inversely proportional to the spectral line width, bP. Since infrared absorption lines are broadened by collisions, the line width can be reduced by decreasing the pressure of the gas being sampled until the line width approaches its Doppler limit, which usually occurs in the pressure range between 10 and 60 mmHg (760 mmHg $\equiv$ 1 atm). Although reducing the pressure also decreases ΔI, since n is proportional to pressure at constant temperature, the resultant spectral narrowing can minimize interferences from nearby spectral lines of other atmospheric species and also reduce the amount of laser frequency tuning required to go from line centre to line edge. Thus, for measurement systems which pump atmospheric gas into a closed optical sample cell, adjustment of the sample's pressure is a valuable experimental variable. The optimum sampling pressure is generally in the range where the pressure broadened line width, bP, equals the Doppler width.

The dependence of infrared absorption features on sampling pressure is illustrated in Figs 8.1 and 8.2 for methane lines in the 3.3 μm wavelength region. Figure 8.1 shows the infrared transmission through a 36 m path at atmospheric pressure (760 mmHg) and at a reduced pressure of 30 mmHg. Interferences from water vapour lines completely dominate the methane multiplet at 3067 cm^{-1} at atmospheric pressure, while at 30 mmHg, they

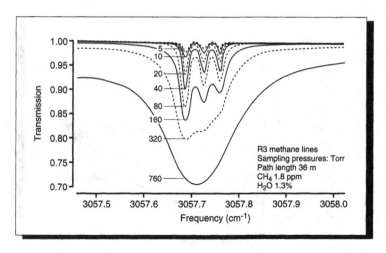

Fig. 8.2 Sampling pressure dependence on $3057\,cm^{-1}$ CH_4 multiplet from 5 to 760 mmHg.

are nearly separated. The methane feature near $3058\,cm^{-1}$ is isolated at both pressures. Figure 8.2 shows the effect of varying the sampling pressure from 5 to 160 mmHg on the $3058\,cm^{-1}$ multiplet.

Systems using open optical paths must work with spectral lines broadened by atmospheric pressure, and sensitive measurements will depend on the existence of individual spectral features free of significant interference by abundant infrared active gases such as H_2O and CO_2. Since H_2O and/or CO_2 absorb infrared radiation strongly between about 2.5–3.3 μm, 4.2–8.0 μm, 15–16 μm and beyond 22 μm, atmospheric pressure systems are most likely to be successful operating in the infrared 'windows' between 0.8 and 2.5 μm, 3.3–4.2 μm, 8–14 μm and 16–22 μm.

The keys to designing a successful TILDAS measurement system are the identification of a suitable, isolated absorption feature for the target species, the availability of a laser tunable over the required frequency range to probe this absorption feature, the provision of sufficient path length to achieve a measurable $\Delta I/I$, and the design of suitable signal processing equipment to distinguish the $\Delta I/I$ due to target gas absorption from background interferences and instrumental noise.

Two typical TILDAS experimental arrangements are shown in Figs 8.3 and 8.4. Figure 8.3 shows an open long path system where a remote retroreflector returns the laser light to a receiving optical system co-located with the laser. This type of system can be used to monitor average concentrations over the sampling path length as well as vertical or horizontal gradients if more than one retroreflector is employed (Zahniser *et al.*, 1992).

Figure 8.4 shows a schematic of a closed cell sampling instrument

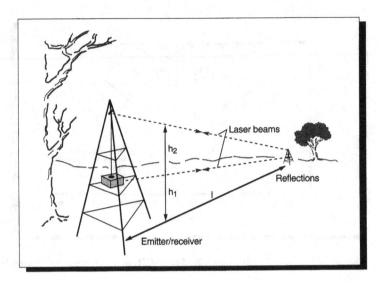

Fig. 8.3 Dual-path retroreflective tunable infrared differential absorption system. Trace gas concentrations are detected simultaneously along two paths using a single laser source.

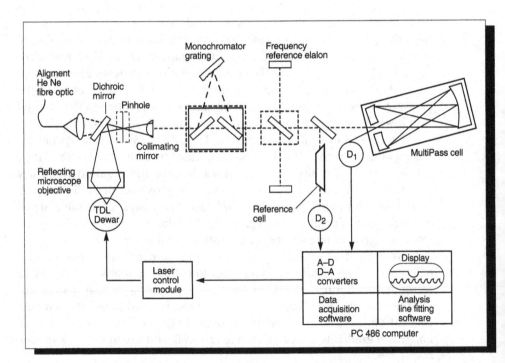

Fig. 8.4 Schematic diagram of closed-path laser absorption system with multiple pass cell.

used to make point source concentration measurements. By keeping the cell volume small (<0.5 l) fast response suitable for eddy correlation flux measurements can be achieved using a modest size vacuum pump. Gradient flux measurements can also be made with the same instrument by sequentially sampling from several vertical points.

The details of various current tunable laser, optical sampling system and signal processing system options will be presented below. In addition the capabilities of several successful biogenic trace gas measurement systems will be presented.

8.2.2 Laser sources
TIRL sources most commonly used for trace gas detection systems fall into three categories:
1 lead salt tunable diode lasers (TDLs) which are available throughout the wave length region from 3.3 to 30 μm;
2 rare gas lasers which operate at selected wave lengths in the same spectral region; and
3 semiconductor diode lasers which operate in the near infrared in selected regions between 0.7 and 2.5 μm.

Lead salt TDLs are the most widely used for atmospheric trace gas monitoring. A review of the current state of this technology may be found in Wall (1991). Examples of applications for atmospheric monitoring by several groups of experimenters may be found in a recent compendium and review article (Schiff, 1991, 1992). Lead salt TDLs have the advantage of wide spectral coverage with very high frequency resolution ($<0.001 \, \text{cm}^{-1}$). Their modest output powers, up to several hundred microwatts, allow the use of multiple pass optics or long atmospheric paths before transmission losses limit the signal to noise ratio of the measurement. Rapid tunability from several kHz to several hundred MHz allows a variety of signal processing techniques to be employed to maximize sensitivity so that fractional absorptions as small as a few parts in 10^6 of incident laser power may be detected.

The main disadvantage of lead salt TDLs is that they operate at temperatures generally below 100 K and thus require cryogenic cooling, usually with liquid nitrogen or closed cycle helium refrigerators. Another inherent disadvantage lies in the laser's mode structure. Since they often operate with multiple frequencies, mode filtering optics are usually necessary to assure spectral purity, and any particular diode's frequency characteristics may change with time. Thus, practical operation and characterization of these devices usually requires a higher degree of operator knowledge and experience than other infrared technologies such as the FTIR devices discussed in the next section.

Rare gas (He, Xe, Ne) lasers as reviewed by Kebabian & Kolb (1993) have selected applications for measuring trace gases by infrared absorption. They overcome the main disadvantages of TDLs but have some substantial limitations of their own. Since they operate at frequencies corresponding to the fixed energy levels of excited atomic states, their spectral output is well characterized and highly reproducible. Thus they are also inherently more economical to manufacture and reproduce. They also operate at room temperature and require no cryogens. However, since they are not widely tunable, they rely on coincidental overlaps between the laser frequency and the trace gas absorption line. Several techniques such as Zeeman splitting or pressure broadening can be used to achieve the degree of fine tuning necessary for a differential measurement. Such overlaps have been successfully exploited for CH_4, CO and N_2O, and although several other trace gases are likely candidates, the method is not as universally applicable as TDL or FTIR methods.

Semiconductor diode lasers fabricated from AlGaAs and InGaAsP type compounds operate in the near infrared from 0.7 to 2 µm and have been developed and mass produced for applications in the communications industry. Their main advantages compared to lead salt diode lasers are that they operate at room temperature, produce higher powers in the 1–15 mW range, and are considerably less expensive because they are mass produced. Unfortunately, the overlap between strong molecular absorption lines for most gases of ecological interest is very limited in this wavelength region. However, these lasers have been used to detect CH_4 (Hovde & Stanton, 1992; Uehara & Tai, 1992), water vapour and CO_2 using weaker overtone transitions. A further disadvantage in using these diodes is that finding a device with the proper wavelength to coincide with a particular molecular absorption can be difficult and time-consuming. Development efforts are underway to extend the range of these devices into the 2–5 µm region which would greatly increase their utility for atmospheric trace gas monitoring. An excellent review of semiconductor diode lasers applied to atomic and molecular spectroscopy is given by Wieman & Hollberg (1991).

8.2.3 Typical optical layout

The optical layout and signal processing schematic for a typical TDL system is shown in Fig. 8.4. The laser diode is housed in either a liquid nitrogen cooled dewar or a closed-cycle helium cryostat. Laser output is collected with a reflecting microscope objective and focused to a 100 µm pinhole with a 15:1 magnification. The infrared beam is combined with a visible alignment beam from a helium–neon laser using a dichroic mirror which reflects the infrared and transmits the visible light. The pinhole, which is mounted on a removable kinematic base, is used to ensure that the two beams are co-aligned at the source. Alignment of the infrared beam through the rest of the optical system is then straightforward.

The beam is collimated with a 2.5 cm diameter spherical mirror. Laser modes may be selected if necessary with a removable pair of mirrors which deflect the beam to a diffraction grating. A removable beam splitter and two flat mirrors spaced 75 cm apart form a simple yet effective etalon fringe pattern which is used for relative frequency calibration. A second beam splitter and a reference gas cell provide absolute frequencies from known line positions. The beam then enters the multipass cell and exits to the main detector.

8.2.4 Multiple pass mirror systems

Multipass absorption cells are a widely used means of providing long optical absorption path lengths in a compact volume. There are several factors to be considered in the evaluation of a multipass cell design, the first of which is the total available pathlength. The signal:noise ratio (SNR) in an absorption measurement increases with the path length, up to a limit where loss from the many reflections becomes important. Another, and often the most important, factor limiting the effective SNR at long path lengths is the appearance of optical interference fringes. These fringes, which arise due to light scattering from the cell mirrors, can have a spectral period close to the frequency width of molecular absorption lines. Thus, fringes tend to obscure the spectral features of interest. Finally, the volume of an absorption cell for a given path length is a limiting factor for applications such as eddy correlation flux measurements where system response time is important.

Multiple pass cells have traditionally used the three-mirror design originally suggested by White (1942). These designs have the advantage of a relatively large collection efficiency which is important when using incoherent light sources as described for FTIR systems in the following section. For laser systems where the light beam can be efficiently collimated, the large apertures available with White-type cells are not advantageous. Furthermore, the overlapping spot pattern of the White cell often leads to more severe interference fringes than with mirror configurations with non-overlapping spot patterns.

We have used both White cells and off-axis resonator or Herriott-type cells (Herriott et al., 1964; Herriott & Schulte, 1965) in our TDL systems, and prefer the latter for ease of alignment and relatively lower level of interference fringes. The Herriott cell consists of two spherical mirrors separated by nearly their radius of curvature. An optical beam is injected through a hole in one mirror in an off-axis direction, and it recirculates a number of times before exiting through the coupling hole. The beam spots fall in elliptical patterns on the mirrors, such that the beams fill a volume within the cell that is describable as a flattened hollow hyper-boloid. The number of passes is adjustable with changes in the mirror separation. The Herriott cell is easy to construct and to align, and is relatively insensitive to mirror misalignments. For a re-entrant pattern, the location of the output spot is independent of the alignment of the mirrors. The only effect of mirror alignment is to change the shape of the pattern of beam spots on the mirrors.

An additional valuable characteristic of the Herriott cell is the pos-sibility of establishing paths which minimize the effects of interference fringes which are due primarily to spillover of light from beam passes neighbouring the coupling hole. Maximizing the number of passes between the input/output beams and the beamspots neighbouring the coupling hole produces high-frequency fringes which are much narrower than molecular absorption lines and therefore are much easier to suppress (McManus & Kebabian, 1990).

A variation on the traditional Herriott cell uses astigmatic mirrors (Herriott & Schulte, 1965) to minimize the cell volume for a given path length. The different horizontal and vertical radii of curvature produce a pattern of beam spots which fill the mirrors as shown in Fig. 8.5, rather than the elliptical paths found in the conventional Herriott cell. The optical path, therefore, more completely fills the optical absorption volume, and thus the astigmatic Herriott cell can have a smaller volume for a given path length than spherical mirror cells.

A cell based on a particularly favourable combination of curvatures to produce 182 passes before exiting has been designed and developed at

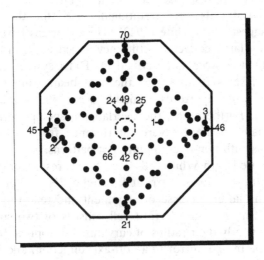

Fig. 8.5 Spot pattern (91 spots) on astigmatic multipass mirror with central coupling hole. Numbers indicate order of passes.

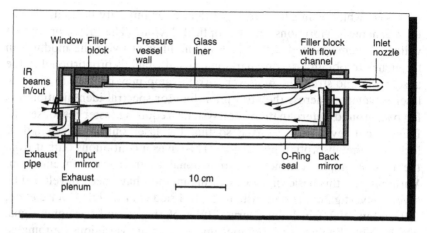

Fig. 8.6 Fast response sampling cell for eddy correlation flux measurements using infrared TDL absorption.

Aerodyne Research, Inc. (ARI) (Kebabian, 1994; McManus *et al.*, 1994). One version has a 55 cm base length which gives a path length of 100 m. With 7.6 cm diameter mirrors, the enclosed volume is 3 l. A smaller base path version of this same design has an overall path length of 36 m and an enclosed volume of 0.3 l. The number of passes between either the entrance or exit beams and the beam spots adjacent to the coupling hole is not less than 48. Thus the dominant fringe period should correspond to 48 times the base path length, or 26.4 m for the larger cell, giving a dominant fringe frequency of $3.8 \times 10^{-4} \mathrm{cm}^{-1}$. This is much narrower than the typical spectral width of molecular lines, and minimizes interference with absorption measurements. The mirrors have a reflectivity of 99.2% from 3 to 10 μm resulting in a 20% transmission of light through the cell. An enclosure designed for fast time response eddy correlation measurements which takes advantage of the high path length:volume ratio using these astigmatic mirrors is shown in Fig. 8.6. System response times of 0.3 s ($1 e^{-1}$) for a 100 m path length have been demonstrated using this cell with a 600 l min^{-1} vacuum pump making it suitable for fast response eddy correlation measurements of atmospheric trace gases. The smaller volume cell (0.3 l) has a 0.06 s ($1 e^{-1}$) response time with a 300 l min^{-1} pump. The shorter path length of 36 m is ideal for trace gases such as N_2O, CH_4 and CO which have mixing ratios greater than 100 ppb.

8.2.5 Signal processing

The most common signal processing technique for TDL systems has traditionally been second harmonic detection. In this method the laser current and therefore the wavelength is modulated at a frequency of

1–30 kHz which is much more rapid than laser intensity fluctuations due to mechanical vibrations in the optical layout. The detector output voltage is demodulated with a lock-in amplifier at twice the modulation frequency to obtain a second harmonic signal which is proportional to the absorption line strength. A slower current ramp may be used to scan the laser wavelength across the absorption line for spectral identification and for background determination. For systems requiring faster response, the DC current level may be fixed so that the laser output wavelength is locked to the absorption line centre. This gives a continuous output from the lock-in amplifier which is proportional to trace gas concentration. Variations on this basic signal processing method have been developed by several investigators (Loewenstein, 1988; Fried *et al.*, 1991; Bomse *et al.*, 1992; May, 1992) including curve fitting of the second harmonic signal which helps eliminate uncertainties due to baseline variations, automated line-locking schemes to compensate for long-term drift in the laser output frequency, and the use of higher harmonics to minimize the effects of baseline curvature.

The main advantage of second harmonic detection is that it provides a convenient method for doing the basic signal processing at frequencies higher than most of the sources of noise in the optical train. Even greater noise reductions may be obtainable from higher frequency modulation, or FM, methods (Cooper & Carlisle, 1988; Werle *et al.*, 1989; Bomse *et al.*, 1992). In these methods the laser frequency is modulated in the mHz range to produce sidebands which are detected using Radio Frequency (RF) mixers. The use of higher frequency modulation can reduce noise sources arising from amplitude fluctuation in the laser itself which decrease with increasing frequency. FM methods have been used to detect fractional absorptions as low a one part in 10^7 of the laser intensity (Carlisle *et al.*, 1989). However, this method has not been widely used in trace gas monitoring since it is more complicated to implement than second harmonic detection and its effectiveness is highly dependent on the quality of the laser diodes. Thus it has been more widely used with semiconductor diodes in the near infrared than with lead salt TDLs in the mid infrared. Furthermore, trace gas detection limits are more often due to baseline uncertainties arising from interference fringes or from other weakly absorbing molecular species rather than from laser intensity noise, so the increased sensitivity possible with FM detection is often not realizable in practice.

The main disadvantage of both frequency modulation techniques is that some of the 'absolute' nature of the infrared absorption method is degraded since the amplitude of second harmonic or FM signal is a function of modulation amplitude and absorption line width in addition to trace gas concentration and path length. Direct absorption methods, whereby the unmodulated laser frequency is swept repeatedly over the absorption line, retain the absolute relationship between line strength and species concentration since the latter is directly related to the area under the absorption line. Although this 'sweep integration' method has been in use for some time (Jennings, 1980), recent developments in fast, inexpensive, dedicated computers and high speed data acquisition hardware have allowed this method to provide similar advantages to second harmonic detection and yet retain the convenience of absolute concentration retrieval. A system developed at ARI (Worsnop *et al.*, 1992) using this method is described in more detail below.

The data acquisition method is based on rapid sweep integration over the full infrared transition line shape. This is accomplished by scanning the laser frequency under computer control and synchronously measuring the transmitted infrared light intensity. This spectral information is analysed in real time with a non-linear least squares fitting routine which returns both the spectral line profile and the absolute laser intensity (baseline). The area under the absorption line, together with the absorption coefficient for the line and the path length, is used to calculate the absolute concentration of the species being observed.

In a typical experiment the laser temperature is held constant while the current through the laser is modulated with a computer-generated sawtooth ramp to sweep the output frequency across the infrared transition. The sawtooth is generated in 30–300 discrete steps with a 300 kHz digital to analogue converter board. Thus the absorption line and baseline are acquired continuously at a rate of 1–10 kHz. During approximately 10% of the duty cycle the laser current is dropped below the lasing threshold to provide a precise measurement of zero light intensity.

A 12-bit analogue to digital converter (ADC) provides a numerical representation of the signal amplitude from the infrared detector. The conversion time for the ADC is 3 μs. The result of each conversion is transferred to the extended memory of a 80486 processor-based computer using direct memory access (DMA). This process is fast enough to maintain easily a continuous 300 kHz data stream with 100% duty cycle and in this regard is far superior to the analogous sweep integration approach using a digital oscilloscope for data acquisition.

A computer program controls the data acquisition process. It monitors the data being written to the circular buffer in extended memory and grabs a block of it for processing whenever one-half of the buffer is ready. Since this process only uses about 7% of the available central processing unit (CPU) time, the program is able to perform several other tasks in real time without losing any data. The most important of these tasks is to signal average the incoming stream of absorption spectra. This is accomplished with a high speed assembly language subroutine. Two channels of data may be sampled with the data points alternating between channels so that the two absorption spectra are interleaved and are therefore effectively simultaneous. Both spectra may be signal averaged for a fixed time (typically a few seconds) and then displayed on the computer screen. The program also executes a non-linear least squares fitting routine to determine the integrated absorption intensity for the absorption spectrum from each channel. The result of the fitting routine is displayed on the computer screen as a function of time so that trends in the trace gas column densities are immediately apparent. Finally, the program stores the fit results on its hard disc for future analysis. The time required to implement each of these tasks is small enough that the program can accomplish them while continuously accumulating data in the background with very little loss of duty cycle. For a 3–5 averaging time the duty cycle exceeds 90% and could be improved further as computer speeds continue to increase.

The non-linear least squares fitting routine is crucial for absolute absorption measurements and for line shape studies. The fitting routine uses the Levenberg–Marquardt approach. The diode laser power spectrum is represented as a slowly varying polynomial of adjustable order; typically a quadratic or cubic polynomial is used. The absorption line shape may be fit to either a Gaussian, pressure broadened Lorentzian, or Voigt profile. The position, width and height of the line are simultaneously fit together with the diode laser polynomial baseline. The absolute accuracy of the area under the peak returned by the fit is a few per cent. The short-term (~1 h) precision can approach one part in one thousand.

For continuous fast response measurements as used in eddy correlation flux methods, a data stream rate of up to 20 Hz is desirable. This sampling rate may be obtained by using the area between the baseline and absorption peak where the baseline is obtained from a linear least squares fit to a second or third order polynomial on either side of peak. While this method is less accurate than the simultaneous line-shape–baseline fitting procedure, it is considerably faster since it does not require an iterative non-linear least squares computation. With the availability of faster processors, the non-linear least squares iterative method will be feasible at a 20 Hz or greater sampling rate. Such a system has been developed at ARI for eddy correlation measurements of CH_4 and N_2O.

8.2.6 Specific systems

A number of research groups have developed differential laser absorption systems for atmospheric trace gas measurements. Several representative systems and their capabilities are highlighted in this section.

Closed-path reduced pressure sampling systems using lead salt TDLs have been developed for tropospheric trace gas measurements including

CO and CH_4 (Sachse *et al.*, 1987), OCS (Fried *et al.*, 1991), H_2O_2, CH_2O and HNO_3 (Schiff, 1991, 1992) from both ground-based and aircraft-based platforms. These systems use multiple pass cells with path lengths from 20 to 100 m, second harmonic detection, and typically detect fractional absorptions as low as 10^{-5} corresponding to trace gas concentrations in the sub-ppbv range. The system developed by Fried *et al.* (1991) is capable of detecting ambient OCS concentrations less than 10 pptv. Isotopic specific detection of trace gases is possible using TDL absorption at reduced pressure since each isotope has a unique infrared absorption spectra. Isotopic specific detection for $^{12}C/^{13}C$ in CO_2 has been demonstrated by Becker *et al.* (1992) and determination of $^{16}O/^{18}O$ ratios in environmental N_2O samples has been reported by Yoshinari *et al.* (Whalen & Yoshinari, 1985; Yoshinari & Whalen, 1985). While these TDL isotopic measurements are much less precise (a few δ to a few tenths δ) compared to the mass spectrometric techniques discussed in Chapter 9, they will often be accurate enough to characterize isotopically gas sources or detect evolution of gases from isotopically spiked systems. Furthermore, many isotopic determinations can be made with minimal sample preparation. N_2O samples, for instance, do not have to be cleared of CO_2 (or vice versa).

A method for closed-path sampling of CH_4 using a rare gas helium–xenon laser has been developed by McManus *et al.* (1989, 1991a). Differential absorption signals are obtained by magnetically tuning the laser line across a nearly coincident methane line around 3.3 μm to obtain a sensitivity of 5 ppbv with a 1 s averaging time. An earlier version of this instrument has been used for tower-based eddy correlation measurements of CH_4 fluxes from the Alaskan tundra (Fan *et al.*, 1992) and a van-mounted instrument has been widely deployed for field measurements of CH_4 fluxes associated with natural gas distribution systems (McManus *et al.*, 1991a,b).

Trace gas flux measurements using open-path absorption at atmospheric pressure using TDLs have been demonstrated for CH_4 and ozone (Anderson & Zahniser, 1991). Representative data showing CH_4 fluxes from a Massachusetts wetland are shown in Fig. 8.7. A similar system is being developed by Cooper *et al.* (1991) for measuring NH_3 fluxes. Open-path flux measurements of CH_4 with a compact, tower-mounted system using a near infrared semiconductor diode laser have been demonstrated by Hovde & Stanton (1992). Open-path methods have the advantage for eddy correlation flux measurements of being capable of truly simultaneous measurement of trace gas mixing ratio and wind field as shown in Fig. 8.7 without the lag time or response time degradation inherent with closed-path sampling systems. The disadvantages are that they impose more stringent tuning and selection requirements on the TDL due

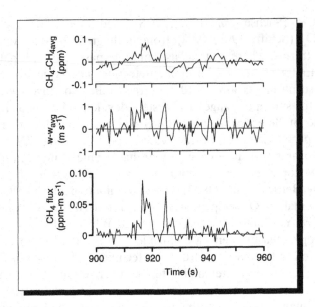

Fig. 8.7 Representative CH_4 flux data from a Massachusetts wetland.

to atmospheric line broadening and resultingly greater interferences from other absorbing species. A further disadvantage is that the instrument itself must be mounted on a tower or that the measurement height is limited to several metres above the optical platform. In the case of CH_4 sampling using near infrared semiconductor diodes (Hovde & Stanton, 1992) the instrument can be sufficiently compact that tower mounting is feasible.

Open-path absorption using long path retroreflective optics has been implemented using lead salt TDLs (Ku et al., 1975; Diehl & Wiesemann, 1987). The long path arrangement is useful for measuring spatially averaged concentrations over extended trace gas sources. A two-leg system for simultaneous measurements at different heights above the ground (see Fig. 8.3) has been developed and used to determine CH_4 gradients over a swamp (Zahniser et al., 1992) with a resolution of 1% of the ambient level in a 2 s integration time and a (total) path length of 240 m. A data set from this instrument in Fig. 8.8 shows the build-up of CH_4 gradients over a Massachusetts wetland before and after sunset. The system is also suitable for CO and N_2O measurements over path lengths up to 1 km. Applications to other gases including NO, CO_2, O_3 and H_2O are also possible.

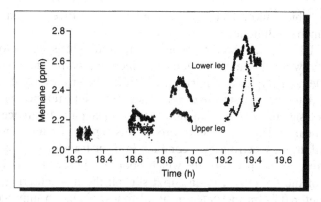

Fig. 8.8 CH$_4$ gradients measured over a Massachusetts wetland (25 October 1991). The gradient increases in response to the temperature inversion and decreased vertical mixing after sunset (18.4 h). Gaps in the data stream are where the optical system required realignment.

8.3 FTIR

8.3.1 Methods overview

FTIR spectrometers can be considered for atmospheric trace gas measurements because of substantial recent progress, both in the mechanical design of the spectrometer itself, and in the various computational aspects which are such a large part of this technique. While many infrared active trace gases can be detected by either TIRL/DA and FTIR techniques, the intrinsic spectral properties of the target species will usually dictate the most sensitive technique. In general, the TIRL/DA spectroscopy methods will be more sensitive than FTIR spectroscopy for trace gas species with relatively widely spaced vibrational/rotational lines, including diatomic species like CO, linear triatomics such as CO_2, N_2O, OCS and CS_2 and some polyatomics hydride species like H_2S, NH_3 and CH_4. FTIR spectroscopy will usually be the superior method for species whose vibrational/rotational lines merge into infrared band spectra at the sampling pressure of interest. Examples include DMS, most NMHCs, and most partially oxygenated and/or halogenated hydrocarbons.

Discussions of the details of FTIR spectrometry can be found in a number of books (Bell, 1972; Ferraro & Basile, 1979; Griffiths & deHaseth, 1986; Ferraro & Krishnan, 1990). Combining these texts with manufacturers' literature will give a clear picture of the capabilities and operating characteristics of currently available systems. Therefore, in this section. We will present only brief descriptions of the principles of the spectrometer and other system components and then describe the advances which allow FTIR to be used as a sensitive, automated field measurement technique.

In an FTIR spectrometer, incoming infrared light is split into the two paths of a two-beam interferometer in which one path contains a moving optical element (typically a mirror). When the beams are combined at an infrared detector, constructive and destructive interference produces a modulated signal which is a function of the optical path difference between the two beams. The digitized record of this signal, the interferogram, is converted into a spectrum by a complex Fourier transform. In contrast to a prism or grating spectrometer which operates by selecting only a small fraction of the light entering its input slit, all the light entering an FTIR spectrometer falls on the detector at all times, giving an improvement in instrument SNR. This is known as the multiplex advantage. FTIR instruments also have a throughput advantage over scanning monochromators, meaning that their light-gathering power is greater for the same spectral resolution.

The other components of a system are an infrared source, an optical system which defines the sampling region between source and spectrometer, and a computer which not only performs the data acquisition and spectrum generation functions mentioned above, but can be programmed to aid in the analysis of the spectra as well. Optical systems can be a single long path, a double long path (allowing source and receiver to be at the same point, with only a retroreflector positioned remotely), a compact multipass cell with an associated sampling system, or an open multipass cell with a variety of base path lengths. A schematic of a single open path FTIR system is shown in Fig. 8.9. An example of a spectrum taken over a long open path is given in Fig. 8.10. Additional system details can be found in several recently published review articles (Grant *et al.*, 1992; Schiff, 1992) and in the articles detailing the examples presented below.

8.3.2 Commercial FTIR instruments

Recognition of a market for industrial process monitoring devices, and the further stimulus of several military remote sensing projects, have resulted in commercially available spectrometers with resolution in the $0.5-1 \, \text{cm}^{-1}$ range which are relatively inexpensive, lightweight (22.5–67.5 kg), and rugged (to the point of being operable on helicopters). Most proof-of-principle experiments published so far have used somewhat larger laboratory spectrometers with resolutions in the $0.1-0.5 \, \text{cm}^{-1}$ range.

The spectral resolution required depends on the nature of the spectra of the molecules which are to be detected (and possibly also on characteristics of interfering molecules as well). How narrow the molecular absorption features are compared to the spectrometer resolution affects

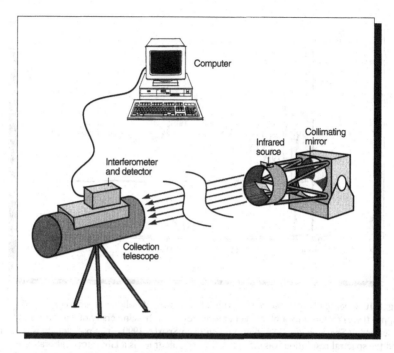

Fig. 8.9 Schematic of single open path FTIR system.

both the sensitivity and the accuracy of the observation. Individual pressure-broadened vibrational/rotational line widths are typically in the $0.05-0.1\,\text{cm}^{-1}$ range. An excellent source of line width, as well as line position, information is the HITRAN database (Rothman *et al.*, 1992); an earlier version of the compilation is also available in useful hardcopy plots (Park *et al.*, 1987). Many molecules, including most containing three or more atoms other than hydrogen, have line spacings of this order or smaller. However, clumps of lines can still generate spectral features with widths of a few cm^{-1}. Increasing spectrometer resolution substantially below the width of the narrowest strong features will not improve sensitivity. Computer spectral databases are becoming an increasingly useful method of assessing the shape and position of candidate molecular bands. An excellent place to start is the recently introduced NIST/EPA database (National Institute of Standards and Technology, 1992) which provides over 5000 spectra at $12\,\text{cm}^{-1}$ resolution. A number of commercial compilations offer more spectra and/or higher resolution (Warr, 1991).

For small molecules with widely separated lines, sensitivity is maximized when the spectrometer resolution width is small compared to the line spacing. However, still higher resolution can be useful in improving

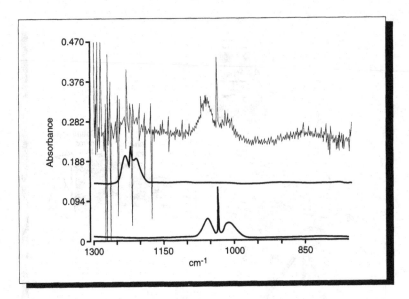

Fig. 8.10 Open-path absorbance spectrum of a mixture of volatile organic compounds (upper trace) released by a plume generating device at the University of Kansas (field tests described in Spartz *et al.*, 1989; figure taken from Spartz, 1990). The path length was 100 m, and the spectral resolution was $0.5\,cm^{-1}$. The middle trace is a laboratory calibration spectrum for acetone, while the bottom trace is the same for methanol. When the methanol model spectrum is subtracted from the field observation, a residual spectral feature can be seen due to ethanol and *n*-butanol.

accuracy. The optical depth or absorbance of a spectral feature, the natural logarithm of the ratio of transmitted light intensity without and with the absorption, is proportional to the absorbing species concentration as long as the observation is made with sufficiently high resolution that the shape of the feature is not distorted. However, the same quantity (an apparent optical depth) obtained when using a spectral resolution which is broader than the true width of the features observed is not proportional to species concentration. Corrections can be applied (and must be applied when subtracting out interfering lines as well), but higher resolution will lessen the magnitude of the corrections and the accompanying errors.

As significant as hardware developments have been, the development of mathematical methods of analysis of spectra with contributions from a number of absorbing species has not only made considerable progress but may be the area in which the greatest advances in detection sensitivity are made in the future. A number of least squares techniques have been developed, which are now becoming available in commercial software packages (Haaland, 1985; Beebe & Kowalski, 1987; McClure, 1987;

Hong-kui *et al.*, 1989; Li-shi & Levine, 1989; Saarinen & Kauppinen, 1991). A key is the proper weighting of spectral regions, so that those regions containing bands of interest control the fitting procedure. Although a number of investigations to date have used single spectral features for quantification, it is clear that both greater sensitivity and greater accuracy will result if many spectral points are fit simultaneously. Best results will be obtained if all standard spectra used in the analysis are obtained using the same instrument making the field measurements. A possible alternative approach in cases in which the compounds of concern are not only chemically but spectrally similar, is to use mathematical techniques to derive a few spectra which are representative of an entire class of spectra, and use these representer spectra to quantify summed concentrations of all members of the class (Frankel, 1984; Gruninger, 1985).

8.3.3 Multipass absorption cells for FTIR spectroscopy

Multipass cells can be divided into two types: (a) closed cells in which the beam paths are contained inside a tube or other housing and the sampled gas must be drawn through inlet and outlet ports; and (b) open cells which sample the path through whatever atmosphere happens to be between the mirrors which define an observation volume. The total path length is the product of a base path length and the number of passes. Clearly, closed cells with large base lengths will have drawbacks in portability and pumping requirements. On the other hand, increasing the number of passes will not result in indefinitely increasing sensitivity: mirror reflectivities are such that eventually transmitted intensity is decreasing faster than depth of absorption is increasing. Typically, this point is reached somewhere in the 40–100 pass range. Therefore, cells with 1.5 m base lengths, as unwieldy as could be tolerated for most applications, are commercially available from several sources and are specified as having total path lengths up to 100 m. A few larger, and many smaller closed-path cells are also sold by FTIR manufacturers and specialists in sampling accessories, and some even larger closed-path systems have been built by research groups.

The base length of an open-path cell is not similarly constrained, but its useful length is still limited by the effects of atmospheric turbulence. Open cells with 25 m base lengths, yielding total path lengths of 1–2 km, have been used successfully. Most cells used with FTIR spectrometers have been based on the three-mirror White design (White, 1942). Light is focused and refocused in two rows of spots across the front mirror, the refocusing being performed alternately by a pair of back mirrors, with the final or exit spot passing the front mirror at the opposite side from the input point. Some cells for particularly long paths include a corner cube reflector at the output point, yielding two more rows of spots across the front mirror (Horn & Pimentel, 1971), with the advantage that passes are better separated by being spread over two dimensions.

8.3.4 Long-path optics for FTIR spectroscopy

Long-path layouts can also be divided into two classes; (a) bistatic (source and receiver sit at opposite ends of the observation path); and (b) mono-static (source and receiver sit side by side). In a monostatic system, the source beam traverses the observation region twice, being directed back to the receiver by a retroreflector which also slightly displaces the beam. Once again, performance is often improved by using optical components whose size is somewhat larger than would be ideal in a field measurement system: both source and spectrometer are typically coupled into telescopes (of Cassegrainian or Newtonian design) with primary mirror diameters in the 25–50 cm range, while a displacing corner cube retroreflector at least 75 cm wide is appropriate (though this is simply an assemblage of plane mirrors). The path length above which further gains in sensitivity are not realized appears to be somewhere between 500 m and 1 km: this is to be expected given the limits discussed above for multipass cells and the possibility that a one or two pass system is even more sensitive to atmospheric turbulence than a multipass system with the same total path but a shorter base path. However, there are certainly situations in which the spatial averaging provided by a long-path system is desirable, just as there are others in which point sampling is most appropriate.

8.3.5 Examples of system performance

In order to estimate detection limits for a particular molecule, one needs to know its absorption strength, the observation path length, and the minimum detectable optical depth for the spectrometer and data analysis procedure to be used. The first quantity can be obtained from the spectral databases mentioned above, while the second is subject to the constraints discussed in the preceding two subsections. The purpose of this subsection is to review selected literature reports of FTIR field measurements to extract estimates of minimum detectable optical depths for a variety of systems, and to give the resulting minimum detectable column densities (expressed in ppmv·m^{-1}, the product of path length and number density expressed as mixing ratio) for a few example molecules.

We begin with NH_3, a strong absorber with fairly sharp spectral features in a spectral region relatively free from atmospheric interferences. Spellicy *et al.* (1991) made measurements over a 500 m open path using 0.5 cm^{-1} spectral resolution, averaging 150 spectra collected in 4 min. They reported a column density detection limit of 1.5 ppmv· m^{-1}. This can be compared with two measurements made using multipass cells, one closed path (Tuazon *et al.*, 1978) and one open path (Biermann *et al.*,

1988). The two cells were similar, the former having a 22.5 m base length and a 1080 m total path, the latter having 25 and 1150 m, respectively. The open-path measurement collected 64 scans in 4.5 min at a resolution of $0.125\,cm^{-1}$, while the closed-path spectra were formed from 40 scans at $0.5\,cm^{-1}$ resolution (if the same instrument had been used, acquiring a $0.125\,cm^{-1}$ scan would take four times as long as a $0.5\,cm^{-1}$ scan). In the open-path measurement the minimum detectable column density was quoted as $1-2\,ppmv\cdot m^{-1}$, while in the earlier closed-path work it was estimated as about $2\,ppmv\cdot m^{-1}$, both quite similar values to that of Spellicy *et al.* Both multipass cell references quoted a minimum detectable optical depth of about 5×10^{-3}. This is consistent with our analysis of several of the other column density detection limits quoted by Spellicy *et al.* as representing minimum detectable optical depths of from 10^{-3} to 10^{-2}, with the range resulting from the varying severity of atmospheric interferences.

All of this is also roughly consistent with the work of Plummer (1991), who presented results of extractive sampling of combustion gases into heated long-path cells with 4 and 5 m paths, along with a detailed listing of peak-to-peak noise levels in various spectral regions. For $1\,cm^{-1}$ resolution and a 6 min accumulation time, the noise level in the NH_3 region is reported to be 1.4×10^{-3} (in optical depth units), leading to a lower detection limit for NH_3 column density of $0.76\,ppmv\cdot m^{-1}$.

Grant *et al.* (1992) published an extensive table of detection limits based on absorption data from a variety of sources, including laser studies as well as FTIR work from two laboratory studies and the combined laboratory and field work of Spartz *et al.* (1989). They quote an underlying minimum detectable optical depth of 10^{-2} at $1000\,cm^{-1}$ (referencing the Spartz *et al.* group and a second group, and assuming a noise level three times smaller than this). However, they assume that this value changes with the spectral region. Comparing detection limits for compounds listed by both Spartz *et al.* and Grant *et al.* (adjusting for different averaging times by the square root of the ratio), we find that the Grant *et al.* limit is lower for only one case, methanol, while for most cases the Spartz *et al.* estimates imply more sensitive detection by an average of a factor of 2. This means minimum detectable optical depths in the 5×10^{-3} range for the 4 min averaging time assumed by Grant *et al.* Methanol is also the only compound shared between Spellicy *et al.* and Spartz *et al.* Again correcting for different averaging times, their minimum detectable column densities are 2.2 and $6.5\,ppmv\cdot m^{-1}$, compared to $2\,ppmv\cdot m^{-1}$ quoted by Grant *et al.* Furthermore, Spartz *et al.* published a long-path spectrum of methanol with an accompanying comment indicating that the minimum detectable optical depth in this spectral region is about 2×10^{-3}, consistent with our understanding of the detection limits of Spellicy *et al.*, and implying that the estimate of Grant *et al.* takes into account the characteristics of this relatively favourable spectral region.

Spartz *et al.* (1989) made extensive long-path observations of a number of compounds, using $0.1\,cm^{-1}$ resolution, collecting 256 scans in about 27 min, in a bistatic configuration with path lengths of 100 m and up. However, the comment quoted above is the closest connection they make between their field work and their table of detection limits, which they had based on laboratory spectra. Another group which has reported field measurements but has not quantified their detection limits is Russwurm *et al.* (1991). They used $0.5\,cm^{-1}$ resolution, collected 256 scans in 4 min and then co-added spectra up to 32 min total collection time, and used 250 and 500 m paths in a monostatic configuration. As evidence for their sensitivity levels, we can take their comment that comparisons between chlorobenzene levels measured by FTIR and canister methods show high variability because chlorobenzene levels are always very close to the detection limits of the FTIR. Those concentration levels are in the 50 to 100 ppbv range. If we assume a detection limit of 10 ppbv over the 500 m path, this implies a minimum detectable column density of $5\,ppmv\cdot m^{-1}$. By comparison, Spartz *et al.* and Grant *et al.* quote 3.1 (adjusted) and $16\,ppmv\cdot m^{-1}$, respectively.

Finally, we can make a comparison with another measurement system by considering a fourth molecule, methyl bromide. Green *et al.* (1991) state that their (as yet unoptimized) system should be able to detect 35 ppbv with satisfactory SNR. They took spectra at $0.25\,cm^{-1}$ resolution, accumulated 256 scans in 6 min, and used a 2.5 m base path length, 140 m total path open cell. Therefore their 35 ppbv translates into a $5\,ppmv\cdot m^{-1}$ column density detection limit. (Grant *et al.*, on the other hand, give a value of $128\,ppmv\cdot m^{-1}$.) Examination of the methyl bromide spectrum associated with the 35 ppbv concentration limit indicates a minimum detectable optical depth of about 3×10^{-3}.

We can draw the conclusion that present FTIR systems are doing well when they approach minimum detectable optical depths of 10^{-3}, and that this level is particularly difficult to achieve in regions with significant atmospheric interferences such as water or CO_2 lines. However, we also expect that further improvements in FTIR design, in thermal source characteristics, in source and collection optics, and especially in data analysis software, will bring minimum detectable optical depths down very close to 10^{-4}. This will result in FTIR detection limits for many trace species of interest in the $1.0–0.1\,ppmv\cdot m^{-1}$ range.

8.4 Electronic state spectroscopy techniques

8.4.1 Overview

The preceding TILDAS and FTIR techniques utilized the vibrational/rotational energy levels of various target gaseous molecules. In principle, it is also possible to use electronic state spectroscopic transitions in the visible (V_{15}) and ultraviolet portions of the electromagnetic spectrum. Several trace species measurement techniques based on electronic energy level spectroscopy have been developed and utilized in the atmosphere. Two techniques, differential optical absorption spectroscopy (DOAS) and laser-induced fluorescence (LIF) rely on the target molecule displaying a visible (700–400 nm), near ultraviolet (400–300 nm) or mid-ultraviolet (300–200 nm) electronic transition from its ground electronic state. Unfortunately, almost none of the biogenically produced gases of primary interest to this review meet this criterion, since they have their first strong electronic transition in the vacuum ultraviolet ($\lambda < 220$ nm) making sensitive spectroscopy in the presence of atmospheric O_2 and N_2 unfeasible.

DOAS is a long-path absorption technique, conceptually similar to the long-path infrared absorption techniques described above, with either a bright lamp or V_{15}/ultraviolet laser as the photon source. Typical DOAS target species with suitable visible or near ultraviolet absorptions include NO_3, NO_2, O_3, SO_2, HONO, CH_2O and OH (Kolb, 1991; Grant et al., 1992; Schiff, 1992).

LIF is a point source measurement technique. It not only requires a strong visible/ultraviolet transition from the target species' ground state but a bound upper state which can reradiate a photon, with or without a wavelength shift. Typical atmospheric species detected using LIF techniques include NO and the hydroxyl radical (OH) (Kolb, 1991; Schiff, 1992).

8.4.2 Photofragmentation/LIF (PF/LIF)

The addition of a second, high powered PF laser to a LIF detection technique will allow sensitive detection of some of the biogenic gases of interest to ecologists. The PF/LIF technique generally uses two or more ultraviolet photons from a high power 'pump' laser to photodissociate the target species of interest. LIF detection is then performed on a radical photofragment with a strong fluorescing transition accessible to a visible/ultraviolet 'probe' laser. This technique has been well developed for the NH_3 molecule by Bradshaw and co-workers at the Georgia Institute of Technology (Schendel et al., 1990).

In the Georgia Tech technique two 193 nm photons (with energy of $h\nu_2$) from an ArF excimer laser photodissociate NH_3 producing an electronically excited NH ($b^1\Sigma^+$) fragment:

$$NH_3 + 2h\nu_1 \xrightarrow{\lambda_1 = 193\,nm} NH(b^1\Sigma^+) + products.$$

A second LIF laser (a doubled neodymium-yttrium aluminium garnet pumped dye laser) probes the photodissociation products at 452 nm to produce NH in the $c^1\Pi$ state:

$$NH(b^1\Sigma^+, v'' = 0) + h\nu_2 \xrightarrow{\lambda_2 = 452\,nm} NH\ (c^1\Pi, v' = 0)$$

which in turn fluoresces at 325 nm to the NH $a\ ^1\Delta$ state:

$$NH(c^1\Pi, v' = 0) \rightarrow NH\ (a\ ^1\Delta, n'' = 0) + h\nu_3, \lambda_3 = 325\,nm.$$

While both optically and spectroscopically complex this method is also both extremely sensitive and specific to NH_3. As utilized by the Georgia group, the PF/LIF detection sensitivity for NH_3 is ~5 pptv for a 1 min integration time in relatively clean air (Schendel *et al.*, 1990). The technique was recently shown to perform very well over a wide range of conditions in an intercomparison study with four more traditional NH_3 measurement techniques (Williams *et al.*, 1992).

The PF/LIF technique has been utilized by the Georgia group and others to detect a number of NO containing trace species such as NO_2, HNO_2 and HNO_3 (Kolb, 1991). In terms of ecologically important biogenic gases beyond NH_3, the Georgia group has proposed PF/LIF schemes for CS_2 and H_2S (Schendel *et al.*, 1990) but has not yet demonstrated these schemes in the field.

8.5 Advanced mass spectrometry

8.5.1 Method overview

The methods discussed above are all based on optical probing of one or more transitions between energy states of the target species (or its photo-fragment). In contrast, mass spectrometry (or mass spectroscopy) ionizes the target species and separates the resulting fragment ions by mass using magnetic and/or electric fields. While mass spectrometry is a nearly universal detector, with moderately high sensitivity for almost all molecular gases, it can suffer from interferences since simple field instruments cannot always distinguish among species with similar molecular weights.

Traditionally, precision mass spectroscopy has required relatively high vacuum, which has severely restricted its use in field studies since sampling from atmospheric pressure to low operating pressures requires large vacuum pumps drawing significant levels of electric power. Furthermore, without high pumping speed, relatively reactive trace species such as NH_3 or trace sulphur species can be lost on sampling system walls leading to imprecise results. Recently, several research groups have addressed these problems and deployed novel mass spectrometric systems in both ground-based and airborne trace gas field studies. While such systems are still quite bulky and are restricted to sites and/or sampling platforms with relatively high power available, they do illustrate significant progress in adapting mass spectroscopy to field conditions.

8.5.2 Atmospheric pressure ionization/mass spectrometry

One approach to couple effectively atmospheric pressure sampling to mass spectrometric detection involves the use of a corona discharge to create target ions which are then pumped through an orifice into a tandem quadrupole mass spectrometer for analysis. Over a decade ago (Hijazi & DeBrou, 1982) demonstrated that this technique is useful for real-time detection of reduced sulphur gases produced in industrial settings. More recently a group at Battelle, Columbus has adopted this technique and demonstrated airborne measurements of species as diverse as DMS, HCOOH, HNO_2 and HNO_3 (Kelly & Kenny, 1991; Kenny et al., 1992). They have investigated several versions of the basic corona discharge ionization scheme, all under the general title of atmospheric pressure chemical ionization (APCI), including the monitoring of both positive and negative ions. Selectivity is achieved by ion pair monitoring, i.e. the detection of selected daughter ions by the third-in-line quadrupole. These ion pairs are created by collisions in the second quadrupole from a specific parent ion selected by the first quadrupole.

The Battelle group has adapted commercial tandem (triple) quadrupole trace atmospheric gas analyser (TAGA) instruments built by SCIEX. This type of instrument is described in more detail by Dawson et al. (1982a,b). Continuous detection limits for reduced sulphur gases at a 1 s response time have been measured to be 2 pptv for DMS, 1 ppbv for H_2S, and ~10 ppbv for OCS, CS_2 and CH_3SH (Kelly & Kenny, 1991). More sensitive detection levels could be achieved with sample preconcentration.

8.5.3 Isotopic dilution gas chromatography/mass spectrometry

The extremely low clean air levels of reduced sulphur gases and their partial oxidation products present a major analytical challenge. In addition to the TAGA/APCI mass spectrometry technique described above, a second advanced mass spectroscopy technique has been developed and demonstrated for monitoring atmospheric sulphur gases by Bandy and co-workers at Drexel University (Driedger et al., 1987; Thornton et al., 1990; Bandy et al., 1992). This technique is gas chromatography/mass spectrometry with isotopically labelled internal standards (ILS).

In this technique, the species specific internal losses arising from exposure to multiple surfaces in the gas chromatography portion of the analysis are calibrated for by injecting known levels of isotopically unique compounds corresponding to the target species being quantified. For instance, if DMS is the target compound, deuterated DMS is used as the internal standard. The mass spectrometer then uses the ratio of the unlabelled atmospheric DMS to the isotopically labelled DMS internal standard to quantify the atmospheric level accurately. Pumping requirements on the mass spectrometer are reduced since only the relevant chromatographic output peaks need to be analysed.

Bandy and co-workers have used this technique for airborne measurements of SO_2, DMS, CS_2, DMSO and $DMSO_2$. Using cryogenic preconcentration and 3 min integration times they have demonstrated detection limits of less than 1 pptv for these species (Bandy *et al.*, 1992).

8.6 Summary

We have reviewed four types of advanced spectroscopic measurement techniques for quantifying trace biogenic gases of ecological and climatological interest. While none of these techniques are capable of precisely measuring all trace gases of interest, they each have appealing characteristics which match the needs of specific field experiments quite well.

Recent advances in tunable diode and tunable rare gas infrared lasers, multipass optical cell technology and associated signal processing electronics make TILDAS techniques very attractive for monitoring infrared active biogenic diatomic (CO), triatomic (CO_2, N_2O, H_2S CS_2, OCS) and hydride polyatomic (NH_3, CH_4) species. In particular, the fast response capabilities demonstrated for this technique make eddy correlation flux measurements feasible for several of these species.

FTIR absorption spectroscopy is best suited for moderately long-path, atmospheric pressure monitoring of multiple species, given its wide intrinsic spectral range. While not as sensitive as laser differential absorption for small molecules with sharp line spectra, it may be more sensitive for larger molecules which display only relatively broad spectral features.

PF/LIF spectroscopy is a highly equipment intensive technology, but may be the only method capable of measuring clean air levels of species like NH_3, CS_2 and H_2S accurately and/or quickly enough for some types of micrometeorological flux measurements.

Finally, advanced mass spectrometric techniques such as TAGA/APCI and gas chromatography and mass spectrometry ILS have provided direct and reliable measurement of key reduced sulphur species at pptv levels, providing viable measurement options to traditional sample concentration/ gas chromatography technology.

Laboratory-based physical and analytical chemists are continuously developing new spectroscopic trace species measurement techniques which can be expected to be extended to field use as they mature. In addition, the technologies discussed in this chapter are undergoing continuous improvements in both their laboratory and field measurement guises. If an advanced spectroscopic measurement technique capable of meeting your needs has not yet been developed, just wait awhile: it may well be on the way.

References

Albritton D.L., Fehsenfeld F.C. & Tuck A.F. (1990) Instrumentation requirements for global atmospheric chemistry. *Science*, **250**, 75–81.

Anderson S.M. & Zahniser M.S. (1991) Open-path tunable diode laser absorption for eddy correlation flux measurements of atmospheric trace gases. *Proceedings of SPIE*, **1433**, 167–178.

Bandy A.R., Thornton D.C., Ridgeway R.G. Jr & Blomquist B.W. (1992) Determination of key sulfur containing compounds in the atmosphere and ocean by gas chromatography/ mass spectrometry and isotopically labeled internal standards. In: Kaye J.A. (ed) *Isotope Effects in Chemical Reactions and Photodissociation Processes*. American Chemical Society, Washington, DC.

Becker J.F., Sauke T.B. & Loewenstein M. (1992) Stable isotope analysis using tunable diode laser spectroscopy. *Applied Optics*, **31**, 1921–1927.

Beebe K.R. & Kowalski B.R. (1987) An introduction to multivariate calibration and analysis. *Analytical Chemistry*, **59**, 1007A–1017A.

Bell R.J. (1972) *Introductory Fourier Transform Spectroscopy*. Academic Press, New York.

Biermann H.W., Tuazon E.C., Winer A.M., Wallington T.J. & Pitts J.N. Jr (1988) Simultaneous absolute measurements of gaseous nitrogen species in urban ambient air by long pathlength infrared and ultraviolet-visible spectroscopy. *Atmospheric Environment*, **22**, 1545–1554.

Bomse D.S., Stanton A.C. & Silver J.S. (1992) Frequency modulation and wavelength modulation spectroscopies: comparison of experimental methods using a lead-salt diode laser. *Applied Optics*, **31**, 718–730.

Carlisle C.B., Cooper D.E. & Preier H. (1989) Quantum noise-limited FM spectroscopy with a lead salt diode laser. *Applied Optics*, **28**, 2507–2576.

Cooper D.E. & Carlisle C.B. (1988) High-sensitivity FM spectroscopy with a lead-salt diode laser. *Optics Letters*, **3**, 719–721.

Cooper D.E., Riris H., van der Laan J.E. (1991) Frequency modulation spectroscopy for chemical sensing of the environment. *Proceedings of SPIE*, **1433**, 120–127.

Dawson P.H., French J.B., Buckley J.A., Douglas D.J. & Simmons D. (1982a) The use of triple quadrupoles for sequential mass spectrometry 1: the instrument parameters. *Journal of Organic Mass Spectrometry*, **17**, 205–211.

Dawson P.H., French J.B., Buckley J.A., Douglas D.J. & Simmons D. (1982b) The use of triple quadrupoles for sequential mass spectrometry 2: a detailed case study. *Journal of Organic Mass Spectrometry*, **17**, 212–219.

Diehl W. & Wiesemann W. (1987) A transportable laser system for remote sensing of road traffic emission. In: Grisar R., Preier H., Schmidtke G. & Restelli G. (eds) *Monitoring of Gaseous Pollutants by Tunable Diode Lasers*, pp. 29–38. Reidel, Dordrecht.

Driedger A.R. III, Thornton D.C., Lalevic M. & Bandy A.R. (1987) Determination of ppt levels of atmospheric sulfur dioxide by isotope dilution gas chromatography/mass spectrometry. *Analytical Chemistry*, **59**, 1196–1200.

Fan S.M., Wofsy S.C., Bakwin P.S. *et al.* (1992) Micrometeorological measurements of CH_4 and CO_2 exchange between the atmosphere and subarctic tundra. *Journal of Geophysical Research*, **97**, 16627–16643.

Ferraro J.R. & Basile L.J. (eds) (1979) *Fourier Transform Infrared Spectroscopy: Applications to Chemical Systems*, vol. 2. Academic Press, New York.

Ferraro J.R. & Krishnan K. (eds) (1990) *Practical Fourier Transform Infrared Spectroscopy: Industrial and Laboratory Chemical Analysis*. Academic Press, San Diego.

Frankel D.S. (1984) Pattern recognition of Fourier transform infrared spectra of organic compounds. *Analytical Chemistry*, **56**, 1011–1014.

Fried A., Drummond J.R., Henry B. & Fox J. (1991) Versatile integrated tunable diode laser system for high precision: application for ambient measurements of OCS. *Applied Optics*, **30**, 1916–1932.

Grant W.B., Kagann R.H. & McClenny W.A. (1992) Optical remote measurement of toxic gases. *Journal of the Air and Waste Management Association*, **42**, 18–30.

Green M., Seiber J.N. & Biermann H.W. (1991) *In situ* measurement of methyl bromide in indoor air using long path FTIR spectroscopy. *Proceedings of SPIE*, **1433**, 270–274.

Griffiths P.R. & deHaseth J.A. (1986) *Fourier Transform Infrared Spectrometry*. Wiley, New York.

Gruninger J. (1985) Generalized singular value decomposition and discriminant analysis. *Optical Engineering*, **24**, 991–995.

Haaland D.M. (1985) Theoretical comparison of classical (K-matrix) and inverse (P-matrix) least-squares methods for quantitative infrared spectroscopy. In: Cameron D.G. & Grasselli J.G. (eds) *Fourier and Computerized Infrared Spectroscopy*, pp. 241–242. Proceedings of SPIE 553.

Herriott D.R. & Schulte H.J. Jr. (1965) Folded optical delay lines. *Applied Optics*, **4**, 883–889.

Herriott D.R., Kogelnik H. & Kompfer R. (1964) Off-axis paths in spherical mirror interferometers. *Applied Optics*, **3**, 523–526.

Hijazi N.H. & Debrou G.B. (1982) Speciation of reduced sulfur compounds by the use of atmospheric pressure ionization/mass spectrometry (API/MS). Transactions of the Technical Section. *Journal of Pulp and Paper Science*, **8**, 100–103.

Hong-kui X., Levine S.P. & D'Arcy J.B. (1989) Iterative least-squares fit procedures for the identification of organic vapor mixtures by Fourier transform infrared spectrophotometry. *Analytical Chemistry*, **61**, 2708–2714.

Horn D. & Pimentel G.C. (1971) 2.5 km low-temperature multiple-reflection cell. *Applied Optics*, **10**, 1892–1898.

Hovde D.C. & Stanton A.C. (1992) Fast response instrumentation for methane flux measurements; an open path near infrared diode laser sensor. *Journal of Atmospheric Chemistry*.

Jennings D.E. (1980) Absolute line strengths in the v_4 band of CH_4; a dual beam diode laser spectrometer with sweep integration. *Applied Optics*, **19**, 2695–2700.

Kebabian P.L. (1994) Off-axis cavity absorption cell with improved tolerance of fabrication errors, US patent. Number 5291265.

Kebabian P.L. & Kolb C.E. (1993) The neutral gas laser: a tool for remote sensing of chemical species by infrared absorption. In: Winegar E.D. & Keith L.H. (eds) *Sampling and Analysis of Airborne Pollutants*, pp. 257–273. Lewis Publishers, Boca Raton, Florida.

Kelly F.J. & Kenny D.V. (1991) Continuous determination of dimethylsulphide at part-per-trillion concentrations in air by atmospheric pressure chemical ionization mass spectrometry. *Atmospheric Environment*, **25A**, 2155–2160.

Kenny D.V., Spicer C.W., Sverdrup G.M., Busness K. & Hannigan R. (1992) *Airborne tandem mass spectrometry for investigation of regional and global environmental issues*. Presented at the 1992 Annual Meetings of the American Waste Management Association.

Kolb C.E. (1991) Instrumentation for Chemical Species Measurements in the Troposphere and Stratosphere. *Reviews of Geophysics Supplement*, 25–36.

Ku R.T., Hinkley E.D. & Sample J.O. (1975) Long-path monitoring of atmospheric carbon monoxide with a tunable diode laser system. *Applied Optics*, **14**, 854–861.

Lenschow D.H. & Hicks B.B. (eds) (1989) *Global Tropospheric Chemistry – Chemical*

Fluxes in the Global Atmosphere. National Center for Atmospheric Research, Boulder, Colorado.

Li-shi Y. & Levine S.P. (1989) Fourier transform infrared least-squares methods for the quantitative analysis of multicomponent mixtures of airborne vapors of industrial hygiene concern. *Analytical Chemistry*, **61**, 677–683.

Loewenstein M. (1988) Diode laser harmonic spectroscopy applied to *in situ* measurements of atmospheric trace molecules. *Journal of Quantitative Spectroscopy and Radiative Transfer*, **40**, 249–256.

May R.D. (1992) Correlation-based technique for automated tunable diode laser scan stabilization. *Review of Scientific Instruments*, **63**, 2922–2926.

McClure G. (ed) (1987) *Computerized Quantitative Infrared Analysis.* ASTM Special Technical Publication No. 934, American Society for Testing Materials, Philadelphia.

McManus J.B. & Kebabian P.L. (1990) Narrow optical interference fringes for certain setup conditions in multipass absorption cells of the Herriott type. *Applied Optics*, **29**, 898–900.

McManus J.B., Kebabian P.L. & Kolb C.E. (1989) Atmospheric methane measurement instrument using a zeeman-split He-Ne laser. *Applied Optics*, **28**, 5016–5023.

McManus J.B., Kebabian P.L. & Kolb C.E. (1991a) *Aerodyne Research Mobile Infrared Methane Monitor, in Measurement of Atmospheric Gases.* Proceedings of SPIE 1433, 330–339.

McManus J.B., Kebabian J.B. & Zahniser M.S. (1994) Astigmatic mirror multiple pass absorption cells for long pathlength spectroscopy. *Applied Optics*, (submitted).

McManus J.B., Kolb C.E. & Crill P.M. *et al.* (1991b) Measuring urban fluxes of methane. *World Resource Review*, **3**, 162–182.

National Institute of Standards and Technology (1992) *NIST/EPA Gas Phase Infrared Database, Standard Reference Data.* NIST, Bldg. 221/Room A320, Gaithersburg, Maryland 20899.

Park J.H., Rothman L.S., Rinsland C.P., Pickett H.M., Richardson D.J. & Namkung J.S. (1987) *Atlas of Absorption Lines from 0 to 17900 cm^{-1}.* NASA Reference Publication No. 1188, National Aeronautics and Space Administration. NTIS, N87-28955.

Plummer G.M. (1991) Field and laboratory studies of Fourier transform infrared (FTIR) spectroscopy in continuous emissions monitoring applications. In: Santoleri J.J. (ed) *Environmental Sensing and Combustion Diagnostics*, pp. 78–89. Proceedings of SPIE, 1434.

Rothman L.S., Gamache R.R. & Tipping R.H. *et al.* (1992) The HITRAN Molecular Database: Editions of 1991 and 1992. *Journal of Quantitative Spectroscopy and Radiative Transfer*, **48**, 469–507.

Russwurm G.M., Kagann R.H., Simpson O.A., McClenny W.A. & Herget W.F. (1991) Long-path FTIR measurements of volatile organic compounds in an industrial setting. *Journal of the Air and Waste Management Association*, **41**, 1062–1066.

Saarinen P. & Kauppinen J. (1991) Multicomponent analysis of FT-IR spectra. *Applied Spectroscopy*, **45**, 953–963.

Sachse G.W., Hill G.F., Wade L.O. & Perry M.G. (1987) Fast-response, high-precision carbon monoxide sensor using a tunable diode laser absorption technique. *Journal of Geophysical Research*, **92**, 2071–2081.

Schendel J.S., Stickel R.E., van Dijk C.A., Sandholm C.T., Davis D.D. & Bradshaw J.D. (1990) Atmospheric ammonia measurement using a VUV/photofragmentation laser-induced fluorescence technique. *Applied Optics*, **29**, 4924–4937.

Schiff H.I. (ed) (1991) *Measurement of Atmospheric Gases*, Proceedings of SPIE 1433, 94–210.

Schiff H.I. (1992) Ground based measurements of atmospheric gases by spectroscopic methods. *Berichte der Bunsengesellschaft*, **96**, 296–306.

Spartz M.L. (1990) Development of a mobile spectrometer laboratory for on-site atmospheric measurement of volatile organic compounds (VOCs) using Fourier transform infrared spectrometry, Thesis, Department of Chemistry, Kansas State University.

Spartz M.L., Witkowski M.R., Fateley J.H. *et al.* (1989) Evaluation of a mobile FT-IR system for rapid VOC determination, Part I: Preliminary qualitative and quantitative calibration results. *American Environmental Laboratory*, **11**, 15–30.

Spellicy R.L., Crow W.L., Draves J.A., Buchholtz W.F. & Herget W.F. (1991) Spectroscopic remote sensing: addressing requirements of the Clean Air Act. *Spectroscopy*, **6**, 24–34.

Thornton D.C., Bandy A.R., Ridgeway R.G., Driedger A.R. III & Lalevic M. (1990) Determination of part-per-trillion levels of atmospheric dimethyl sulphide by isotopic dilution gas chromatography/mass spectrometry. *Journal of Atmospheric Chemistry*, **11**, 299–308.

Tuazon E.C., Graham R.A., Winer A.M., Easton R.R., Pitts J.N. Jr & Hanst P.L. (1978) A kilometer pathlength Fourier-transform infrared system for the study of trace pollutants in ambient and synthetic atmospheres. *Atmospheric Environment*, **12**, 865–875.

Uehara K. & Tai H. (1992) Remote detection of methane with a 1.66-μm diode laser. *Applied Optics*, **31**, 809–814.

Wall D.L. (1991) Advances in tunable diode laser technology for atmospheric monitoring applications. In: Schiff H. (ed) *Measurement of Atmospheric Gases*, pp. 94–103. Proceedings of SPIE 1433.

Warr W.A. (1991) Spectral databases. *Chemometrics and Intelligent Laboratory Systems*, **10**, 279–292.

Werle P. & Slemr F. (1991) Signal-to-noise ratio analysis in laser absorption spectrometers using optical multipass cells. *Applied Optics*, **30**, 430–434.

Werle P., Slemr F., Gehrtz M. & Bräuchle C. (1989) Quantum-limited FM-spectroscopy with a lead-salt diode laser. *Applied Physics B*, **49**, 99–108.

Whalen M. & Yoshinari T. (1985) Oxygen isotope ratios in N_2O from different environments. *Nature*, **313**, 780–782.

White J.U. (1942) Long optical paths of large aperture. *Journal of the Optical Society of America*, **32**, 285–288.

Wieman C.E. & Hollberg L. (1991) Using diode lasers for atomic physics. *Review of Scientific Instruments*, **62**, 1–20.

Williams E.J., Sandholm S.T., Bradshaw J.D. *et al.* (1992) An intercomparison of five ammonia measurement techniques. *Journal of Geophysical Research*, **97**, 11591–11611.

Worsnop D.W., Nelson D.D. & Zahniser M.S. (1992) Chemical kinetic studies of atmospheric reactions using tunable diode laser spectroscopy. In: Schiff H. & Platt V. (eds) *Optical Methods in Atmospheric Chemistry*, pp. 18–33. Proceedings of SPIE, **1715**.

Yoshinari T. & Whalen M. (1985) Oxygen isotope ratios in N_2O from nitrification at a wastewater treatment facility. *Nature*, **317**, 349–350.

Zahniser M.S., McManus J.B., Nelson D.D. & Kebabian P. (1992) *A Tunable Diode Laser-based Gradient Sensor for Atmospheric Trace Gas Surface Flux Measurements*. Aerodyne Technical Report ARI-RR-953.

CHAPTER 9

Use of isotopes and tracers in the study of emission and consumption of trace gases in terrestrial environments

S.E. TRUMBORE

9.1 Introduction

This chapter is primarily devoted to techniques using direct measurement of stable and radioactive isotopes of carbon, nitrogen, oxygen and hydrogen in trace gases (here CO_2, CH_4, CO and oxides of nitrogen), and in substrates such as plants, soils and sediments. Isotopic tracers take advantage of the fact that isotopes of an element, while having the same general chemical properties, will undergo mass-dependent fractionation during physical, biological or chemical transformations. Thus isotopic data may be used to infer the dominant mechanisms controlling the production/consumption and emission/deposition of trace gases in terrestrial ecosytems. The global importance of a particular process or ecosystem in a trace gas budget may be deduced by comparing its isotopic 'signature' to the average tropospheric isotopic composition. Temporal records of tropospheric trace gas isotopic abundance recorded in plant cellulose, lake sediments, corals or determined from archived air samples, may be used to deduce changes in cycling of carbon, nitrogen, oxygen and hydrogen in the past, and are key to interpreting present observations of isotopic abundance.

The final part of this chapter will focus on the use of non-isotopic tracers, both deliberate and naturally occurring, to quantify the role of gas exchange between the soil atmosphere and overlying air in determining trace gas fluxes. A better understanding of the physical controls of trace gas exchange may be used to ascertain the causes of observed temporal and spatial variability in trace gas emission/deposition within an ecosystem. Such an understanding will help us to predict future climate- or land-use change effects on trace gas fluxes from ecosystems.

Any review is necessarily incomplete. The use of isotopically enriched tracers in studies of biogeochemical processes are discussed elsewhere (see Chapter 10). Sulphur isotopes will not be discussed here. The reader is referred to several excellent recent books which cover the use of isotopes in ecology, including Rundel *et al.* (1989), Coleman & Fry (1991) and Schimel (1993). More general discussions of the fundamentals of isotope applications may be found in Rankama (1954), Craig *et al.* (1964), Hoefs (1980), Galimov (1985) and Faure (1986).

9.2 Measurement of isotopes of carbon, nitrogen, oxygen and hydrogen in trace gases and substrates

9.2.1 Isotopic abundances

The isotopic abundances of the elements which make up the most commonly measured trace gases (CO_2, CH_4, CO, N_2O, NO, NH_3, H_2O) are summarized in Table 9.1. Stable isotopic abundances are commonly expressed as the ratio of the rare (heavy) isotope to the abundant (light) one (for example, $^{13}C/^{12}C$). As the difference in isotopic ratios between two samples may be measured more precisely than the absolute ratios, stable isotope data are generally reported as the deviation in the ratio of rare to common isotope in the sample (R_x) from that of a commonly accepted standard (R_{std}):

$$\delta_x = \left[\frac{R_x - R_{std}}{R_{std}}\right]1000. \tag{9.1}$$

When the δ value is negative (i.e. the ratio of the rare (heavy) to common isotope is less than that of the standard) it is often referred to as 'depleted' (in the heavy isotope) or isotopically 'light'. In contrast, a reservoir with a positive delta value may be referred to as 'enriched' or 'heavy'. These terms are also commonly employed when comparing the isotopic compositions of two reservoirs. For example, plant tissues produced during photosynthesis (with $\delta^{13}C$ of roughly $-25‰$) are referred to as depleted, or lighter, or more negative than the precursor atmospheric CO_2 (roughly $-8‰$).

Table 9.1 identifies the standards commonly used in reporting stable isotope data for hydrogen, carbon and oxygen. Many of these original standards have been used up and are no longer available. In practice, each stable isotope laboratory maintains a set of working standards which have been cross-calibrated with standards used in other laboratories, and eventually are relatable to the primary reporting standard.

The two radioisotopes identified in Table 9.1, tritium (^3H) and radiocarbon (^{14}C), are both far less abundant than the stable isotopes. Reporting of these data are either in the form of activity (decays min^{-1} unit $mass^{-1}$ of substance measured), or as deviation in isotopic ratio from that of a standard, similar to the stable isotope expression. The standard differs from those for stable isotopes in that it must be fixed in both isotopic content and time; for ^{14}C the ultimate standard is 1895 wood, now defined as 95% of the activity of an oxalic acid standard in 1950 (for a review of radiocarbon methods see Mook, 1980).

Table 9.1 Isotopic abundance and half-lives of isotopes discussed in this chapter. The units in which isotopic abundances are expressed are also given for each isotope. For information on stable isotope standards, see Craig (1961) for standard mean ocean water (SMOW); Craig (1957) for Pee Dee Belemnite (carbonate) (PDB), and Mariotti (1984) for N_2

Element	Mass	Abundance (%) (half-life)	Units of expression
H (hydrogen)	1	99.985	
D (deuterium)	2	0.0148	$\delta D = \left(\dfrac{\frac{D}{H}_{sam}}{\frac{D}{H}_{std}} - 1 \right) * 1000;\ std = SMOW$
T (tritium)	3	10^{-16} (12.33 y)	TU (tritium unit): 1 TU = 1 atom T per 10^{18} atoms H, or 7.1 dpm l^{-1} H_2O
C (carbon)	12	98.89	
	13	1.11	$\delta^{13}C = \left(\dfrac{\frac{^{13}C}{^{12}C}_{sam}}{\frac{^{13}C}{^{12}C}_{std}} - 1 \right) * 1000;$ std = PDB (carbonate)
	14	$<10^{-10}$ (5730 y)	$fM = \dfrac{\frac{^{14}C}{^{12}C}_{sam}}{\frac{^{14}C}{^{12}C}_{std}}{}^{*};\ \Delta^{14}C = (fM - 1)^{*}\ 1000$ std = activity of oxalic acid in 1950 13.56 dpm gC^{-1} * corrected to $\delta^{13}C$ of $-25‰$
N (nitrogen)	14	99.63	
	15	0.366	$\delta^{15}N = \left(\dfrac{\frac{^{15}N}{^{14}N}_{sam}}{\frac{^{15}N}{^{14}N}_{std}} - 1 \right) * 1000$ std = atmospheric N_2
O (oxygen)	16	99.76	
	17	0.038	
	18	0.204	$\delta^{18}O = \left(\dfrac{\frac{^{18}O}{^{16}O}_{sam}}{\frac{^{18}O}{^{16}O}_{std}} - 1 \right) * 1000;\ std = SMOW$

9.2.2 Fractionation of isotopes: equilibrium and kinetic

Variations in isotopic abundance are caused in two ways. First, the strength of chemical bonds formed by different isotopes will vary according to their mass. Thus, at chemical equilibrium, two coexisting phases or compounds will contain different abundances of an isotope. This is referred to as 'equilibrium' fractionation. An example is the observed fractionation of ^{13}C and ^{18}O between calcium carbonate and the solution of H_2O and CO_3^{2-} from which it is precipitated. The degree to which heavier isotopes are concentrated in the solid phase is expressed using a fractionation factor, α, equal to the ratio of $^{13}C/^{12}C$ or $^{18}O/^{16}O$ in $CaCO_3$ divided by that of the dissolved carbonate or H_2O. The fractionation factor is dependent on temperature and (in the case of $CaCO_3$ and seawater) on the ionic strength of the solution (see discussions in Rankama, 1954; Craig et al., 1964; Faure, 1986). The temperature effect has been used in measurements of calcium carbonate shells precipitated by marine organisms to infer palaeotemperatures (Craig, 1965; Emiliani, 1966).

Isotopes react and diffuse at different rates, leading to 'kinetic' fractionation of isotopes between reactants and products in irreversible reactions, or along diffusion gradients. Both physical (e.g. diffusion) and biological (e.g. microbial metabolism and photosynthesis) processes cause kinetic isotope fractionation. An example of the influence of fractionation during diffusion may be found in comparing the $\delta^{13}C$ of CO_2 in soil atmospheres to the $\delta^{13}C$ of CO_2 produced in the soil. Because $^{12}CO_2$ diffuses out of the soil faster than $^{13}CO_2$, the soil atmosphere CO_2 is enriched in the heavy isotope by up to 4.4‰ over the isotopic composition of CO_2 produced (Cerling, 1991; Hesterberg & Siegenthaler, 1991; Aravena et al., 1992).

The net fractionation by a chemical or biological kinetic process depends on the fraction of the initial chemical reservoir consumed during the reaction. For example, if CH_4 oxidizing bacteria consume 100% of the CH_4 present in a jar experiment, the CO_2 produced from CH_4 oxidation will have a $\delta^{13}C$ value identical to that of the original CH_4. If the CH_4 in the jar is continuously replaced, such that only a fraction of the CH_4 passing through the system ever gets oxidized, the resulting CO_2 will have a ^{13}C value much more negative (i.e. depleted in the heavy isotope) than the substrate CH_4.

The distribution of a radioisotope within a reservoir is determined both by mass-dependent fractionation, which affects the initial isotope ratio, and by subsequent radioactive decay. In order to isolate the effects of radioactive decay, measurements of radioisotopes are often corrected for isotope fractionation effects when they are reported. For example, because the mass difference between ^{14}C and ^{12}C is twice that between

[13]C and ^{12}C, ^{14}C is assumed be depleted or enriched twice as much as [13]C (for more see Craig, 1954, 1961). [14]C data are thus reported not only with respect to the ^{14}C/^{12}C value of oxalic acid, but are corrected to a common value of $\delta^{13}C$ (−25‰) (Stuiver & Polach, 1977).

9.2.3 Production and radioactive decay of ^{14}C, [3]H

Radiocarbon (^{14}C) and tritium (^{3}H) are cosmogenic isotopes, produced by collisions involving neutrons resulting from the interaction of cosmic rays with gases in the Earth's outer atmosphere. [14]C is formed primarily by the [14]N(n, p)[14]C reaction, and ^{3}H by $^{14}N + n \Rightarrow {}^{12}C + {}^{3}H$. [14]C decays back to ^{14}N with a half-life of 5730 y, while ^{3}H decays to [3]He with a half-life of 12.3 y. Most of the production of cosmogenic isotopes takes place in the stratosphere (Lal & Peters, 1967; Lingenfelter, 1963). The subsequent global distributions of ^{14}C and ^{3}H reflect the different chemistries of carbon and hydrogen. [14]C is oxidized to [14]CO in the order of hours, then more slowly (order of months) to $^{14}CO_2$ (Lingenfelter, 1963). The distribution of [14]CO, which has an atmospheric lifetime shorter than the time needed for the atmosphere to mix completely, reflects atmospheric exchange and removal processes. In contrast, $^{14}CO_2$ has an atmospheric residence time of years, and its distribution reflects the exchanges of carbon between atmosphere, biosphere, ocean and sediment reservoirs. If the production rate of ^{14}C is constant (about 2 atom $cm^{-2}s^{-1}$; Lal & Peters, 1967), the steady-state ^{14}C burden in each of these reservoirs will reflect the balance between exchange with atmospheric CO_2 and loss by radioactive decay. At present, roughly 90% of the total ^{14}C burden is in the oceans, 8% in the biosphere and soils, and 2% in the atmosphere (Broecker & Peng, 1982). Isolation of a carbon reservoir from exchange with atmospheric $^{14}CO_2$ results in a decrease in the ^{14}C/^{12}C ratio in the reservoir over time, as ^{14}C decays. In closed, homogeneous systems, such as unaltered plant matter, the ^{14}C/^{12}C value is a measure of the time since death of the organism. In heterogeneous reservoirs which are continuously exchanging carbon with the atmosphere, such as soil or sedimentary organic matter, the ^{14}C content will reflect the mixture of materials with both very long and very short residence times. In these heterogeneous systems, it is particularly important to distinguish between the average age of a carbon atom in the reservoir (given by the decrease in ^{14}C/^{12}C in the reservoir below that of fresh carbon inputs), and the residence time of a carbon atom in the reservoir (given by the reservoir size divided by the input flux). For reservoirs such as soil organic matter, in which a majority of the carbon added annually is rapidly decomposed, but a small portion accumulates slowly over long time scales, the average age will be greater than the average residence time for carbon (Balesdent et al., 1987).

Just as the distribution of ^{14}C reflects the carbon cycle, ^{3}H is controlled by the water cycle. Tritium is oxidized in the atmosphere to water vapour, then rapidly removed in rainfall. Evaporation recycles some ^3H back into water vapour. Tritium has not yet been extensively used in the study of trace gases, but is a useful tracer of global hydrological process (Broecker & Peng, 1982; Zaucker & Broecker, 1992).

The interpretation of $^{14}C/^{12}C$ and tritium/hydrogen values depends in part on the assumption that the production and global distribution of ^{14}C and tritium have remained constant through time. In fact, variations in $^{14}C/^{12}C$ of the order of 10% in the atmosphere are seen over the past 10 000 y, recorded in the $^{14}C/^{12}C$ content of known-age tree ring cellulose (Lerman et al., 1970; Stuiver & Quay, 1981). These changes have been linked to both variations in solar activity (and therefore production rate; Beer et al., 1988) and to changes in the distribution of radiocarbon among ocean, atmosphere and biosphere reservoirs (Stuiver et al., 1991).

The natural abundance of ^{14}C and ^3H was perturbed by atmospheric testing of thermonuclear weapons, which peaked before ending with the implementation of the Nuclear Test Ban Treaty in 1963. Figure 9.1 shows the time history of ^{14}C in the atmosphere measured in the last 30 years (data from Burchuladze et al., 1989; Levin et al., 1989; Manning et al., 1989). The bomb production of ^{14}C during 1961–1963 approximately doubled the atmospheric burden of ^{14}C (to about $\Delta^{14}C = +900‰$ in the Northern Hemisphere), while the ^3H content of precipitation increased from about 25 tritium unit (TU) before 1960 (see Table 9.1 for definition of TU) to a maximum of 2200 TU in the Northern Hemisphere in 1963

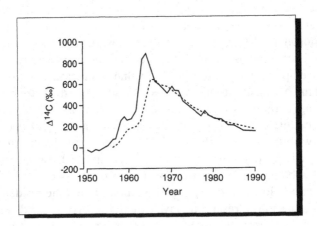

Fig. 9.1 Time history of ^{14}C in atmospheric CO_2 based on measurements made in the Northern (Burchuladze et al., 1989) and Southern (Manning et al., 1989) Hemispheres. ^{14}C is given in the Δ notation (see Table 9.1); a $\Delta^{14}C$ of +1000‰ indicates a doubling of the atmospheric ^{14}C burden. The peak in ^{14}C is lower (about 600‰) in the Southern Hemisphere (Manning et al., 1989). (—— North; --- south).

(Gat, 1980). While the increased production of these isotopes caused by atmospheric weapons testing has complicated their use as measures of age or residence time in contemporary samples, it has provided a global isotopic tracer experiment which has been particularly useful for following the exchange of carbon between atmospheric, oceanic and terrestrial reservoirs on annual to decadal timescales (Broecker & Walton, 1959; Broecker & Peng, 1982).

9.3 Methodology — isotopic measurements

Table 9.2 summarizes the methods used to isolate and quantify the isotopic contents of several trace gases. Methods for measurement of isotope abundances in trace gases require careful consideration of all

Table 9.2 Methods of measurement and minimum sample size requirements for isotopes in commonly measured trace gases. The measurement techniques are identified as: MS, stable isotope mass spectrometry; AMS, accelerator mass spectrometry; TDL, tunable diode laser. The required size to make a measurement from tropospheric air is calculated in the far right-hand column. For references related to methodology, see text

Isotope	Measured as	Method	Minimum sample	Precision	Species/ mixing ratio	Volume required for measurement
^{13}C	CO_2	MS	5 µmol	0.02–0.05‰	CO_2 (355 ppmv) CH_4 (1.7 ppmv) CO (120 ppbv) substrate (10% C)	0.3 l 651 (0.065 m³) 9001 (0.9 m³) 0.6 mg
	CO_2 in air	TDL	?	4‰	Measured *in situ*	
^{14}C	Graphite	AMS	40 µmol* (500 µg C)	0.5–1% Modern	CO_2 (370 ppmv) CH_4 (1.7 ppmv) CO (120 ppbv) substrate (10% C)	2.4 l 5201 (0.52 m³) 72001 (7.2 m³) 4.8 mg
	CO_2 or benzene	Decay counting	0.25 mol (3 g C)	0.2–1% Modern	6000–10 000 times values for AMS above	
^{18}O	CO_2	MS	5 µmol	0.02–0.05‰	Same as for ^{13}C	
	N_2O	TDL	0.27 mmol	1‰	N_2O (0.35 ppmv 3001 (0.3 m³)	
D	H_2	MS	0.45 µmol H_2O	1‰	H_2O CH_4	8 µl 0.651
^{15}N	N_2	MS	6 µmol 0.6–1 mg N	0.1‰ 1%	N_2O (0.35 ppmv) NO_3^{2-} (dissolved) NO_3^{2-} (particulate) NO_x in combustion exhaust only (ppm levels)	0.4 l depends on concentration

* For the AMS ^{14}C measurements, samples up to 20 times smaller are measureable, but uncertainties due to contamination during sample processing increase dramatically. For example, Brenninkmeijer *et al.* (1991) measured the ^{14}C content CO using only 10001 of air by diluting the sample with ^{14}C-free CO_2.

phases of sample collection and treatment. Typically four steps are involved:

1 collection of a sufficient quantity of air;

2 isolation and purification of the trace gas of interest from the bulk sample;

3 conversion of the trace gas into a measurable form; and

4 measurement.

All of these processes have the potential to fractionate isotopes, and therefore must be tested carefully to ensure the accuracy of the final result.

Many of the isotopes are measured using mass spectrometry. This technique relies on the principle that charged particles accelerated through an electric field will be deflected by a magnetic field. For two isotopes with the same charge state, the isotope with heavier mass will be deflected to a lesser degree than the lighter isotope. Detectors are arranged downstream of the magnet to collect the ions associated with a given mass. For stable isotope measurements, the inlet system to the mass spectrometer switches rapidly back and forth between the standard gas and the unknown. The δ value is calculated as the difference in integrated current collected in the detectors between sample and standard gases (McKinney *et al.*, 1950).

Radioactive isotopes have in the past been measured primarily by measuring the activity of the sample (decays min^{-1}). For both radiocarbon and tritium, however, the half-life of the isotope is long enough that often prohibitively large quantities of sample are needed to obtain sufficient activity for detection by decay counting. More sensitive methods have been developed to measure both isotopes. Today, ^{14}C is commonly measured by accelerator mass spectrometry (AMS; see reviews by Elmore & Phillips, 1987; Finkel & Suter, 1993). This technique accelerates carbon ions sputtered from a graphite target to high energies prior to introducing it into a magnetic field. Because of the high energies (millions of electronvolts, as opposed to thousands for standard stable isotope mass spectrometry), individual atoms of the rare isotope may be detected. The advantages of AMS include the more than 10 000-fold reduction in samples size required for measurement (see Table 9.2), and the more rapid throughput of samples, rather than improvements in the accuracy or precision of AMS measurements over decay counting methods. Low levels of tritium are more easily measured by allowing tritium to decay to ^{3}He, then determining the $^{3}He\!:^{4}He$ ratio with a mass spectrometer (Tolstikhin & Kamenskiy, 1969).

Recently, measurements of isotopic abundance using tunable diode laser optical techniques have been developed (e.g. Wahlen & Yoshinari,

1985; Becker *et al.*, 1992). These techniques rely on small differences in energy levels for rotational and vibrational states in gaseous molecules, and may be made directly on air samples without the need for purification. While not yet as widely used or as precise as mass spectrometry for isotope measurements, rapid advances in laser optical methods hold promise for the future.

9.3.1 Sample collection

As demonstrated in Table 9.2, large quantities of air must be processed to produce a sample of sufficient size for the isotopic measurement of most trace gases. The simplest method of sample collection is to compress the air sample into a gas cylinder for return to the laboratory for extraction, purification and analysis. Possible complications include contamination of the sample by the pump or compressor, reaction of the gas of interest with the container and alteration of the sample during collection. For example, the presence of strong pressure gradients or capillary flow may cause depletion of the heavy isotope in the collected sample. Even small bore syringes improperly used may create differences of up to several per millilitre in $^{13}CO_2$ (Alan Townsend, Stanford University, personal communication, 1992). Sample storage must be thoroughly tested to ensure that there is no contamination through reaction of the gas with the container walls. For example, any H_2O in the sample container can exchange oxygen isotopes with CO_2 (Francey & Tans, 1987), thus affecting the measured oxygen isotope ratio. Special care should be taken with reactive gases like CO or non-methane hydrocarbons (NMHC), which may react with the metal of the canister wall (e.g. Lowe *et al.*, 1991; Brenninkmeijer *et al.*, 1992).

Chemical trapping offers the advantage of reduced bulk in samples to be transported from the field. CO_2 is commonly extracted and concentrated in the field, either through taking advantage of the high solubility of CO_2 in bases like NaOH, or trapping on a molecular sieve (Tyler *et al.*, 1988; Lowe *et al.*, 1991; Bauer *et al.*, 1992). NO_x can be trapped from combustion exhaust streams by bubbling through KOH (Moore, 1977), or equilibrating collected gas with NaOH plus H_2O_2 (Heaton, 1990). Particulate NO_3^- and gas phase HNO_3 are collected on filters (Freyer, 1991). Chemical trapping must be tested carefully, as large isotopic fractionation may result from trapping with less than 100% efficiency. In addition, contaminants within the trapping medium (e.g. CO_2 in NaOH), may cause large procedural blanks.

Samples of organic substrates, such as soils and sediments, should be stored dry or refrigerated to reduce microbial activity. Special care must be taken to homogenize samples before combustion for isotopic measurements. Procedures designed to extract organic matter from soils and sediments should be employed with care, as extracted and unextracted materials may differ dramatically in isotopic composition.

9.3.2 Sample purification

Purification methods rely on chemical or cryogenic trapping techniques, usually in vacuum lines, and vary according to the type and concentration of the trace gas. In some cases, the gas of interest must be quantitatively converted to another gas species (e.g. CO and CH_4 to CO_2, or NO_3^- to N_2) for measurement. Table 9.2 summarizes some of this information; details for commonly measured trace gases are given here. Any purification method must be tested with known amounts of the trace gas (of known isotopic content) to ensure that no fractionation or contamination is occurring during sample purification.

CO_2

The concentration of CO_2 in the free troposphere in 1992 was about 355 ppmv. CO_2 concentrations in soil atmospheres can be much greater, up to several per cent. Thus, isotopic abundances of ^{13}C, ^{14}C and ^{18}O may be measured in CO_2 extracted from between 0.3 and 2.41 of air (see Table 9.2).

CO_2 may be concentrated from air by either chemical or cryogenic trapping methods. These methods may also be used to remove CO_2 quantitatively from air samples in which measurement of CO or CH_4 is desired. CO_2 may be trapped cryogenically at liquid nitrogen (LN_2) temperatures ($-196°C$). Vacuum line pressures should be kept low (below about 200 mmHg) to ensure efficient trapping and to avoid oxygen condensation. Water vapour is usually removed upstream of the LN_2 or chemical traps using a trap maintained at $-76°C$ by a dry ice/alcohol slurry.

Because N_2O condenses in a LN_2 trap along with CO_2, care must be taken to ensure that CO_2 is separated from N_2O for stable isotope measurements ($^{14}N_2{}^{16}O$ has the same mass as $^{12}C^{16}O^{16}O$). One approach to solving this problem has been to reduce the trapped N_2O to N_2 with hot Cu wire in the vacuum line. Alternatively, the N_2O content of the gas may be measured by gas chromatography on an aliquot of sample, or directly by mass spectrometer, and a correction factor calculated (e.g. Craig & Keeling, 1963; Mook & Jongsma, 1987; Friedli & Slegenthaler, 1988). The currently accepted correction factors for measurements of atmospheric CO_2 are $+0.23‰$ in $\delta^{13}C$ and $+0.33‰$ in $\delta^{18}O$ (Mook & Jongsma, 1987).

For ^{14}C measurement by AMS, purified CO_2 must be converted to a solid graphite target (although the use of CO_2 gas is being developed;

Middleton *et al.*, 1989). This is easily accomplished by reducing the CO_2 with either H_2 (Vogel *et al.*, 1984, 1987; Lowe & Judd, 1987) or zinc (Vogel, 1992) and either Fe or Co powder as catalysts.

CH_4

CH_4 concentrations in the troposphere in 1992 were about 1.7 ppmv. The volume of air needed for isotopic measurement of ^{13}C was thus about 200 times that required for the measurement of CO_2 (see Table 9.2). Two methods have been used to isolate and combust CH_4 for isotopic analysis. The method of Lowe *et al.* (1991), adapted from Stevens & Rust (1982), removes CO_2 and water vapour by cryogenic trapping (LN_2 temperature). The gas stream is then passed through a column of Schutze reagent (I_2O_5), which quantitatively converts CO to CO_2, but does not affect CH_4. The CO_2 produced from CO is trapped at liquid nitrogen temperatures, and the remaining gas, including CH_4, is combusted to CO_2 and H_2O (using a heated Pt or CuO wire catalyst) for measurement or conversion to graphite AMS targets. Wahlen *et al.* (1989) report a different method. After removal of H_2O, CO_2 and N_2O in LN_2 traps, the gas stream is passed over activated charcoal, on which CH_4 and other gases are trapped. The gases (including Kr, CH_4 and others) desorbed on heating the charcoal trap are separated chromatographically, and trapped individually for analysis. Purified CH_4 is combusted in a Pt-coated combustion tube at 600°C. The chromatographic separation technique offers the advantage of separating NMHCs (e.g. ethane) which are combusted in Schutze reagent (Tyler, 1986). The amount of CO_2 contributed by NMHCs, however, is negligible compared to that from CH_4 in many instances (Tyler, 1986).

To measure the deuterium/hydrogen ratios in CH_4, the H_2O formed during combustion of CH_4 must be quantitatively converted to H_2. This is accomplished either by equilibrating the evolved H_2O with H_2 of known isotopic content in the presence of a Pt catalyst, and measuring the change in the H_2 isotopic content (Horita *et al.*, 1989; Coplen *et al.*, 1992), or by reducing the H_2O directly to H_2 using zinc as a reductant (Coleman *et al.*, 1982; Kendall & Coplen, 1985). Tanweer (1990) points out the importance of properly cleaning the zinc used to obtain good results with the latter procedure.

CO

The CO concentration in ambient air is highly variable in space and time, but ranges between about 50 and 200 ppbv. The basic method used to isolate and oxidize CO to CO_2 was developed by Stevens & Krout (1972). After removal of CO_2 and N_2O using molecular sieves and LN_2 traps (Stevens & Krout, 1972), or with NaOH absorbant traps (Volz *et al.*,

1981), Schutze reagent is used to quantitatively convert CO to CO_2 for ^{14}CO and ^{13}CO analyses (Stevens & Krout, 1972; Brenninkmeijer *et al.*, 1992). Volz *et al.* (1981) used a hot Pt catalyst to combust CO to CO_2 instead of the Schutze reagent. The hot Pt will also combust any CH_4 present in the sample, which could cause an overestimation of ^{14}CO concentration.

N_2O

N_2O is cryogenically trapped from a gas stream under the same conditions as CO_2. CO_2 may be subsequently removed by passing over ascarite (NaOH-coated asbestos). The N_2O passing through the trap is further purified using gas chromatography. N_2O must be reduced to N_2 for the mass spectrometric measurement. Kim & Craig (1990) reduce N_2O by exhaustive reaction with carbon rods wrapped with Pt wire at 700°C. Using this method the ^{18}O in N_2O could be measured by quantitatively converting the oxygen in N_2O to CO_2. Another method reacts the purified N_2O with hot Cu wire (Yoshida, 1988; Ueda *et al.*, 1991), but ^{18}O is not measured.

^{18}O in N_2O was measured with high resolution infrared spectroscopy using tunable diode lasers (Wahlen & Yoshinari, 1985).

Other nitrogen oxides

Though methods vary, the preparation of N_2 from NO_3^-, HNO_3, NH_3 and NO_x share several common features. NO_x collected by trapping in basic solution (NaOH or KOH) will be in the form of NO_3^- and NO_2^- (Heaton, 1990). NO_3^- trapped on filters must be extracted into solution (Freyer, 1991). NO_3^- and NO_2^- in the sample are then reduced to NH_3 using Devarda's reagent (Zn with a coating of Cu). The NH_3 is purified through distillation, and reduced to N_2 by reaction with NaOBr (Moore, 1977; Heaton, 1990) or LiOBr (Freyer, 1991) in a vacuum system.

Measurement of substrates

The isotopic composition of carbon, nitrogen and oxygen in organic matter substrates is determined by combustion of the sample, followed by cryogenic purification of the resulting CO_2, N_2 and H_2O. For example, both the dried and ground organic matter and CuO wire are sealed into an evacuated quartz tube and combusted at 900°C (Buchanan & Corcoran, 1959). Ag or Cu wire may also be added, the Ag to remove chlorine and sulphur, the Cu to ensure reduction of nitrogen to N_2. Kendall & Grim (1990) describe how this method may be optimized for isotopic measurements of N_2.

Recently, mass spectrometers which use high temperature combustion/ gas chromatographic inlet systems for the measurement of stable isotopic composition of carbon and nitrogen in organic matter have become commercially available. These machines provide the advantage of speed in sample processing, but are currently less precise than many older methods.

9.4 Isotopic variation during production of trace gas species

9.4.1 General isotopic contents of ecosystem components

The isotopic composition of atmospheric trace gases produced in terrestrial ecosystems will depend on several factors, including the composition of the substrate compound(s) from which the gas is produced, biological fractionation during production, any reactions involving the trace gas following its formation, and any fractionating process (such as diffusion) involved in transfer of the gas from soil or plant to the troposphere.

Figure 9.2 shows the isotopic distribution of ^{13}C and ^{15}N among common reservoirs which may be substrates in trace gas formation. One well-studied cause of variations in carbon isotopes is the discrimination against ^{13}C during photosynthesis by plants (see reviews by Deines, 1980; O'Leary, 1981; Farquhar et al., 1982; Tieszen & Boutton, 1989). Atmospheric CO_2 in the late 1980s had a $\delta^{13}C$ value of $-7.8‰$ (Keeling et al., 1989), while the $\delta^{13}C$ of leaves and woody tissues of plants with a C3 photosynthetic pathway (used by most terrestrial plant species) range from -20 to $-32‰$. Plants with C4 photosynthetic pathways (most importantly, tropical and subtropical grasses), have $\delta^{13}C$ values between -17 and $-9‰$ (Smith & Epstein, 1971; Deines, 1980). The difference in ^{13}C between photosynthetic pathways is due to the difference in biochemical fractionation. The variability in $\delta^{13}C$ values observed for whole plants falling within each photosyntheic pathway type arises from a combination of:

1 temperature dependence of the biochemical fractionation;

2 the magnitude of the diffusion gradient for CO_2 between intracellular leaf water and air outside the leaf; or

3 variations in the ^{13}C content of local atmospheric CO_2 (Keeling, 1958).

In addition to the variations in ^{13}C observed for the whole plant, differences in $\delta^{13}C$ value are observed for different classes of compounds within the plant. For example, reduced components (such as lignin) are depleted in ^{13}C relative to cellulose and water-soluble components (Deines, 1980; Benner et al., 1987; Nadelhoffer & Fry, 1988). Thus roots, with high lignin contents, may be expected to be more depleted in ^{13}C than leaves for the same plant (e.g. Nadelhoffer & Fry, 1988). Animals tend to reflect the ^{13}C content of their diet to within about 1‰, though differences between organs and biochemical components within an in-

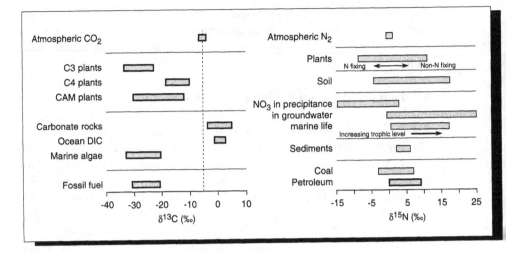

Fig. 9.2 ^{13}C content of major carbon reservoirs, modified from Deines (1980), and ^{15}N content of major nitrogen reservoirs.

dividual are observed (DeNiro & Epstein, 1978; Tieszen & Boutton, 1989).

Plant tissues also vary in nitrogen isotopic content, in relation to the source of plant nitrogen (atmospheric N_2 vs. soil inorganic nitrogen; Delwiche et al., 1979). The fractionation factor associated with nitrogen fixation is not well determined (Schearer & Kohl, 1989), but is small relative to the difference in $\delta^{15}N$ between atmospheric nitrogen (0‰) and organic nitrogen in soils (generally in the range of +5 to +17‰, Fig. 9.2). Like the carbon isotopes, nitrogen isotopes may be unequally distributed among plant components, although systematic differences are not yet well understood (Schearer & Kohl, 1989). The ^{15}N content of animals tends to be enriched in comparison with the food they consume (Ambrose & DeNiro, 1987). Thus as trophic level increases, the ^{15}N content of the animal increases (Minegawa & Wada, 1984).

Differences in $\delta^{13}C$ and $\delta^{15}N$ values are observed between soil organic matter and the fresh plant matter from which it is ultimately derived. These differences may be due to one or a combination of several causes, summarized by Nadelhoffer & Fry (1988):

1 overall discrimination between heavy and light isotopes during decomposition;

2 differential preservation of compounds with lighter or heavier ^{13}C and ^{15}N;

3 changes in the ^{13}C and ^{15}N of plant inputs with time; or

4 selective translocation of depleted or enriched organic matter from litter to soil layers.

In their study of ^{13}C and ^{15}N changes in an upland forest, Nadelhoffer & Fry (1988) concluded that the enrichment of ^{13}C and ^{15}N generally observed in soils over litter input values where no vegetation change has occurred is due to general discrimination during decomposition of organic matter. They point out that selective preservation of lignin components would cause a depletion of ^{13}C, rather than the observed enrichment. The decrease in ^{13}C in atmospheric CO_2 caused by combustion of fossil fuels (see discussion, below) may contribute to some of the present observed differences in ^{13}C between soil organic matter and fresh plant litter.

9.4.2 Determining pathways of CH₄ production and oxidation using stable isotopes

Wetlands are the largest natural source of CH_4 to the atmosphere (Cicerone & Oremland, 1988). The average ^{13}C content of CH_4 over all wetland environments from which data are available is about −58‰ (Tyler, 1991), and varies from a low of −80‰ (Lansdown et al., 1992) to a high of −40‰ (Chanton et al., 1989). Seasonal variations of the isotopic content of CH_4 have been observed in both coastal marine (Martens et al., 1986) and freshwater (Chanton & Martens, 1988) environments. The $\delta^{13}C$ and δD values of CH_4 produced in wetlands are determined by:

1 the isotopic content of substrate material (organic matter and coexisting water);

2 the process by which CH_4 is produced (fermentation vs. reduction); and

3 the dominant mechanism by which CH_4 is transported from the site of production to the atmosphere (ebullition, diffusion or transport through plants).

The effect of slow transport is twofold in that enrichment of the heavy isotope may arise both because of diffusion and because of the increased time that CH_4 is subjected to CH_4 oxidation.

The majority of biogenic CH_4 production occurs by two different pathways: (a) cleavage of acetate to produce CO_2 + CH_4 (fermentation); and (b) reduction of CO_2 with H_2 (McCarty, 1964). Both are microbial processes which strongly discriminate in favour of the lighter isotopes, producing CH_4 which is depleted with respect to its substrate with a range from −50 to −110‰. Whiticar et al. (1986) describe a method to distinguish the dominant pathway for CH_4 production by comparing the $\delta^{13}C$ and δD in CH_4 to that of coexisting CO_2 and H_2O (Fig. 9.3). The depletion of ^{13}C is greater during CO_2 reduction than for acetate fermentation, while CH_4 formed by the transfer of a methyl group during acetate fermentation is more depleted in deuterium than that formed by CO_2 reduction (Whiticar et al., 1986). Whiticar et al. noted that CH_4 produced in marine environments has lighter $\delta^{13}C$ (−110 to −60‰) and

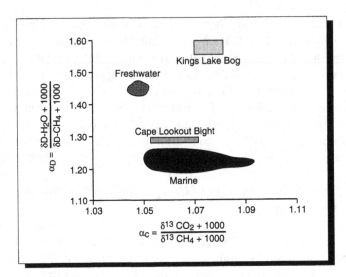

Fig. 9.3 Plot of ^{13}C of coexisting CH_4 and CO_2 versus deuterium in CH_4 and pore water, modified from Whiticar *et al.* (1986). The isotope discriminations are expressed as (Whiticar *et al.*, 1986):

$$\alpha_c = \frac{\delta^{13}C_{CH_4} + 1000}{\delta^{13}C_{CO_2} + 1000}, \quad \text{and} \quad \alpha_H = \frac{\delta^{13}D_{CH_4} + 1000}{\delta^{13}D_{H_2O} + 1000}$$

In Whiticar's original work, marine systems (dominated by CO_2 reduction) are clearly separated from freshwater systems (dominanted by acetate fermentation), and fell within the regions shown on the plot. Additional data from Lansdown *et al.* (1992) and Martens *et al.* (1986) show that fermentation and CO_2 reduction may be observed in both freshwater and marine environments, and thus the isotope systematics may fall outside of the two end member conditions.

heavier δD (-250 to $-170‰$) than that produced in freshwater environments ($\delta^{13}C = -65$ to $-50‰$, $\delta D = -400$ to $-250‰$). Thus CO_2 reduction has been thought to be the dominant pathway for CH_4 production in marine systems, while fermentation was thought to be the dominant source of CH_4 in freshwater environments. Subsequent investigations have shown that the CH_4 produced in both systems may show evidence of either or both production pathways. All of the CH_4 in the freshwater bog studied by Lansdown *et al.* (1992) was produced by CO_2 reduction, while there is evidence of acetate fermentation at Cape Lookout Bight, a coastal marine site (Martens *et al.*, 1986). Blair *et al.* (1987) and Blair & Carter (1992) further investigated the isotopic fractionation during the formation and consumption of acetate at Cape Lookout Bight.

Microbial CH_4 oxidation, like fermentation and reduction, discriminates in favour of the light isotope, in this case acting to enrich the unoxidized CH_4 in the heavier isotopes (Alperin *et al.*, 1988). Thus CH_4

which is quickly lost from sediments in which it was produced (e.g. by ebullition), is presumably least affected by CH_4 oxidation, and therefore retains more negative ^{13}C values (Martens et al., 1986; Chanton & Martens, 1988). Because the isotopic enrichment associated with CH_4 oxidation complicates the interpretation of the CH_4 production pathway from $\delta^{13}C$ and δD values, care must be taken to ensure that CH_4 collected reflects in situ production (e.g. stirred sediment or bubbles). CH_4 lost from the sediment primarily by diffusion or transport through plants (Chanton et al., 1988, 1989) is usually enriched in heavier isotopes compared to CH_4 collected from bubbles. Loss of CH_4 by diffusion may cause enrichment of the residual CH_4 of up to 19‰ because the lighter isotope will diffuse faster (see Eqn. 9.5, below). Alternatively, enrichment of the heavier isotopes resulting from CH_4 oxidation may occur in the plant rooting zone. Chanton et al. (1989, 1992) discuss the effects of plant enhanced transport on the isotopic content of CH_4.

Seasonal variations in both CH_4 flux and isotopic content are observed at both marine and freshwater sites (Martens et al., 1986; Chanton & Martens, 1988). Lansdown et al. (1992) observed no seasonal change in isotopic content of CH_4 from an acidic bog, although CH_4 fluxes varied about 300-fold between summer and winter. Explanations for seasonal variations in isotopic content of CH_4 include a change in the dominant CH_4 production mechanism (Martens et al., 1986) and a change in the transport pathway for CH_4, resulting in more or less influence of CH_4 oxidation (Chanton & Martens, 1988).

9.4.3 Dynamics of soil organic matter (SOM)
The isotopic content of SOM and sedimentary organic carbon may be compared to that of trace gases produced from it to give an idea of the degree to which substrates are labile or refractory. For example, CH_4 produced in wetlands (Wahlen et al., 1989), and CO_2 produced in soils (Dörr & Münnich, 1986), are both enriched in ^{14}C compared to bulk organic matter substrates. This is due to the heterogeneous nature of SOM and sedimentary organic matter, which is a mixture of both ^{14}C-depleted, refractory and ^{14}C-enriched labile components. While the ^{14}C data from evolved gases show the preferential oxidation of more labile components, they often show significant ^{14}C depletion compared to the contemporary atmosphere, indicating that some 'old' components are metabolized.

Quantification of the amount of labile and refractory organic matter in soils and sediments may help place limits on the potential fluxes of trace gases from a given ecosystem, and provide better understanding of the role of substrates in determining the amount and isotopic composition of trace gas fluxes. Separation of bulk organic matter into constituent labile and refractory pools has been attempted in several ways, including: monitoring the rate of loss of organic matter in soils on conversion to agriculture, observing the rate of loss of ^{14}C labelled organic matter added to soils (Jenkinson & Raynor, 1977; Jenkinson et al., 1991) using the change in ^{13}C in SOM with time after conversion of C3 dominant vegetation to C4 (Balesdent et al., 1987; Cerri et al., 1991); observations of the rate of increase of ^{14}C in soil organic matter during the last 30 y due to incorporation of bomb ^{14}C (O'Brien & Stout, 1978; Harkness et al., 1986; Trumbore et al., 1989; Trumbore, 1993). In addition, observations of the ^{14}C content of respired CO_2 and/or CH_4 may provide stringent tests for models of soil and sedimentary organic matter dynamics.

9.5 Extrapolation of isotopic and flux data to global scales

The average concentrations and isotopic contents of tropospheric CH_4 and CO_2 reflect the sizes and isotopic signatures of their major sources and sinks. In this sense, the isotopes act as integrators which may be used to constrain the importance of various ecosystems and biogeochemical processes as sources and sinks at the global scale.

9.5.1 CO₂ budget

Whereas ^{13}C provides information about the distribution of carbon among atmosphere, biosphere and ocean reservoirs, ^{14}C primarily reflects the influence of CO_2 derived burning of (^{14}C-free) fossil fuels. The ^{13}C contents of various carbon reservoirs are shown in Fig. 9.2. Biospheric (predominantly C3 vegetation) and fossil fuel sources are both depleted with respect to atmospheric CO_2 by about 20‰. Carbon dissolved in seawater is enriched in ^{13}C compared to the atmosphere. Thus, changes in the apportionment of carbon among oceanic, biospheric and fossil fuel sources will be reflected as a change in the ^{13}C content of atmospheric CO_2. Proxy records of the ^{14}C and ^{13}C record in atmospheric CO_2 preserved in tree ring and corn cellulose have been used to deconvolve the roles of biospheric and fossil fuels as sources of CO_2 since 1900 (Stuiver & Quay, 1981; Peng et al., 1983; Peng, 1985; Siegenthaler & Oeschger, 1987; Keeling et al., 1989). An example of the use of isotopic records to constrain the magnitude of fossil fuel vs. biosphere contributions to observed CO_2 increases is described in Fig. 9.4 (from Peng et al., 1983).

Hypotheses explaining glacial–interglacial differences in carbon cycling are constrained by data on ^{13}C in glacial CO_2 trapped in ice cores (Friedli et al., 1986; Leuenberger et al., 1992) and in C4 plant remains (Marino et al., 1992). Both measurements show that the $\delta^{13}C$ value of atmospheric CO_2 was lower during the last glacial interval than during the subsequent interglacial period.

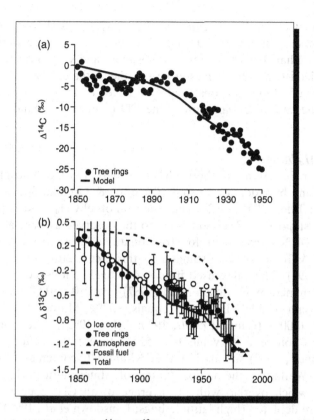

Fig. 9.4 The deconvolution of the ^{14}C and ^{13}C records recorded in tree rings (after Peng *et al.*, 1983). (a) The decrease in ^{14}C in atmospheric CO_2 recorded by tree rings (data from Stuiver & Quay, 1981) is primarily due to dilution of atmospheric [14]CO_2 by fossil fuel emissions of ^{14}C-free CO_2. The ^{14}C decrease thus provides an estimate of the total amount of fossil fuel burned which remains as CO_2 in the atmosphere (model curve from Peng *et al.*, 1983). (b) Once the fossil fuel input is known, the decrease in ^{13}C due to fossil fuel emissions alone may be calculated. The data points summarize ^{13}C measurements in tree rings (Freyer, 1979) and ice cores (Friedli *et al.*, 1986), presented as the difference ($\Delta\delta$[13]C) from preindustrial ^{13}C values. The tree ring record shows increased ^{13}C depletion over and above that predicted from the fossil fuel curve, particularly during the early part of the century. This is attributed to a net transfer of isotopically light carbon from the biosphere and soils to the atmosphere. [13]C records in tree rings (C3 plants) may be strongly affected by plant water use efficiency (Farquhar *et al.*, 1982), and by incorporation of ^{13}C-depleted, soil respired CO_2. Thus this deconvolution exercise must be viewed with some scepticism.

^{18}O in CO_2 and deuterium in CH_4 may be used to constrain the latitudinal distribution of CO_2 and CH_4 sources. The ^{18}O content of CO_2 is set by exchange with H_2O in the biosphere and soils (Francey & Tans, 1987; Friedli *et al.*, 1987; Hesterberg & Siegenthaler, 1991). Latitudinal variations of ^{18}O in CO_2 observed by Francey & Tans (1987) thus reflect in part the latitudinal variation of the ^{18}O content of groundwater (Gat, 1980).

9.5.2 CH₄ budget

The major sink for atmospheric CH_4 is oxidation in the atmosphere, with a lesser sink by oxidation in soils (Cicerone & Oremland, 1988; Fung *et al.*, 1991; Tyler, 1991, 1992). Thus the sum of all CH_4 sources (at steady-state) must have a ^{13}C content equal to the observed average atmospheric $^{13}CH_4$ ($-47‰$), corrected for fractionation during oxidation by OH (Stevens & Rust, 1982). Estimates of this factor differ in the literature (Tyler, 1991). The value used in global budgets published prior to 1990 was determined by Davidson *et al.* (1987) as 1.010 ± 0.007, which infers that the sum of all CH_4 sources equals $-57‰$. More recent values are 1.0054 ± 0.0009 (Cantrell *et al.*, 1990; Tyler, 1992), which increase the integrated source $\delta^{13}C$ value to $-52‰$. Figure 9.5 (after Stevens & Engelkemeir, 1988) summarizes the ^{13}C and ^{14}C contents of major CH_4 sources. Wetlands, the major source of atmospheric CH_4, have ^{13}C contents which overlap with the average of all CH_4 sources, although both more depleted (high latitude bogs; Lansdown *et al.*, 1992) and more enriched (tropical floating grassmats; Chanton *et al.*, 1989) sources have been measured. Burning of C4 vegetation (tropical and subtropical grasslands) may be an increasing source of isotopically light CH_4 (Lassey *et al.*, 1993). Monitoring of the ^{13}C content of atmospheric CH_4 at remote sites shows interhemispheric differences and decreasing $\delta^{13}C$ values in the late 1980s (Fung *et al.*, 1991; Lowe *et al.*, 1991; Quay *et al.*, 1991).

The ^{14}C content of tropospheric CH_4 in the late 1980s was roughly 122 pmc ($+220‰$), in the Northern Hemisphere (Wahlen *et al.*, 1989; Quay *et al.*, 1991), and 120 pmc ($+200‰$) in the Southern Hemisphere (Manning *et al.*, 1990). The ^{14}C record in atmospheric CH_4 is complicated by the increasing importance of $^{14}CH_4$ produced by pressurized water nuclear reactors (Kunz, 1985; Wahlen *et al.*, 1989; Levin, quoted in Manning *et al.*, 1989), which now are causing a global increase in $^{14}CH_4$ values (Fung *et al.*, 1991). Based on data from the late 1980s, which were corrected for the pressurized water reactor (PWR) source, 20% (Wahlen *et al.*, 1989; Quay *et al.*, 1991) to 30% (Manning *et al.*, 1989) of the global source of CH_4 in the late 1980s was attributable to fossil fuel inputs. These values are larger (by 5–15%) than the estimates of fossil fuel

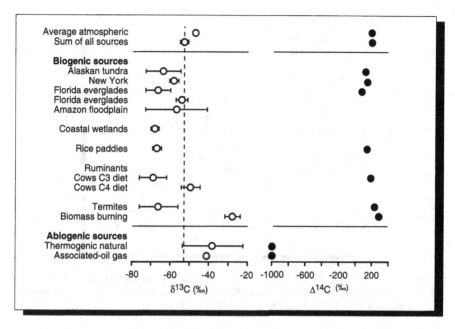

Fig. 9.5 Summary of the ¹³C and ¹⁴C contents of sources of atmospheric methane. (Adapted from Stevens & Engelkmeier, 1988, and from tables in Tyler, 1991 and Lansdown *et al.*, 1992.)

inputs based on emissions data (Cicerone & Oremland, 1988). One possible reconciliation is achieved if additional sources of ¹⁴C-depleted CH_4 (such as from wetlands or CH_4 hydrates) exist, though to date no such sources have been found (Wahlen *et al.*, 1989; Tyler, 1992). For recent, detailed reviews of the global CH_4 budget, the reader is referred to Tyler (1991, 1992) and Fung *et al.* (1991).

The δD value of CH_4 reflects the isotopic content of water where the CH_4 was formed (Wahlen *et al.*, 1990), although this signal is complicated due to changes in the ratio of fermentation to reduction sources for CH_4 (see Fig. 9.3). Tyler (1992) discusses the interpretation of δD in CH_4 in more detail.

9.5.3 ¹⁵N and the origins of nitrogen in precipitation

The large number of species of atmospheric nitrogen, combined with their low concentrations and sometimes high reactivity, have limited the use of isotopes in the study of the atmospheric nitrogen cycle. ¹⁵N and ¹⁸O isotopes offer promise for enhancing our understanding of the nitrogen cycle, though due to analytical difficulties, relatively few studies have been done to date. Figure 9.6 summarizes the published δ¹⁵N values

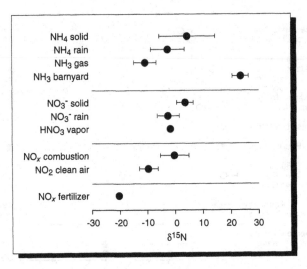

Fig. 9.6 Summary of ^{15}N measurements in atmospheric nitrogen species. (After Moore, 1977.) Data for NH_3 and NH_4^+ are from Wood (1977); data for particulate and rain NO_3^-, combustion NO_x and fertilizer estimates from Freyer (1991).

of some of the nitrogen species in the atmosphere. Moore (1977) provided quantitative estimates of the influence of source contributions and atmospheric modifications of $\delta^{15}N$ for several nitrogen species. Freyer (1991) reported an extensive record of the seasonal variability of ^{15}N in rain and particulate NO_3^-, and HNO_3 vapour. These investigators propose that the primary source of NO_3^- in precipitation is from oxidation of other NO_x species, rather than NH_3. Sources of NO_x from combustion vary in their isotopic content (Heaton, 1990), but average about 0‰, reflecting the primary source of NO_x in burning (N_2 in air). No data are available for biosphere burning, though some inherited organic nitrogen should influence the isotopic content of NO_x produced during burning. Fertilizer and biospheric NO_x sources are assumed to be depleted in ^{15}N relative to atmospheric values (Freyer, 1991). Freyer (1991) offers the interpretation that seasonal differences in the ^{15}N content of precipitation reflect changes in the relative influences of natural and anthropogenic NO_x sources between summer and winter.

N_2O

The relatively sparse measurements of ^{15}N in N_2O suggest that dentrification and nitrification sources of N_2O will have different isotopic signatures (Yoshida, 1988). Studies of ^{18}O in N_2O have shown that different sources also have different isotopic signatures (Wahlen &

Yoshinari, 1985; Kim & Craig, 1990), thus offering hope of using combined ^{15}N and ^{18}O to help interpret the atmospheric N_2O budget. Kim and Craig (1993) have published an attempt to constrain the global N_2O budget using stable isotopes.

9.5.4 Global OH distributions deduced from ^{14}CO

The majority of ^{14}CO is produced in the stratosphere and upper troposphere from interaction of cosmic ray-produced neutrons and ^{14}N. Minor amounts of ^{14}CO are produced by pressurized water reactors (mentioned earlier) and by recycling of carbon through the biosphere. Burning of wood, for example, produces CO with approximately Modern ^{14}C content (one ^{14}C atom in 10^{12} carbon atoms; Volz et al., 1982; Brenninkmeijer et al., 1992; Mak et al., 1992). Although biomass burning is a large contributor to the global sources of ^{12}CO, its low ^{14}C content makes it less important in the ^{14}CO budget. The major loss mechanism for both ^{12}CO and ^{14}CO is by reaction with OH radical to produce CO_2. In this regard, the use of ^{14}CO has been proposed to measure of the global concentration of OH radical. The production rate of ^{14}C is known (Lal & Peters, 1967; Lingenfelter, 1963). Thus, if the atmospheric burden of ^{14}CO is known (assuming ^{14}CO is in steady-state), the rate of destruction of ^{14}CO is also known, and the concentration of OH may be deduced (Volz et al., 1982; Brenninkmeijer et al., 1992; Mak et al., 1992). In practice, the distribution of ^{14}CO and OH differ enough vertically and zonally that their interpretation would benefit from the use of a full three-dimensional atmospheric transport model (Mak et al., 1992). Interpretation of this data is still largely limited by the narrow range of ^{14}CO measurements available (Mak et al., 1992).

During oxidation of CO to CO_2 by the Schutze reagent, the original oxygen in the CO is retained (Stevens & Krout, 1972), and the second oxygen derives from the reagent. Thus ^{18}O in the final CO_2 may be measured, and back-corrected for the ^{18}O added by the I_2O_5 reagent to estimate the $C^{18}O$ content (Stevens & Krout, 1972). No geochemical interpretation of these data has yet been made.

9.6 Use of deliberate tracers and ^{222}Rn to determine soil gas exchange rates

The previous sections have concentrated on using measurements of stable and radioactive isotopes at their natural abundances to understand the sources and sinks of trace gases on local and global scales. The discussion in the remaining section shifts to measurements using non-isotopic tracers to understand physical controls on trace gas fluxes. The primary physical control on trace gas exchanges is the rate of gas exchange between the soil atmosphere and overlying air.

Fluxes of trace gases into and out of soils vary spatially and temporally. Often these fluxes are correlated with factors such as temperature or soil moisture. These physical variables may influence trace gas movements through their effects on biological activity (rate of trace gas production/ consumption). In addition, however, temperature and moisture conditions in the soil affect the rate of exchange between the soil atmosphere and overlying air, which will also influence the measured trace gas flux. In order to separate biological and physical controls on trace gas fluxes, gas exchange rates must be determined over a range of soil conditions.

Two methods for determining soil gas exchange rates will be discussed here. The first method uses the loss of a deliberate a tracer following its addition to the head space of a flux chamber to deduce soil gas exchange rates (Trumbore, 1988; Rolston *et al.*, 1991). This method is most suitable for studying gases with highest production and consumption in the upper 10 or 20 cm of the soil, as that is the depth to which a tracer commonly penetrates during an experiment. The second method measures the deficit of ^{222}Rn in the soil atmosphere near the surface to estimate gas exchange rates, which are generally rates averaged over the upper 50 cm or so of soil. This method is most applicable for gases, such as CO_2, with sources and sinks extend deeper into the soil.

9.6.1 Diffusion of gases from soil

The concentration of a trace gas in the soil air space represents a balance between biological activity (production and consumption) and the rate at which the soil air exchanges with the overlying atmosphere. The rate of soil gas exchange is often parameterized by assuming that diffusion is the dominant transport process (for reviews, see Nazaroff, 1922; Glinski & Stepnowski, 1985). An effective diffusion coefficient (D_{eff}) is defined, which is ultimately related to soil physical properties (porosity, air-filled pore space, tortuosity) and to the diffusivity of the trace gas of interest in air (D_0). Fick's first law may be used to describe the concentration (C) of a gas in the soil air space:

$$\frac{dC}{dt} = D_{eff} \frac{d^2 C}{d^2 z} + Q(z), \tag{9.2}$$

where $Q(z)$ is the depth-dependent production rate for the gas C. At steady-state $dC/dt = 0$ and losses through diffusion equal production (minus consumption) in the layer. The flux through the surface then equals $D_{eff}^* dC/dz)_{z \Rightarrow 0}$, assuming no production at the interface. If D_{eff} and the concentration profile are known, and the profile is at steady-state, Eqn. 9.2 may be solved for production as a function of soil depth.

Two methods have been employed to determine soil effective diffusivity in the field. The first uses a deliberately introduced tracer gas to follow exchange. The second uses the steady-state concentration profile of ^{222}Rn in the soil atmosphere to derive D_{eff}. In both cases, the tracer is an inert gas, whose distribution is affected only by physical transport, rather than by biological processes in the soil.

9.6.2 Deliberate tracer measurement

The determination of D_{eff} using a deliberate tracer injected into the head space of a flux chamber has been described in Trumbore (1988) and in Rolston et al. (1991). Both of these studies used man-made gases, such as sulphur hexafluoride (Trumbore, 1988) or chlorofluorocarbons (CClF$_3$; Rolston et al., 1991). The advantages of SF$_6$ are its low solubility, and the extreme sensitivity with which it can be detected using an electron capture detector (Wanninkhof et al., 1985). The chlorofluorocarbon used by Rolston et al. (1991) was detected using a flame ionization detector, which simultaneously detected CO$_2$ and CH$_4$ for concentration measurements. Similar methods have followed the decrease through time of a gas injected into the soil (O$_2$, Lai et al., 1976; N$_2$O, Jellick & Schnabel, 1986). The solution of the diffusion equation (9.2 above) for a semi-infinite plane (zero initial concentration) in contact with a well-stirred reservoir with length a, constant effective diffusivity D_{eff}, and initial concentration C_0 is adapted from Carlslaw & Jaeger (1959, p. 306):

$$C(z, t) = C_0 \exp\left(\frac{z}{a} + \frac{D_{eff}t}{a^2}\right) \text{erfc}\left(\frac{z}{2\sqrt{D_{eff}t}} + \frac{\sqrt{D_{eff}t}}{a}\right) \tag{9.3}$$

where C is the concentration of the gas, z is the distance from the interface, t is the time since tracer injection, and D_{eff} is the effective diffusivity. The concentration of tracer in the well-stirred reservoir, $C_s t$ (the box head space) at any time may be found by substituting $z = 0$ into Eqn. 9.3:

$$C_s(t) = C_0 \exp\left(\frac{D_{eff}t}{a^2}\right) \text{erfc}\left(\frac{\sqrt{D_{eff}t}}{a}\right). \tag{9.4}$$

Rolston et al. (1991) used Eqn. 9.4 to deduce D_{eff} from the loss rate of a tracer introduced into a chamber head space. Trumbore (1988) followed concentration changes over time in both the head space and the soil atmosphere, using a specially constructed chamber base (closed to 25 cm depth in the soil) to prevent loss of the sulphur hexafluoride tracer. (Eqns 9.3 and 9.4 are valid only for cases in which the soil can be considered one-dimensional; that is, no tracer is lost laterally once in the soil atmosphere.) Figure 9.7 shows data obtained from a typical tracer experiment (Trumbore, 1988). The effective diffusivity determined in the experiment was $0.015–0.020\,\text{cm}^2\,\text{s}^{-1}$.

One of the advantages of the deliberate tracer method is that D_{eff} may

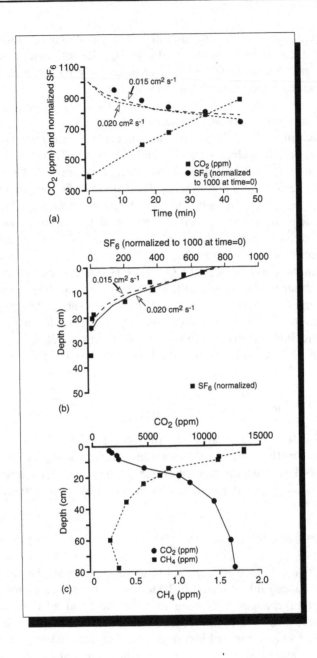

(a)

(b)

(c)

be determined simultaneously with other measurements. Fluxes were also measured for the two gases by monitoring the concentrations of CO_2 and CH_4 in the head space with time. Figure 9.7 also shows the depth profiles of CO_2 and CH_4 measured in the sample samples used for the tracer injection experiment. Assuming steady-state for both profiles, we may calculate the flux of CO_2 and CH_4 from $D_{eff} \times dC/dz)_{z \Rightarrow 0}$. The D_{eff} must first be corrected for mass-dependent differences in the rate of diffusion of CO_2, CH_4 and SF_6:

$$\frac{D_{A,air}}{D_{B,air}} = \frac{\sqrt{\dfrac{m_{air} + m_A}{m_{air} m_A}}}{\sqrt{\dfrac{m_{air} + m_B}{m_{air} m_B}}} \tag{9.5}$$

(Jost, 1960) where $D_{A,air}$ and $D_{B,air}$ are the diffusivities of gases A and B in air; m_{air} is the molecular weight of air (28 g/mole), m_A and m_B are the molecular weights of A and B. Using this relation, CO_2 diffuses 17% faster and CH_4 54% faster than sulphur hexafluoride (SF_6) at STP. Equation 9.5 is also used to calculate isotopic fractionation for diffusing gases, discussed earlier.

The diffusive fluxes (from $D_{eff}{}^* dC/dt$) calculated for the experiment shown in Fig. 9.7 are $0.93-1.24 \times 10^{14}$ molecules cm^{-2}s^{-1} for CO_2 and -2.72 to -3.6×10^7 molecules cm^{-2}s^{-1} for CH_4 (negative flux is into the soil). The fluxes measured concurrently by following the change in concentration of the gases in the head space air with time were 1.5×10^{14} molecules cm^{-2}s^{-1} for CO_2 and -3.3×10^7 molecules cm^{-2}s^{-1} for CH_4 (Trumbore, 1988).

If we assume the D_{eff} is constant with depth in the soil profile, we may calculate the production as a function of depth in the soil by assuming steady-state and solving the equation for $Q(z)$. To a first approximation, $Q(z)$ will equal the second derivative of curves fit to the CO_2 and CH_4 data. Thus a second order polynomial fit to the data will give constant production rates with depth, while an exponential best-fit curve implies

---

Fig. 9.7 (*opposite*) Example of an experiment to determine D_{eff} from a purposeful tracer experiment using SF_6. (From Trumbore, 1988.) A modified flux chamber was used for the measurement, adapted from a design by Seiler & Conrad (1981). The chamber base extends 25 cm into the soil, to prevent exchange of gases in the soil atmosphere inside and outside of the box. The experiment was conducted in a small woodland adjacent to the Lamont-Doherty Geological Observatory in New York. The tracer was injected into the box head space at time 0. (a) Shows the subsequent decrease in SF_6 in the box head space with time (normalized so that the initial spike concentration is 1000). (b) Shows the penetration of the tracer into the soil air space measured 45 min after the spike addition; (——) and (———) represent the solution to Eqns 9.3 and 9.4 in the text for a D_{eff} of 0.015 and 0.020 cm^{2}s^{-1}, respectively (for SF_6 in air). (c) Shows the CO_2 and CH_4 measured in the soil atmosphere during the SF_6 tracer experiment (time = 45 min after spike addition).

exponential distribution of source or sink strength in the soil. This approach provides a method for estimating the distribution and magnitudes of trace gas sources and sinks within the soil profile. To be more than an estimate, the underlying assumptions of steady-state (no change in concentration profile with time) and constant D_{eff}, must be examined rigorously.

9.6.3 ^{222}Rn

A second method for estimating D_{eff} is through the distribution of radon (^{222}Rn) in the soil atmosphere. ^{222}Rn is produced from the decay of ^{226}Ra, which is present in mineral lattices of soils and rocks. ^{222}Rn which escapes into the soil atmosphere has one of two fates, either radioactive decay (half-life = 3.8 days), or loss to the atmosphere. The deficit of ^{222}Rn in the upper layer of the soil below the activity supported by ^{226}Ra decay thus may be used as a measure of the soil gas exchange rate. The following discussion of the process is adapted from Broecker (1965) and Dörr & Münnich (1987, 1990).

Equation 9.2 may be adapted to reflect the additional loss of ^{222}Rn by decay:

$$\frac{dC}{dt} = D_{eff}\frac{d^2C}{d^2z} + Q(z) - \lambda C, \tag{9.6}$$

where λ is the decay constant of ^{222}Rn, $2.1 \times 10^{-6}\,s^{-1}$, and z and t are depth and time variables. Assuming D_{eff} and Q are constant with depth, and that the ^{222}Rn profile reflects a steady state condition ($dC/dt = 0$), together with boundary conditions that $C(0) = 0$ and $dC/dz_{zfi\infty} = 0$, the solution to Eqn. 9.6 is:

$$C(z) = \frac{Q}{\lambda}\left[1 - \exp\left(-z\sqrt{\frac{\lambda}{D_{eff}}}\right)\right]. \tag{9.7}$$

Examples of the potential uses of independently determining both soil gas exchange rates and soil gas fluxes are demonstrated in the recent literature on controls of CH_4 consumption by soils. Born et al. (1990) and Whalen et al. (1992) both show that soils consuming the most CH_4 have the highest soil gas diffusivity (measured in both cases using ^{222}Rn). Thus valuable process information is obtained which may be useful in estimating the importance of CH_4 oxidation in soils as a sink for CH_4 on global scales.

9.7 Conclusions

Isotopic and tracer techniques provide powerful methods with which to understand ecosystem trace gas exchanges. The isotopic content of trace gases produced in terrestrial ecosystems is a reflection of substrate isotopic content, the pathways leading to trace gas production, and the processes mediating transport of the trace gas to the atmosphere. For example, CH_4 produced by acetate fermentation in a salt marsh will have $\delta^{13}C$ and δD values determined by its substrate isotopic content and kinetic isotope fractionations (as predicted by Whiticar *et al.*, 1986). If the CH_4 produced is bubbled quickly from the sediment, its isotopic content will remain essentially unchanged. If it remains in the sediment, CH_4 oxidation will remove isotopically light carbon, and the remaining CH_4 will be enriched in ^{13}C. Once in the atmosphere, the global average trace gas isotopic content will depend on the balance with other trace gas sources and on chemical reactions. Factors influencing the isotopic content of CH_4 include the balance of global CH_4 sources, and the fractionation associated with oxidation by OH and in soils.

Non-isotopic tracers are useful in developing a process-level understanding of the controls on atmosphere–ecosystem exchanges, including gas exchange. The same soil physical properties which influence the rate of exchange of soil air with the overlying atmosphere will also influence the amount and type of biological activity in the soil. For those investigating the fluxes of various trace gases from soils, it is desirable to separate and quantify physical and biological controls on the net flux. The methods described here for quantifying physical controls on gas exchange are inexpensive additions to any field campaign. The deliberate tracer method may be performed simultaneously with trace gas fluxes, using the same flux chambers, and should be included in any field campaign aimed at a predictive understanding of ecosystem trace gas exchange.

Many challenges remain in the use of isotopes and tracers to study trace gas fluxes into and out of ecosystems. Several applications are limited by the quality, cost or availability of isotopic measurements. One recent advance in measurement technology which has improved isotopic measurements is the development of AMS for ^{14}C measurements. The improved sample throughput of AMS makes possible global monitoring of the temporal trends for ^{14}C in atmospheric CO_2, CH_4 and CO. The observed trends will be complicated by the production of ^{14}C in pressurized water reactors; special efforts are needed to improve quantification of this source. AMS may also provide clues as to the importance of fossil fuel combustion compared to biospheric sources in the global budgets of other carbon-containing gases, such as ethylene, formaldehyde and several NMHCs. The greatest difficulty in making ^{14}C measurements of these gases, which have mixing ratios in the ppb to ppt range, will be in the development of non-contaminating, non-fractionating procedures to concentrate, purify and convert these gases into measurable form of sufficient quantity. Due to the difficulty in purification and the relatively small dynamic range, the potential for increasing our understanding of atmospheric nitrogen cycling through ^{15}N measurements, is still not fully realized. In particular, the combination of using nitrogen and oxygen isotopes for studying the N_2O budget requires further work. Finally, the potential of new technology using laser optic measurements of stable isotope ratios, though as yet imprecise compared to mass spectrometric measurements, shows great promise for increased ease and availability of isotopic measurements in the future.

Acknowledgements

I thank S. Tyler, J. Chanton, S. Whalen and B. Deck for helpful discussions, and P. Crill, D. DesMarias and P. Matson for thoughtful reviews.

References

Alperin M.J., Reeburgh W.S. & Whiticar M.J. (1988) Carbon and hydrogen isotope fractionation resulting from methane oxidation. *Global Biogeochemical Cycles*, **2**, 279–288.

Ambrose S.H. & DeNiro M.J. (1987) Bone nitrogen isotopic composition and climate. *Nature*, **325**, 201.

Aravena R., Schiff S.L., Trumbore S.E. & Elgood R. (1992) Evaluating dissolved inorganic carbon cycling in a forested lake watershed using carbon isotopes. *Radiocarbon*, **34**, 615–626.

Balesdent J., Mariotti A. & Guillet B. (1987) Natural ^{13}C abundance as a tracer for studies of soil organic matter dynamics. *Soil Biology and Biochemistry*, **19**, 25–30.

Bauer J.E., Williams P.M. & Druffel E.R.M. (1992) Recovery of sub-milligram quantities of carbon dioxide from gas streams by molecular sieve for subsequent determination of isotopic natural abundance. *Analytical Chemistry*, **64**, 824–827.

Becker J.F., Sauke T.B. & Lowenstien M. (1992) Stable isotope analysis using tunable diode laser spectroscopy. *Applied Optics*, **31**, 1921–1927.

Beer J., Siegenthaler U., Bonani G. *et al.* (1988) Information on past solar activity and geomagnetism from ^{10}Be in the Camp Century ice core. *Nature*, **331**, 675–679.

Benner R., Fogel M.L., Sprague E.K. & Hodson R.E. (1987) Depletion of ^{13}C in the lignin fractions of plants: Implications for stable carbon isotope studies. *Nature*, **329**, 708–710.

Blair N.E. & Carter W.D. (1992) The carbon isotope biogeochemistry of acetate from a methanogenic marine sediment. *Geochimicaet Cosmochimica Acta*, **56**, 1247–1258.

Blair N.E., Martens C.S. & DesMarias D. (1987) Natural abundances of carbon isotopes in acetate from a coastal marine sediment. *Science*, **236**, 66–68.

Born M., Dörr H. & Levin I. (1990) Methane consumption in aerated soils of the temperate zone. *Tellus*, **42B**, 2–8.

Brenninkmeijer C.A.M., Manning M.R., Lowe D.C. *et al.* (1992) Interhemispheric asymmetry in OH abundance inferred from measurements of atmospheric ^{14}CO. *Nature*, **356**, 50–52.

Broecker W.S. (1965) The application of natural radon to problems in ocean circulation. In: Ichiye T. (ed) *Symposium on Diffusion in Oceans and Freshwaters*. Lamont Geological Observatory, New York.

Broecker W.S. & Peng T.-H. (1982) *Tracers in the Sea*, pp. 500–568. Eldigio Press, New York.

Broecker W.S. & Walton A. (1959) The geochemistry of ^{14}C in the freshwater systems. *Geochimica et Cosmochimica Acta*, **16**, 15–38.

Buchanan D.L. & Corcoran B.J. (1959) Sealed-tube combustions for the determination of carbon-13 and total carbon. *Analytical Chemistry*, **31**, 1635–1637.

Burchuladze A.A., Chud'y M., Eristavi I.V. *et al.* (1989) Anthropogenic 14C variations in atmospheric CO_2 and wines. *Radiocarbon*, **31**, 771–777.

Cantrell C.A., Shetter R.E., McDaniel A.H. *et al.* (1990) Carbon kinetic isotope effect in the oxidation of methane by hydroxyl radicals. *Journal of Geophysical Research*, **95**, 22455–22462.

Carlslaw H.S. & Jaeger J.C. (1959) *Conduction of Heat in Solids*. Clarendon Press, Oxford.

Cerling T.R. (1991) Carbon dioxide in the atmosphere: evidence from Cenozoic and Mesozoic paleosols. *American Journal of Science*, **291**, 377–400.

Cerri C.C., Eduardo B.P. & Piccolo M.C. (1991) Use of stable isotopes in soil organic matter studies. In: *Stable Isotopes in Plant Nutrition, Soil Fertility and Environmental Studies*, Symposium Proceedings, pp. 247–259. International Atomic Energy Agency, Vienna.

Chanton J.P. & Martens C.S. (1988) Seasonal variations in ebullitive flux and carbon isotopic composition of methane in a tidal freshwater estuary. *Global Biogeochemical Cycles*, **2**, 289–298.

Chanton J., Crill P., Bartlett K. & Martens C.S. (1989) Amazon capims (floating grassmats): a source of ^{13}C enriched methane to the troposphere. *Geophysical Research Letters*, **16**, 799–802.

Chanton J.P., Martens C.S., Kelley C.A., Crill P.M. & Showers W.J. (1992) Methane transport mechanisms and isotopic fractionation in emergent macrophytes of an Alaskan tundra lake. *Journal of Geophysical Research*, **97**, 16681–16688.

Chanton J.P., Pauly G.G., Martens C.S., Blair N.E. & Dacey J.W.H. (1988) Carbon isotopic composition of methane in Florida Everglades soils and fractionation during its transport to the troposphere. *Global Biogeochemical Cycles*, **2**, 245–252.

Cicerone R.J. & Oremland R.S. (1988) Biogeochemical aspects of atmospheric methane. *Global Biogeochemical Cycles*, **2**, 299–328.

Coleman D.C. & Fry B. (1991) *Carbon isotope techniques*, San Diego, Academic Press, 274 pp.

Coleman M.L., Shepherd T.J., Durham J.J., Rouse J.E. & Moore G.R. (1982) *Analytical Chemistry*, **54**, 993–995.

Coplen T.B., Wilman J.D. & Chen J. (1992) Improvements in the gaseous hydrogen water equilibration technique for hydrolgen isotope ratio analysis. *Analytical Chemistry*, **63**, 910–912.

Craig H. (1954) Carbon-13 in plants and the relationships between carbon-13 and carbon-14 in nature. *Journal of Geology*, **62**, 115–149.

Craig H. (1957) Isotopic standards for carbon and oxygen and correction factors for mass spectrometric analysis of carbon dioxide. *Geochimica Cosmochimica Acta*, **12**, 133–149.

Craig H. (1961) Mass-spectrometer analysis of radiocarbon standards. *Radiocarbon*, **3**, 1–3.

Craig H. (1965) The measurement of oxygen isotope paleotemperatures. In: *Stable Isotopes in Oceanographic Studies and Paleotemperatures*, pp. 1–24. Consiglio Nazionale delle Richerche, Laboratorio di Geologia Nucleare, Pisa.

Craig H. & Keeling C.D. (1963) The effects of atmospheric N_2O on the measured isotopic concentration of atmospheric CO_2. *Geochimica et Cosmochimica Acta*, **27**, 549–551.

Craig H. & Lal D. (1961) The production rate of natural tritium. *Tellus*, **13**, 85–105.

Craig H., Miller S.L. & Wasserberg G.J. (1964) *Isotopic and Cosmic Chemistry*. North-Holland, Amsterdam.

Davidson J.A., Cantrell C.A., Tyler S.C. *et al.* (1987) Carbon kinetic isotope effect in the reaction of CH_4 with HO. *Journal of Geophysical Research*, **92**, 2195–2199.

Delwiche C.C., Zinke P.J., Johnson C.M. & Virginia R.A. (1979) Nitrogen isotope distribution as a presumptive indicator of nitrogen fixation. *Botanical Gazette*, **140**, 565–569.

DeNiro M.J. & Epstein S. (1978) Influence of diet on the distribution of carbon isotope ratios in animals. *Geochimica et Cosmochimica Acta*, **42**, 495–506.

Deines P. (1980) The isotopic composition of reduced organic carbon. In: Fritz P. & Fontes

J. (eds) *Handbook of Environmental Isotope Geochemistry*, vol. 1a, pp. 329–406. Elsevier, Amsterdam.

Dörr H. & Münnich K.O. (1986) Annual variations of the ^{14}C content of soil CO_2. *Radiocarbon*, **28**, 338–345.

Dörr H. & Münnich K.O. (1987) Annual variations in soil respiration in selected areas of the temperate zone. *Tellus*, **39B**, 114–121.

Dörr H. & Münnich K.O. (1990) ^{222}Rn flux and soil air concentration profiles in West-Germany. Soil ^{222}Rn as tracer for gas transport in the unsaturated soil zone. *Tellus*, **42B**, 20–28.

Elmore D. & Phillips F. (1987) Accelerator Mass Spectrometry for measurement of long-lived isotopes. *Science*, **236**, 543–550.

Emiliani C. (1966) Isotopic Paleotemperatures. *Science*, **154**, 851–857.

Farquhar G.D., O'Leary M.H. & Berry J.A. (1982) On the relationship between carbon isotope discrimination and the intercellular carbon dioxide concentration in leaves. *Australian Journal of Plant Physiology*, **9**, 121–137.

Faure G. (1986) *The Principle of Isotope Geology*, 2nd edn., pp. 386–521. John Wiley and Sons, New York.

Finkel R. & Suter M. (1994) *AMS in the Earth Sciences: Techniques and Applications*, in press.

Francey R.J. & Tans P.P. (1987) Latitudinal variation in oxygen-18 of atmospheric CO_2. *Nature*, **327**, 495–497.

Freyer H.D. (1991) Seasonal variation of $^{15}N/^{14}N$ ratios in atmospheric nitrate species. *Tellus*, **43B**, 30–44.

Freidli H., Lötsher H., Oeschger U., Siegenthaler. & Stauffer B. (1986) Ice core record of the $^{13}C/^{12}C$ ratio of atmospheric CO_2 in the past two centuries. *Nature*, **324**, 237–238.

Friedli H., Siegenthaler U., Rauber D. & Oeschger H. (1987) Measurements of $^{13}C/^{12}C$ and $^{18}O/^{16}O$ ratios of tropospheric carbon dioxide over Switzerland. *Tellus*, **39B**, 80–88.

Friedli H. & Siegenthaler U. (1986) Influence of N_2O on isotope analyses of CO_2 and mass-spectrometric determination of N_2O in air samples. *Tellus*, **40B**, 129–133.

Freyer H.D. (1979) On the ^{13}C record in tree tings. Part I. ^{13}C variations in northern hemispheric trees during the past 150 years. *Tellus*, **31B**, 124–137.

Fung I., John J., Lerner J. *et al.* (1991) 3-Dimensional model synthesis of the global methane cycle. *Journal of Geophysical Research*, **96**, 13033–13065.

Galimov E.M. (1985) *The Biological Fractionation of Isotopes*. Academic Press, New York.

Gat J.R. (1980) The isotopes of hydrogen and oxygen in precipitation. In: Fritz P. & Fontes J. (eds) *Handbook of Environmental Isotope Geochemistry*, vol. 1a, pp. 21–47. Elsevier, Amsterdam.

Glinski J. & Stipniewski W. (1985) Gas transport in the soil environment. In: *Soil Aeration and its Role for Plants*. CRC press, Boca Raton, Florida.

Harkness D.D., Harrison A.F. & Bacon P.J. (1986) The temporal distribution of 'bomb' ^{14}C in a forest soil. *Radiocarbon*, **28**, 328–337.

Heaton T.H.E. (1990) $^{15}N/^{14}N$ ratios of NO_x from vehicle engines and coal-fired power stations. *Tellus*, **42B**, 304–307.

Hesterberg R. & Siegenthaler U. (1991) Production and stable isotopic composition of CO_2 in a soil near Bern, Switzerland. *Tellus*, **43B**, 197–205.

Hoefs J. (1980) *Stable Isotope Geochemistry*. Springer-Verlag, New York.

Horita J., Ueda A., Mizukami K., Takatori I. (1989) Automatic delta-D and delta-O-18 analysis of multiwater samples using H_2-water and CO_2-water equilibration methods with a common equilibration set-up, *Applied Radiation and isotopes – International Journal of Radiation Applications and Instrumentation*, Part A, **40**, 801–805.

Jellick G.J. & Schnabel R.R. (1986) Evaluation of a field method for determining the gas diffusion coefficient in soils. *Soil Science Society of America Journal*, **50**, 18–23.

Jenkinson D.J. & Raynor J.H. (1977) The turnover of soil organic matter in some of the

Rothamsted classical experiments. *Soil Science*, **123**, 298–305.

Jenkinson D.J., Adams D.E. & Wild A. (1991) Model estimates of CO_2 emissions from soil in response to global warming. *Nature*, **351**, 304–306.

Jost W. (1960) *Diffusion in Solids, Liquids, Gases.* Academic Press, New York.

Keeling C.D. (1958) The concentration and isotopic abundances of atmospheric carbon dioxide in rural areas. *Geochimica et Cosmochimica Acta*, **13**, 322–334.

Keeling C.D., Bacastow R.B., Carter A.F. *et al.* (1989) A three-dimensional model of atmospheric CO_2 transport based on observed winds: 1. Analysis of observational data. In: Peterson P.H. (ed) *Aspects of Climate Variability in the Pacific and Western Americas*, pp. 165–236. Geophysical Monograph 55, American Geophysical Union, Washington, DC.

Kendall C. & Coplen T.B. (1985) Multisample conversion of H_2O to H_2 by zinc for stable isotope determination. *Analytical Chemistry*, **57**, 1437–1440.

Kendall C. & Grim E. (1990) Combustion tube method for measurement of nitrogen isotope ratios using calcium oxide for total removal of carbon dioxide and water. *Analytical Chemistry*, **62**, 526–529.

Kim K.-R. & Craig H. (1990) Two-isotope characterization of N_2O in the Pacific ocean and constrains on its origin in deep water. *Nature*, **347**, 58–61.

Kim K.R. & Craig H. (1993) Nitrogen-15 and oxygen-18 characteristics of nitrous oxide – a global perspective. *Science*, **262**, 1855–1857.

Kunz C. (1985) Carbon-14 discharge at three light-water reactors. *Health Phys*, **49**, 25–35.

Lai S.-H., Tiedge J.M. & Erickson A.E. (1976) *In situ* measurement of gas diffusion coefficients in soils. *Soil Science Society of America Journal*, **40**, 3–6.

Lal D. & Peters B. (1967) Cosmic-ray produced radioactivity in the Earth. In: Sitte K. (ed) *Handbuch der Physik*, pp. 551–612. Springer-Verlag, Berlin **46**.

Lansdown J.M., Quay P.D. & King S.L. (1992) CH_4 production via CO_2 reduction in a temperate bog: a source of ^{13}C-depleted CO_2. *Geochimica et Cosmochimica Acta*, **56**, 3493–3503.

Lassey K.R., Lowe D.C., Brenninkmeijer C.A.M. & Gomez A.J. (1993) Atmospheric methane and its carbon isotopes in the southern hemisphere – their time series and an instructive model. *Chemosphere*, **26**, 95–109.

Lerman J.C., Mook W.G. & Vogel J.C. (1970) ^{14}C in tree rings from different localities. In: Olsson I.U. (ed) *Radiocarbon Variations and Absolute Chronology*, pp. 275–299. Almquist and Wiksell, Stockholm.

Leuenberger M., Siegenthaler U. & Langway C.C. (1992) Carbon isotope composition of atmospheric CO_2 during the last ice age from an Antarctic Ice Core. *Nature*, **357**, 488–490.

Levin I., Kromer B., Schoch-Fischer H. *et al.* (1989) 25 years of tropospheric ^{14}C observations in central Europe. *Radiocarbon*, **27**, 1–19.

Lingenfelter R.E. (1963) Production of carbon-14 by cosmic-ray neutrons. *Reviews of Geophysics*, **1**, 35–55.

Lowe D.C. & Judd W.J. (1987) Graphite preparation for radiocarbon dating by accelerator mass spectrometry. *Nuclear Instrumental Methods*, **B28**, 113–116.

Lowe D.C., Brenninkmeijer C.A., Tyler S.C. & Dlugkencky E.J. (1991) Determination of the isotopic composition of atmospheric methane and its application in the Antarctic. *Journal of Geophysical Research*, **96**, 15455–15467.

Mak J.E., Brenninkmeijer C.A.M. & Manning M.R. (1992) Evidence for a missing carbon monoxide sink based on tropospheric measurements of ^{14}CO. *Geophysical Research Letters*, **19**, 1467–1470.

Manning M.R., Lowe D.C., Melhuish W.H., Sparks R.J., Wallace G., Brenninkmeijer C.A.M. & McGill R.C. (1989) The use of radiocarbon measurements in atmospheric studies. *Radiocarbon*, **32**, 37–58.

Marino B.D. & McElroy M.B. (1991) Isotopic composition of atmospheric CO_2 inferred from carbon in C4 plant cellulose. *Nature*, **349**, 127–131.

Marino B.D., McElroy M.B., Salawitch R.J. & Spaulding SW.G. (1992) Glacial to interglacial variations in the carbon isotopic composition of atmospheric CO_2. *Nature*, **357**, 461–466.

Mariotti A. (1984) Natural ^{15}N abundance measurements and atmospheric nitrogen standard calibration. *Nature*, **311**, 251–252.

Martens C.S., Blair N., Green C.D. & Des Marias D.J. (1986) Seasonal variations in the stable carbon isotopic signature of biogenic methane in a coastal sediment. *Science*, **233**, 1300–1303.

McCarty P.L. (1964) The methane fermentation. In: Heukelekian H. & Dondero N.C. (eds) *Principles and Applications in Aquatic Microbiology*, pp. 91–108. J. Wiley and Sons, New York.

McKinney C.R., McRea J.M., Epstein S., Allen H.A. & Urey H.C. (1950) Improvements in mass spectrometers for the measurement of small differences in isotopic abundance ratios. *Review of Scientific Instruments*, **21**, 724–730.

Middleton R., Klein J. & Fink D. (1989) A CO_2 negative ion source for ^{14}C dating. *Nuclear Instrumental Methods*, **B43**, 231–239.

Minegawa M. & Wada E. (1984) Stepwise enrichment of ^{15}N along food chains: further evidence and the relation between $\delta^{15}N$ and animal age. *Geochimica et Cosmochimica Acta*, **48**, 549–555.

Mook W.G. (1980) *The Principles and Applications of Radiocarbon Dating*. Elsevier, Holland.

Mook W.G. & Jongsma J. (1987) Measurement of the N_2O correction for $^{13}C/^{12}C$ ratios of atmospheric CO_2 by removal of N_2O. *Tellus*, **39B**, 96–99.

Moore H. (1977) The isotopic composition of ammonia, nitrogen dioxide and nitrate in the atmosphere. *Atmospheric Environment*, **11**, 1239–1243.

Nadelhoffer K.J. & Fry B. (1988) Controls on natural N-15 and C-13 abundances in forest soil organic matter. *Soil Science Society of America Journal*, **52**, 1633–1640.

O'Brien B.J. & Stout J.D. (1978) Movement and turnover of soil organic matter as indicated by carbon isotopic measurements. *Soil Biology and Biochemistry*, **10**, 309–317.

O'Leary M.H. (1981) Carbon isotope fractionation in plants. *Phytochemistry*, **20**, 553–567.

Peng T.-H. (1985) Atmospheric CO_2 variations based on the tree-ring ^{13}C record. In: Sundquist E.T. & Broecker W.S. (eds) *The Carbon Cycle and Atmospheric CO_2: Natural Variations Archean to Present*, pp. 123–131. Geophysical monograph 32, American Geophysical Union, Washington, DC.

Peng T.-H., Broecker W.S., Freyer H.D. & Trumbore S.E. (1983) A deconvolution of the tree ring based $\delta^{13}C$ record. *Journal of Geophysical Research*, **88**, 3609–3620.

Quay P.D., King S.L., Stutsman J. *et al.* (1991) Carbon isotopic composition of atmospheric CH_4: fossil fuel and biomass burning source strengths. *Global Biogeochemical Cycles*, **5**, 25–47.

Rankama K. (1954) *Isotope Geology*. McGraw-Hill, New York.

Rolston D.E., Glauz R.D., Grundmann G.L. & Louie D.T. (1991) Evaluation of an *in situ* method for measurement of gas diffusivity in surface soils. *Soil Science Society of America Journal*, **55**, 1536–1542.

Rundel P.W., Ehleringer J.H. & Nagy K.A. (eds) (1989) *Stable Isotopes in Ecological Research*. Springer-Verlag, Berlin.

Schearer G. & Kohl D.H. (1989) Estimates of N_2 fixation in ecosystems: the need for and basis of the ^{15}N natural abundance method. In: Rundel P.W., Ehleringer J.H. & Nagy K.A. (eds) *Stable Isotopes in Ecological Research*. Springer-Verlag, Berlin. pp. 342–374

Schimel D.S. (1993) *Isotope Techniques in Plant, Soil and Marine Biology*, p. 119, Academic Press, San Diego.

Seiler W. & Conrad R. (1981) Field measurements of natural and fertilizer-induced N_2O release rates from soils. *APCA Journal*, **31**, 767–772.

Siegenthaler U. & Oeschger H. (1987) Biospheric CO_2 emissions during the past 200 years reconstructed by deconvolution of ice core data. *Tellus*, **39B**, 140–154.

Smith B.N. & Epstein S. (1971) Two categories of $^{13}C/^{12}C$ ratios for higher plants. *Plant Physiology*, **47**, 380–384.

Stevens C.M. & Engelkemeir A. (1988) Stable carbon isotopic composition of methane from some natural and anthropogenic sources. *Journal of Geophysical Research*, **93**, 725–733.

Stevens C.M. & Krout L. (1972) Method for the determination of the concentration of and of the carbon and oxygen isotopic composition of atmospheric carbon monoxide. *International Journal of Mass Spectrometry and Ion Physics*, **8**, 265–275.

Stevens C.M. & Rust F.E. (1982) The carbon isotopic composition of atmospheric methane. *Journal of Geophysical Research*, **87**, 4879–4882.

Stuiver M. & Polach H. (1977) Reporting of ^{14}C data. *Radiocarbon*, **19**, 355–363.

Stuiver M. & Quay P.D. (1981) Atmospheric ^{14}C changes resulting from fossil fuel release and cosmic ray flux variability. *Earth and Planetary Science Letters*, **53**, 349–362.

Stuiver M. & Quay P.D. (1984) $^{13}C/^{12}C$ ratios in tree-rings and the transfer of biospheric carbon to the atmosphere. *Journal of Geophysical Research*, **89**, 11371–11748.

Stuiver M., Braziunas T.F., Becker B. & Kromer B. (1991) Climatic, solar, oceanic and geomagnetic influences on Late-Glacial and Holocene atmospheric $^{14}C/^{12}C$ change. *Quaternary Research*, **35**, 1–24.

Tanweer A. (1990) Importance of clean metallic zinc for hydrogen isotope analysis. *Analytical Chemistry*, **62**, 2158–2160.

Tieszen L.L. & Boutton T.W. (1989) Stable carbon isotopes in terrestrial ecosystem research. In: Rundel P.W., Ehleringer J.H. & Nagy K.A. (eds) *Stable Isotopes in Ecological Research*, pp. 167–195. Springer-Verlag, Berlin.

Tolstikhin I.N. & Kamenskiy I.L. (1969) Determination of groundwater ages by the T-^{3}He method. *Geochemistry International*, **6**, 810–811.

Trumbore S.E. (1988) *Carbon cycling and gas exchange in soils*, PhD thesis. Columbia University, New York.

Trumbore S.E. (1993) Comparison of carbon dynamics in temperate and tropical soils using radiocarbon measurements. *Global Biogeochemical Cycles*, **7**, 275–290.

Trumbore S.E., Vogel J.S. & Southon J.R. (1989) AMS ^{14}C measurements of fractionated soil organic matter: an approach to deciphering the soil carbon cycle. *Radiocarbon*, **31**, 644–654.

Tyler S.C. (1986) Stable carbon isotope ratios in atmospheric methane and some of its sources. *Journal of Geophysical Research*, **91**, 13232–13238.

Tyler S.C. (1991) The global methane budget. In: Rogers J.E. & Whitman W.B. (eds) *Microbial Production and Consumption of Greenhouse Gases: Methane, Nitrogen Oxides and Halomethanes*, pp. 7–38. American Society for Microbiology, Washington, DC.

Tyler S.C. (1992) Kinetic isotope effects and their use in studying atmospheric trace species, Case study, $CH_4 + OH$. In: Kaye J.A. (ed) *Isotope Effects in Gas-Phase Chemistry*, pp. 390–408. American Chemical Society, Washington, DC.

Tyler S.C., Zimmerman P.R., Cumberbatch C. *et al.* (1988) Measurements and interpretations of $\delta^{13}C$ measurements of methane from termites, rice paddies and wetlands in Kenya. *Global Biogeochemical Cycles*, **2**, 349–355.

Ueda S., Ogura N. & Wada E. (1991) Nitrogen stable isotope ratio of groundwater N_2O. *Geophysical Research Letters*, **18**, 1449–1452.

Vogel J.S. (1992) A rapid method for preparation of biomedical targets for AMS. *Radiocarbon*, **34**, 344–350.

Vogel J.S., Nelson D.E. & Southon J.R. (1987) ^{14}C background levels in an accelerator mass spectrometry system. *Radiocarbon*, **29**, 323–333.

Vogel J.S., Southon J.R., Nelson D.E. & Brown T.A. (1984) Performance of catalytically condensed carbon for use in AMS. *Nuclear Instrumental Methods*, **B5**, 284–293.

Volz A., Ehhalt D. & Derwent R.G. (1981) Seasonal and latitudinal variation of [14]CO and the tropospheric concentration of OH radicals. *Journal of Geophysical Research*, **86**, 5163–5171.

Volz A., Ehhalt D.H. & Derwent R. (1982) Seasonal and latitudinal variation of [14]CO and the tropospheric concentration of OH radicals. *Journal of Geophysical Research*, **86**, 5163–5171.

Wahlen M. & Yoshinari T. (1985) Oxygen isotope ratios in N_2O from different environments. *Nature*, **313**, 780–782.

Wahlen M., Tanaka N., Deck B. *et al.* (1990) δD in CH_4: additional constraints for a global CH_4 budget. *EOS*, **71**(43), 1249.

Wahlen M., Tanaka N., Henry R. *et al.* (1989) Carbon-14 in methane sources and in atmospheric methane: the contribution from fossil carbon. *Science*, **245**, 286–290.

Wanningkhof R., Ledwell J.R. & Broecker W.S. (1985) Gas exchange-wind speed relationship measured with sulfur hexafluoride on a lake. *Science*, **227**, 1224–1226.

Whalen S.C., Reeburgh W.S. & Barber V.A. (1992) Oxidation of methane in boreal forest soils: a comparison of seven measures. *Biogeochemistry*, **16**, 181–211.

Whiticar M.J., Faber E. & Schoell M. (1986) Biogenic methane formation in marine and freshwater environments: CO_2 reduction vs. acetate formation – isotope evidence. *Geochimica et Cosmochimica Acta*, **50**, 693–709.

Yoshida N. (1988) ^{15}N-depleted N_2O as a product of nitrification. *Nature*, **335**, 528–530.

Zaucker F. & Broecker W.S. (1992) The influence of atmospheric moisture transportation on the fresh water balance of the Atlantic drainage basin – general circulation model simulations and observations. *Journal of Geophysical Research*, **97**, 2765–2773.

CHAPTER 10

Microbial processes of production and consumption of nitric oxide, nitrous oxide and methane

E.A. DAVIDSON & J.P. SCHIMEL

10.1 Introduction

Soil microorganisms produce and consume NO, N_2O and CH_4. Emissions of these gases from soils result from the net effects of oxidative and reductive microbial metabolism. Identifying these microbial processes and the conditions under which they occur should help explain spatial and temporal variation in rates of gaseous emissions and should aid efforts to extrapolate results of field studies to larger scales of space and time. The objective of this paper is to describe methods and procedures that have proven useful for identifying soil sources and sinks of NO, N_2O and CH_4, and for identifying soil factors that regulate rates of microbial production and consumption of these gases.

10.2 Identification of functional groups of microorganisms

Our emphasis on functional groups rather than species of microorganisms reflects the common modes of metabolism that exist among many groups of microbes. For most studies of trace gases from a biogeochemical perspective, it is not important to know whether the agent of N_2O production, for example, is *Pseudomonas fluorescens* or *Alcaligenes* spp. However, knowing when and where N_2O is produced by heterotrophic bacteria rather than by chemoautotrophic bacteria may help identify the factors that control rates of production. We shall define the important functional groups and discuss how they may be identified. Several reviews are available that provide more thorough descriptions of these microbial processes (Firestone, 1982; Large, 1983; Haynes, 1986; Tiedje, 1988; Bédard & Knowles, 1989; Conrad, 1989; Firestone & Davidson, 1989; Galchenko *et al.*, 1989; Topp & Hanson, 1991; Hutchinson & Davidson, 1993; Schimel *et al.*, 1993).

10.2.1 Definition of functional groups and processes

Denitrification

The term denitrification has been used to encompass a number of microbial and abiological processes, but the more narrow definition of Firestone & Davidson (1989) is used here: a form of anaerobic respiration in

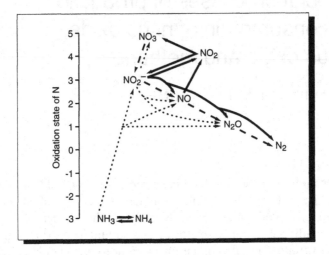

Fig. 10.1 Biological and abiological processes of production and consumption of NO and N_2O (Davidson, 1991). (· · · · nitrifying bacteria; – – dentrifying bacteria; — abiological reactions.)

bacteria which couples reduction of nitrogen oxides to electron transport phosphorylation. Denitrification is a heterotrophic process that requires a source of reductant (usually organic carbon). Bacteria from many genera can denitrify; most are aerobes that use O_2 as a terminal electron acceptor when it is available and switch to nitrogen oxides when O_2 becomes scarce. Hence, denitrification generally occurs in the absence of O_2, or at least at low partial pressures of O_2. Denitrification can result in the production of NO, N_2O and N_2 (Fig. 10.1). As NO and N_2O are oxides that can be further reduced, they can also be consumed by denitrification.

Nitrifying bacteria

Nitrification is defined as the biological oxidation of reduced forms of nitrogen to NO_2^- and NO_3^-. Most nitrification is probably carried out by chemoautotrophic bacteria. These bacteria obtain their energy from this oxidation and use the energy to fix CO_2 to organic carbon. Several genera of these bacteria oxidize NH_4^+ to NO_2^- and one known genus oxidizes NO_2^- to NO_3^-. Nitrifying bacteria produce NO and N_2O as a byproduct of NH_4^+ oxidation (Fig. 10.1). We do not know which of several possible biochemical pathways of NO production by nitrifying bacteria is most important. The production of N_2O most likely results from reduction of NO_2^- by NH_4^+ oxidizers (Poth & Focht, 1985).

Many heterotrophic bacteria and fungi can oxidize organically bound nitrogen and/or NH_4^+; this process is called heterotrophic nitrification.

These heterotrophs do not appear to obtain energy from this process, so the advantage they gain from carrying out this oxidation is unclear (Killham, 1986). Production of NO and N_2O can occur as a byproduct of heterotrophic nitrification (Papen *et al.*, 1989), although the significance of this source has not been demonstrated.

Chemodenitrification

This term does not describe a group of organisms, but rather a group of abiological reactions that result in production of NO, NO_2 and N_2O (Fig. 10.1). Self-decomposition of HNO_2 yields NO and NO_2, and reactions of HNO_2 with phenolic groups of soil organic matter can produce both NO and N_2O (Nelson, 1982). Soil NO_2^- can be involved in chemodenitrification because of the equilibrium between NO_2^- and HNO_2. Chemodenitrification is pH-dependent and occurs at significant rates only below pH 5. However, acidity measured on a bulk soil sample may not accurately reflect the acidity at soil microsites where the reactants may be concentrated. Hence, chemodenitrification and the interaction of chemodenitrification and nitrification have been hypothesized for soils with acidity above pH 5 (Davidson, 1992).

CH_4 production

CH_4 is produced almost exclusively by methanogenic bacteria, which are a group of the archaeobacteria. They are strictly anaerobic and produce CH_4 by two dominant pathways, cleaving acetate to CO_2 and CH_4, and reducing CO_2 with H_2 (Whitman *et al.*, 1992). As different organisms are capable of using different substrates, the overall controls on methanogenesis may vary with the pathway responsible. Other substrates such as methylamines may also be used (Yarrington & Wynn-Williams, 1985; Whitman *et al.*, 1992) but these are generally minor in terrestrial systems.

Methanogenesis is the terminal step in a complex food web. Simple substrates (sugars, amino acids, etc.) are released directly by plants or from soil organic matter breakdown. These substrates are then fermented to acetate and H_2, which are used by methanogens. The different fermentation pathways make it possible for some compounds to act as carbon sources for CH_4, while others are fermented to H_2 and act only as energy sources.

Because methanogenesis is obligately anaerobic it generally only occurs at high rates in systems which are continuously water saturated and are rich in organic matter; thus soil sources include natural wetlands and rice paddies. While CH_4 production can occur in upland soils, it probably only occurs in pulses after rain events that saturate the soil for an extended period; upland soils are more commonly sinks for CH_4 (Schimel *et al.*, 1993).

CH_4 oxidation

CH_4 is oxidized in soils primarily by two enzymes: (a) CH_4 monooxygenase, found in methanotrophic bacteria; and (b) NH_3 monooxygenase, found in nitrifying bacteria (Bédard & Knowles, 1989). NH_3 monooxygenase is capable of oxidizing CH_4 because CH_4 and NH_3 are similar in shape and size (tetrahedra with 31.1 vs. 29.7 nm Van der Waals radii for CH_4 and NH_3 respectively; Weast, 1976). While nitrifiers can oxidize CH_4, they are unable to grow on it (Bédard & Knowles, 1989).

Methanotrophs are a diverse group of aerobic bacteria that oxidize CH_4 through methanol, formaldehyde, and formic acid to CO_2. Methanotrophs require CH_4 for growth, being unable to use other substrates. Other organisms that can use compounds containing methyl groups (e.g. methylamine) are called methylotrophs; most of these probably cannot use CH_4 (Lidstrom, 1992).

10.2.2 Identification of sources and sinks of NO and N_2O

Three approaches to deduce sources and sinks of NO and N_2O will be discussed: (a) selective inhibitors; (b) addition of substrates; and (c) sterilization.

Inhibitors

NH_3 monooxygenase, the enzyme in chemoautotrophic nitrifying bacteria that oxidizes NH_4^+ (actually, NH_4^+ is first deprotonated and NH_3 is oxidized), is very susceptible to a variety of inhibitors. Two of the most useful nitrification inhibitors that irreversibly bind to NH_3 monooxygenase are nitrapyrin (2-chloro-6-(trichloromethyl)-pyridine; Rogers & Ashworth, 1982) and C_2H_2 (Hynes & Knowles, 1982).

Nitrapyrin. Nitrapyrin is a potent inhibitor of nitrification, and it does not appear to have any direct effect on denitrification (although inhibiting nitrification may eventually lead to NO_3^- depletion, thus indirectly affecting rates of denitrification). A disadvantage of nitrapyrin is that the powder must be mixed with a soil sample or added dissolved in solution, so that either mixing or wetting of the soil is required. In laboratory studies of mixed, wetted soil samples from a grassland, addition of nitrapyrin inhibited production of NO and N_2O, indicating that nitrification was the primary source of both gases (Tortoso & Hutchinson, 1990). Nitrapyrin is often not completely effective as a nitrification inhibitor in soils rich in organic matter (Chancy & Kamprath, 1987), such as many forest soils, but it is widely used and generally effective in agricultural soils. Its effectiveness declines after several days or weeks as it is hydrolysed.

C_2H_2. In contrast to nitrapyrin, C_2H_2 is added as a gas and can diffuse throughout an intact soil core, provided that the core is not so wet that gaseous diffusion is restricted. Several designs for incubation vessels and for circulating gases have been developed to improve distribution of C_2H_2 within a soil core (see review by Tiedje et al., 1989). Using C_2H_2 with intact soil cores avoids artifacts of enhanced substrate availability and aeration caused by soil mixing. Two disadvantages of C_2H_2 are that it can be used as a carbon source by heterotrophs (Haider et al., 1983; Terry & Duxbury, 1985), and at high concentrations it inhibits reduction of N_2O by denitrifying bacteria (Yoshinari et al., 1977). Significant consumption of C_2H_2, and hence a decline in its effectiveness, generally occurs only after exposure of a week or more, so the first problem can be avoided by using short-term incubations (1 day or less) and by not repeating treatments on the same soil sample.

The second complication, non-specificity of C_2H_2, can be used to advantage to distinguish between nitrifying and denitrifying sources and sinks of NO and N_2O in laboratory incubations of soil samples (Davidson et al., 1986; Klemedtsson et al., 1988; Davidson, 1992; Davidson et al., 1993b). Nitrifying bacteria are inhibited by only 10 Pa C_2H_2 (Berg et al., 1982; Robertson & Tiedje, 1987), whereas 100 Pa to 10 kPa C_2H_2 is needed to inhibit N_2O reduction by denitrifiers (Ryden et al., 1979). The difference in NO or N_2O emissions between 0 and 10 Pa C_2H_2 treatments therefore indicates the contribution of NO or N_2O production by nitrifying bacteria. The difference between the 10 Pa and 10 kPa treatments then provides an estimate of N_2O consumption (reduction) by denitrifying bacteria (Fig. 10.2).

The 10 Pa C_2H_2 treatment must be used with either a gas circulation system for soil cores (Robertson & Tiedje, 1987) or mixed soils in a static incubation (Davidson, 1992). A partial pressure this low makes it difficult to obtain uniform distribution in an intact static core. Using a slightly higher C_2H_2 concentration might ensure inhibition of nitrification, but might also inhibit N_2O reduction by denitrifiers.

Studies of intact cores with only the high (10 kPa) C_2H_2 treatment are also useful. When N_2O reduction to N_2 by denitrifiers is insignificant, as it often is when soil water content is below field capacity, the 10 kPa treatment either has no effect, or it decreases N_2O production by inhibiting nitrification. In these cases, the 10 kPa C_2H_2 treatment inhibits nitrification and provides an estimate of NO and N_2O production from denitrification, while the difference between the 0 and 10 kPa treatments provides a minimum estimate of the nitrification source. This is a minimum estimate for N_2O production by nitrifiers, however, because some reduction of N_2O by denitrifiers might occur at wet microsites even in relatively dry soil. Inhibition of N_2O reductase by 10 kPa C_2H_2 can

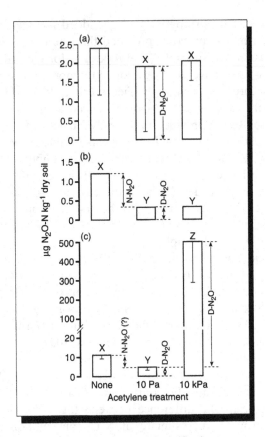

Fig. 10.2 Distinguishing between nitrification and denitrification as sources of N_2O, using incubations with three levels of C_2H_2. Soil samples were collected from (a) a well-drained area near a stream, (b) a well-drained midslope area, and (c) a poorly drained area near a stream of a hardwood forest in North Carolina (Davidson *et al.*, 1986). Means and standard deviations are given for triplicate 24-h incubations. Within each panel, means labelled by the same letter (X, Y or Z) are not significantly different (LSD, $\alpha = 0.05$). Production of N_2O and N_2 from nitrification and denitrification are calculated from subtraction of treatment means as indicated, where the means are statistically significantly different.

increase N_2O production by denitrifiers, partially offsetting the inhibition of N_2O production by nitrifiers.

Acetylene does not affect reduction of NO, so 10 Pa and 10 kPa treatments should give similar results and are redundant for studies of NO sources (unless incomplete diffusion of the C_2H_2 at 10 Pa results in incomplete inhibition of nitrification). The luminol-based method of NO detection used by Davidson (1992) is sensitive to C_2H_2, so C_2H_2 must be applied as a pretreatment and then vented before beginning incubations and analysing head space gas samples.

C_2H_2 has also been used in field studies, both in field incubations of intact soil cores and directly injected beneath the soil surface (see review of these methods by Tiedje *et al.*, 1989). This approach usually is used to quantify total denitrification by inhibiting reduction of N_2O to N_2 with C_2H_2, so that production of N_2O provides a measure of both the 'normal' rate of N_2O production plus the amount that would have been reduced to N_2 had C_2H_2 not been present (NO is often ignored in these studies). As in laboratory incubations, if C_2H_2 addition to soils in the field reduces N_2O emissions, then it can be concluded that nitrification is a contributing source of N_2O. If emissions of N_2O increase when C_2H_2 is added, then reduction of N_2O to N_2 by denitrifiers is significant, but the source of emitted N_2O could be from either nitrifiers or denitrifiers (and may result from both).

C_2H_2 from a cylinder often contains acetone and other contaminants that can be removed by passing it through a sulphuric acid train (Tiedje, 1982). C_2H_2 generated from the reaction of calcium carbide with water contains fewer contaminants, and none that are substrates or inhibitors of denitrification at the concentrations present (Tiedje *et al.*, 1989).

Oxygen. Oxygen is required for nitrification and it generally inhibits denitrification. The complex effects of O_2 on the relative proportions of end-products of nitrification and denitrification are discussed below (see p. 344). Here we focus on the use of O_2 as an inhibitor of denitrification.

Robertson & Tiedje (1987) used a combination of C_2H_2 to inhibit nitrification and very high O_2 partial pressures to inhibit denitrification in intact soil cores from two Michigan forests. They observed a significant source of N_2O that was inhibited by neither C_2H_2 nor O_2. The source may have been heterotrophic nitrifiers (Papen *et al.*, 1989) or a number of nitrate and nitrite respirers that can produce N_2O (Tiedje, 1988). Soil emissions of NO and N_2O are often low in the acid temperate and boreal forest soils where these other microbial processes are most likely to be active (Bowden, 1986; Davidson, 1991; Williams *et al.*, 1992). These functional groups have not been described in greater detail, because their role in NO and N_2O production is thought to be relatively unimportant on a global scale (Firestone & Davidson, 1989). These potential sources have not been well studied, and this assumption has not been tested by field studies.

Substrate addition

The substrate for nitrification is NH_4^+, whereas the substrate for denitrification is usually NO_3^-. It follows that an increase in NO and N_2O production following application of NH_4^+ (or urea) might indicate that nitrifying bacteria are the source, while an increase following NO_3^-

application would indicate denitrification as the source. Two problems arise from this interpretation. First, when NH_4^+ or urea application stimulates nitrification, NO_3^- is produced, which may then also stimulate denitrification. Hence, increased emissions of NO or N_2O following application of NH_4^+ or urea could result from either nitrification or denitrification. Second, while NO_3^- application can stimulate denitrification (Johansson et al., 1988; Keller et al., 1988; Livingston et al., 1988; Bakwin et al., 1990), the process stimulated in the short term may not be the one that produces most of the NO and N_2O in untreated soils over the long term. Nitrification and denitrification can occur simultaneously at different microsites within the soil. Nitrification could be an important source of NO and N_2O, even though the denitrifying bacteria respond to episodes of added NO_3^-.

Sterilization
A simple approach to investigating the importance of abiological sources of NO and N_2O is to compare production of these gases in sterile and non-sterile soils. However, all sterilization treatments have artifactual effects that can affect NO and N_2O production. Autoclaving disrupts soil colloids, causes disappearance of soil NO_2^-, and can result in substantial N_2O production. Robertson & Tiedje (1987) found that N_2O in autoclaved soil cores had to be vented before beginning incubations. Davidson (1992) found that a grassland soil that contained NO_2^- and produced NO lost its NO_2^- and NO production when autoclaved, but NO production was restored at very high rates when sterile NO_2^- was added back to autoclaved soil. These results demonstrated a large capacity for chemodenitrification, but the actual importance of chemodenitrification *in situ* could not be determined.

Gamma irradiation also sterilizes soil, but it has the opposite effect to autoclaving with respect to NO_2^-. Depending on the dose, gamma irradiation can cause radiolysis of NO_3^- to NO_2^- and subsequent production of NO_2 and NO (Cawse & Cornfield, 1972) which must be vented before beginning incubations.

The results of sterilization must be interpreted with caution regarding the importance of chemodenitrification as a source of NO in soils. On the other hand, sterilization by autoclaving has been used effectively to demonstrate the absence of an abiological sink for NO and N_2O (Remde et al., 1989; Davidson, 1992) and the absence of an abiotic source of N_2O (Robertson & Tiedje, 1987; Davidson, 1992).

10.2.3 Identification of sources and sinks of CH_4

Sources

Two aspects of determining the source of CH_4 have been considered: (a) the ultimate source of the carbon and energy (recent plant inputs vs. old peat); and (b) the pathway of CH_4 production (acetate fermentation vs. CO_2 reduction).

Dating with ^{14}C can be used to distinguish between recent and old sources of carbon during CH_4 production. This approach requires the use of an accelerator mass spectrometer to analyse small sample masses (Chanton *et al.*, 1988). This technique has been used to show that plant-derived carbon is the dominant source pool for CH_4 in the Everglades (Chanton *et al.*, 1988) and in the Yukon–Kuskokwim Delta region of Alaska (Martens *et al.*, 1991).

Pathways

The pathway dominating CH_4 production can be examined by two approaches: (a) specific substrate additions; and (b) the natural abundance of ^{13}C in CH_4 produced. Measuring the effect of adding unlabelled compounds can identify which substrates provide energy for methanogens, either directly or through the food web. However, the specific source of carbon is not idenitified by this method, because the added substrate may be fermented by other organisms to H_2, which is then used by the methanogens to reduce CO_2. To solve this problem, a NaOH trap has been included in the incubation flask to remove CO_2 from the head space, so that the only source of carbon is the added substrate, and an increase in CH_4 production should be observed only if the substrate can be used directly (Yarrington & Wynn-Williams, 1985). A more reliable test to determine whether a compound is used for carbon or energy is to add ^{14}C-labelled substrates and to monitor the production of CH_4 and [14]CH_4 (Yarrington & Wynn-Williams, 1985).

Natural abundance isotope methods: [13]*C,* [14]*C.* All of the processes affecting CH_4 (production, consumption and transport) fractionate between ^{12}C and ^{13}C, and between hydrogen and deuterium, always selecting for the lighter isotopes (Chanton & Dacey, 1991). Enrichments of ^{13}C and deuterium in organic matter and CH_4 are always lower than the reference standards, producing negative $\delta^{13}C$ and δD values. Selection for the lighter isotopes produces more negative values, while processes that fractionate less have a product with a greater (less negative) $\delta^{13}C$ or δD.

Methanogenesis produces CH_4 with a lower $\delta^{13}C$ than the substrate from which it was produced. Different methanogenic pathways, however, fractionate differently. The acetoclastic pathway fractionates less than the H_2/CO_2 pathway (Krzycki *et al.*, 1987), providing an approach for identifying the *in situ* pathway for methanogenesis. Lansdown *et al.* (1991) used this approach to show that the bulk of the CH_4 produced in an acid peat bog was via CO_2 reduction, while Martens *et al.* (1986) showed that in a coastal sediment, acetate was the dominant CH_4 source during the summer but that the source varied seasonally.

CH_4 oxidation also fractionates carbon and hydrogen isotopes, leaving behind CH_4 enriched in ^{13}C and deuterium. This fractionation was used to show that CH_4 oxidation is enhanced by aerenchymous plant roots penetrating into the anaerobic layer in the Florida Everglades (Burke *et al.*, 1988). Plant transport can also fractionate carbon isotopes; CH_4 in plants and released to the atmosphere is often $\gg 10$‰ lighter than CH_4 bubbles in the rooting zone (Chanton *et al.*, 1988; Chanton & Dacey, 1991). For these reasons it is preferable to collect gas with soil sampling probes, which minimize the possibility of oxidation and transport, rather than with flux chambers.

Sinks

No simple method exists for determining whether methanotrophs or nitrifiers are primarily responsible for CH_4 oxidation. Moreover, we do not know which of the many species of methanotrophs are dominant. In fact, the K_m values for isolated cultures of methanotrophs are too high to account for CH_4 consumption and growth at atmospheric CH_4 concentrations (Conrad, 1984; Topp & Hansen, 1991). We do not know whether unidentified methanotrophs are generally responsible for the process in nature or whether the organisms isolated have isoenzymes with lower K_m values that are not expressed in culture conditions.

Specific inhibitors are a common way of analysing metabolic pathways in microbial ecology (Oremland & Capone, 1988), and their use has been applied in several cases to the study of CH_4 consumption. Methyl fluoride is a particularly useful inhibitor of CH_4 oxidation, as it is a water-soluble gas (Oremland & Culbertson, 1992a,b). One difficulty has been that many of the compounds that inhibit CH_4 oxidation, such as C_2H_2, nitrapyrin and methyl fluoride also inhibit NH_4^+ oxidation and cannot therefore be used to distinguish between these groups of organisms (Oremland & Capone, 1988; Megraw & Knowles, 1990). Picolinic acid, however, blocks NH_4^+ oxidation by methanotrophs but not by nitrifiers and therefore has potential as a diagnostic inhibitor (Salvas & Taylor, 1984; Megraw & Knowles, 1990).

Another promising method is the ratio of $CO:CH_4$ oxidation rates. Both nitrifiers and methanotrophs can oxidize these compounds, but the relative rates are different (Jones *et al.*, 1984; Bédard & Knowles, 1989). Nitrifiers have a high affinity for CO relative to CH_4, while methanotrophs have the reverse affinities. A high ratio of $CO:CH_4$ oxidation rates is therefore indicative of nitrifier activity, while a low ratio suggests methanotrophs. This techniques has been used by Steudler and co-workers (P.A. Steudler, personal communication), in which they used ^{14}C-labelled CO and CH_4 to measure their oxidation rates in a range of soils. They found that the ratios of $CO:CH_4$ oxidation varied widely between soils, indicating that both nitrifiers and methanotrophs may be important CH_4 oxidizers in soils.

A rough approach to distinguishing whether nitrifiers or meth-anotrophs are responsible for observed oxidation of CH_4 would be to determine relative activities of these two groups of organisms. Assays for nitrification potentials and CH_4 oxidation potentials are described in the next section which provide indices of nitrifier and methanotroph populations. If nitrification potentials are low, then methanotrophs would most likely be responsible for CH_4 consumption. Conversely if meth-anotroph populations are low, then nitrifiers would probably be re-sponsible for CH_4 oxidation.

10.3 Laboratory assays for potential (V_{max}) rates of microbial activity

Laboratory assays can determine the enzymatic capacities of existing soil microbial populations to carry out nitrification (Belser, 1979; Belser & Mays, 1980), denitrification (Tiedje et al., 1989), methanogenesis (Svensson, 1984; Yavitt et al., 1988), and CH_4 oxidation (King, 1990; Nesbit & Breitenbeck, 1992). These assays usually involve short-term (i.e. 1 day or less) incubations in which substrates are added in excess to soil samples that are gently shaken or slurried to avoid diffusional limit-ations. Temperature and acidity are also often controlled. The limiting factor under these defined incubation conditions is assumed to be the amount of enzyme present in the soil sample capable of carrying out the specific process measured. Incubations are usually short to minimize *de novo* enzyme synthesis, and for the denitrification assay, an inhibitor of protein synthesis is added. The utility of these assays varies, described below.

10.3.1 Denitrification

Denitrifying enzymes can persist in the soil for long periods, apparently without being used, so results of the dentrification enzyme assay may reflect denitrification that occurred several months earlier as well as recent activity. Smith & Parsons (1985) found significant denitrifying enzyme capacity after 2 months storage of dry soils. Persistent denitrifying enzymes may be an adaptation enabling denitrifiers to switch quickly from O_2 to NO_3^- for use as a terminal electron acceptor in respiration when soil water content increases abruptly (Rudaz *et al.*, 1991). Groffman & Tiedje (1989) found a significant correlation between denitrifying enzyme activities and annual estimates of denitrification obtained from field studies across several temperate forest sites. Hence, this assay may provide a simple means of making qualitative comparisons of the relative importance of denitrification across sites integrated over annual or longer time periods. A step-by-step description of the method is given by Tiedje *et al.* (1989).

10.3.2 Nitrification

Chemoautotrophic nitrifying bacteria are obligate aerobes that obtain all of their energy from nitrification. Unlike denitrification enzymes that may persist unused for long periods while denitrifiers use aerobic metabolism, nitrification enzymes are the only means for nitrifiers to obtain energy. If substrate availability or some other factor limits nitrification, the nitrifiers eventually starve and the enzymatic capacity of the soil declines (Davidson *et al.*, 1990). The nitrifying enzyme assay thus provides an indication of recent capacity of the soil to nitrify and may be strongly affected by seasonal variation as well as differences among sites. Because nitrification is not only a direct source of NO and N_2O, but it is also the source of NO_3^- used by denitrifiers to produce NO and N_2O, variation in nitrifying enzyme capacity may be a good indicator of the potential for NO and N_2O production. A step-by-step description of the procedure is given by Hart *et al.* (1994).

10.3.3 CH₄ oxidation

The most effective method for measuring CH_4 oxidation potentials uses unsaturated soils, rather than slurries, to facilitate CH_4 diffusion into the soil (Bender & Conrad, 1992; Schimel *et al.*, 1993). A method that we are using involves static incubations of hand mixed soils (5 g) in sealed jars. The soils are adjusted to the optimal moisture content for CH_4 oxidation, and CH_4 is added to the head space to ensure that the oxidation rate is at V_{max}. Values vary for different soils; we have found that 30% of saturation and 2000–4000 ppm CH_4 are reasonable starting points. CH_4 oxidation activity is determined by periodically sampling head space CH_4 over several days.

CH₄ oxidation rates have also been measured by adding $^{14}CH_4$ to a sample in a sealed incubation jar and measuring the decrease in $^{14}CH_4$ and the increase in $^{14}CO_2$ (Whalen & Reeburgh, 1990; Yavitt *et al.*, 1990). Following the labelled CH_4, rather than merely changes in CH_4 concentration, allows measurement of the gross rather than the net rate of consumption. This also allows simultaneous measurement of production and consumption, as the gross rate of production is equal to the difference between gross consumption and net change in CH_4 concentration.

10.3.4 Methanogenesis

CH_4 production potential is most commonly measured using anaerobic slurries (Svensson, 1984; Yarrington & Wynn-Williams, 1985; Yavitt *et al.*, 1988). In this method, soils or sediments are slurried in water with or without additional substrates, the head space is purged of oxygen, and the samples are incubated anaerobically, either shaken or static. CH_4 production is measured by periodically sampling the head space for CH_4 concentration over an extended period of time (often several weeks). To avoid CH_4 build-up from inhibiting production it may be necessary to purge the head space periodically (Valentine *et al.*, 1993). One important component of using this assay is to limit the sample's exposure to air. This can be done by taking intact cores, transporting them sealed, and starting the anaerobic slurries as soon as possible after opening up the cores. If possible, the samples should be processed under a nitrogen atmosphere.

Anaerobic slurries are regularly used to estimate field rates of CH_4 production (Yavitt *et al.*, 1988; King, 1990). This raises the concern that the disturbance involved can alter the CH_4 production rates. Slurrying can alter substrate supply and ensuring anaerobiosis may maximize CH_4 production while eliminating CH_4 consumption. Substate diffusion may be enhanced by slurrying, but probably not excessively in wetland soils. Altering redox by making the slurry anaerobic could also alter rates, but O_2 concentrations and redox usually decrease rapidly below the surface of the water table. Making soils completely anaerobic may therefore not be an unreasonable simulation of soils well below the water table. For soils sampled from the transition zone between highly reduced soils and aerobic surface soils anaerobic slurries may substantially overestimate production rates (King, 1990).

10.4 Related biological processes

Emissions of trace gases result from complex interactions of carbon and nitrogen cycles. Emissions of NO and N_2O are highly sensitive to nitrogen availability, while CH_4 production is most sensitive to carbon availability. Hence, measures of carbon and nitrogen dynamics may help reveal factors that control microbial production and consumption of carbon and nitrogen trace gases.

10.4.1 Nitrogen availability in the soil

Unfortunately, no single assay, incubation or index has been identified that consistently predicts soil nitrogen availability across wide ranges of soil types, ecosystems and scales of investigation. A number of these techniques can be useful to help elucidate factors that control these microbial processes, but the types of comparisons and the scale of the study affects the utility of each technique.

Soil inorganic nitrogen

Several processes, including production and assimilation by micro-organisms, plant uptake and leaching, affect concentrations of inorganic nitrogen in soil. Soil extracts provide only a 'snapshot' of NH_4^+ and NO_3^- concentrations and tell little of the rate at which these substrates are produced and consumed. In a California grassland, concentrations of soil NH_4^+ and NO_3^- were low (1–7 μg nitrogen g $^{-1}$ dry soil) but the rate of nitrification ranged from about 0.6 to 3.5 μg nitrogen g^{-1} dry soil per day, indicating that the pools of soil NH_4^+ and NO_3^- turned over about every 2 days (Davidson *et al.*, 1990). Hence, low concentrations of nitrogen substrates and products do not necessarily indicate low activities of the microbial processes.

Although numerous examples exist in the literature of significant correlations between either NH_4^+ or NO_3^- concentrations and pro-duction of NO or N_2O, the relationships are very site specific, and no consistent trend across studies has emerged. At a coarser scale, however, soil NO_3^- may be a useful indicator of soils with excess nitrogen relative to carbon. For example, Williams & Fehsenfeld (1991) showed that soil NO_3^- was not a good predictor of NO emissions within any given site, but when fertilized croplands were compared with grasslands and forests, soil NO_3^- correlated well with NO emissions over several orders of magnitude. Within each site, modest differences in topographic gradients or distribution of crop and forest residues may have been more important contributors to observed variability, but the large-scale differences between ecosystem types was reflected by accumulation of soil NO_3^-. When nitrogen availability exceeds carbon availability, as it often does in tilled and fertilized croplands, then nitrogen accumulates as NO_3^-, and nitrogen losses as NO and N_2O may also be high. Although this cor-relation shows the importance of carbon and nitrogen availability, it does not indicate the NO source. Emissions of NO and accumulation of NO_3^- could covary with the rate of NO_3^- production (nitrification), with NO_3^- availability for denitrification, or both.

Concentrations of soil NO_2^- may also be relevant to emissions of NO. Two studies from seasonally dry tropical biomes, savannas of Venezuela (Johansson & Sanhueza, 1988) and forests of Mexico (Davidson *et al.*, 1991b), have shown accumulation of soil NO_2^- during the dry season, which may be related to large emissions of NO by either biological denitrification or chemodenitrification when the dry soil is first wetted.

Net nitrogen mineralization and net nitrification
Net nitrogen mineralization is defined as the change in soil NH_4^+ and NO_3^- during an incubation of soil in the absence of plant roots. Net nitrification is defined as the change in soil NO_3^- during such incubations. Typically, intact soil cores placed in polyethylene bags are incubated in the field (Eno, 1960), or mixed soils samples are incubated in the laboratory. Subsamples are analysed for inorganic nitrogen before and after the incubation period. A step-by-step description of the method is given by Hart *et al.* (1994).

The modifying term 'net' emphasizes that soil microorganisms carry out several processes during these incubations that affect inorganic nitrogen concentrations, including mineralization of organic nitrogen to inorganic nitrogen, nitrification, assimilation of both NH_4^+ and NO_3^-, and possibly denitrification. Hence the rate of change in NH_4^+ and NO_3^- concentrations results from the net effects of all of these processes.

With so many potentially confounding factors, it is not surprising that net rates have not proven universally effective in predicting soil emissions of NO and N_2O. Nevertheless, some important relationships have been revealed with these techniques. Robertson & Tiedje (1984) reported that net nitrification correlated with rates of denitrification in forest soils of Michigan, indicating that potential NO_3^- production is more important than is NO_3^- pool size as a controller of denitrification. Matson & Vitousek (1990) showed a significant correlation between net mineralization and N_2O emissions among several humid tropical forest sites, indicating that variation in nitrogen fertility within this biome is related to N_2O emissions. However, sites that had been converted from forest to pasture had higher N_2O emissions than would have been expected from their net mineralization rates. Moreover, we cannot explain why net mineralization seems to be a good predictor of nitrogen gas emissions in some studies, while net nitrification works better for others. Despite these limitations, net rates often reveal differences among sites in nitrogen availability that appear related to nitrogen gas emissions. Measures of net mineralization and net nitrification are relatively simple and inexpensive and have been used widely.

The interactions of nitrogen cycling with CH_4 dynamics are more poorly understood, but appear to be important. In rice paddies, nitrogen fertilization increases CH_4 efflux, but this effect may be due primarily to increased plant productivity and below-ground inputs of carbon (Lindau *et al.*, 1991). CH_4 oxidation was inhibited by high levels of NH_4^+ availability in previously fertilized soils (Steudler *et al.*, 1989; Mosier *et al.*, 1991). The mechanism involved may be competitive inhibition at the active site of either CH_4 or NH_3 monooxygenase (Bédard & Knowles, 1989), but this has not been validated (Schimel *et al.*, 1993). Studies examining the interaction of nitrogen cycling dynamics and both CH_4 production and consumption are badly needed.

Gross nitrogen mineralization and gross nitrification

Gross rates are defined as the actual rates at which organic nitrogen is mineralized to NH_4^+ (gross mineralization) and at which NH_4^+ is oxidized to NO_3^- (gross nitrification; where heterotrophic nitrification is important, oxidation of organic nitrogen to NO_3^- is also included). Estimates are obtained from analysis of ^{15}N pool dilution (Davidson *et al.*, 1991a). A step-by-step description of a soil core method using a 1-day incubation is given by Hart *et al.* (1994), although other methods exist and should be considered if longer incubations are desired (Nason & Myrold, 1991; Wessel & Tietema, 1992).

The value of measuring gross rates of nitrogen mineralization and nitrification has not yet been fully explored for studies of trace gases. Davidson *et al.* (1993b) demonstrated that soil emissions of NO and N_2O were only about 0.1% of the rate of gross nitrification in a seasonally dry forest in Mexico. By measuring gross rather than net rates, the effects of confounding factors in measures of net rates are avoided, and rates of NO and N_2O emissions can be compared specifically to rates of mineralization and nitrification. The trade off is that the ^{15}N pool dilution technique is far more cumbersome, time consuming, expensive and prone to experimental error than is the incubation for net mineralization and net nitrification.

Estimates of gross rates of microbial assimilation of NH_4^+ and NO_3^- can also be obtained from the ^{15}N pool dilution technique (Davidson *et al.*, 1991a; Nason & Myrold, 1991; Wessel & Tietema, 1992; Hart *et al.*, 1993). Hence, the strength of the microbial assimilatory sink for inorganic nitrogen can be measured directly. A strong microbial sink for inorganic nitrogen should represent a competing fate for inorganic nitrogen that precludes significant production of nitrogen gases. These measurements have not yet been made at enough sites to test this hypothesis.

10.4.2 Soil respiration

Soil respiration provides a valuable bioassay of carbon availability to microbes and of total microbial activity. Respiration measurements can include field and laboratory measurements. Field measures of total soil respiration include root respiration, while laboratory measurements generally involve disturbing the soil, which introduces other artifacts. Although root and microbial respiration cannot be reliably differentiated with current techniques of field measurements (Singh & Grupta, 1977), total soil respiration provides a rough index of soil microbial activity and carbon availability. Application of this index to studies of NO, N_2O and

CH_4 has not been pursued deeply. Davidson *et al.* (1993b) used measures of CO_2 flux to demonstrate that microbial activity began immediately following wetting of very dry soil, and that simultaneous emissions of NO and N_2O were probably biological. Mechanistic models that relate microbial dynamics of nitrogen and carbon to fluxes of trace gases may use soil emissions of CO_2 to help parameterize the models (see Chapter 11, this volume).

Carbon availability can be a control on denitrification activity (Burford & Bremner, 1975; Robertson & Tiedje, 1984) and is a critical control on the rate of methanogenesis (Schimel *et al.*, 1993). Therefore, an estimate of available carbon may be useful in studies on CH_4 and N_2O dynamics. There are several methods for estimating carbon availability, but soil respiration may be the most useful and simplest (Davidson *et al.*, 1987).

10.5 Field measurements of related chemical and physical properties of the soil

Abiotic factors have diverse effects on trace gas emissions, and appropriate measures of them are necessary to understand gas dynamics fully. The most important factor controlling NO, N_2O and CH_4 emissions is soil water content, but temperature and acidity may have important effects as well.

10.5.1 Soil water content and gaseous diffusivity

Direct measures of soil diffusivity using tracers such as sulphur hexaflouride (SF_6) or radon are discussed in Chapter 9, so this discussion will focus on the appropriate expressions of soil wetness as they relate to gaseous diffusion. An excellent discussion of expressions of soil water is given by Papendick & Campbell (1981), which will be reviewed here only briefly. Gravimetric soil water content (θ_g) and volumetric soil water content (θ_v) have units of $g\,H_2O\,g^{-1}$ oven-dry soil and $cm^3\,H_2O\,cm^{-3}$ soil, respectively, and are measures of the water that volatilizes when soil is dried at 105°C. Matric potential (Ψ_m) is a measure of the energy of water in contact with the soil matrix relative to the energy of free water. Expressed in units of pressure (MPa), Ψ_m is affected by total soil water content, organic matter and soil texture. Soil Ψ_m is often used to indicate drought stress experienced by plant roots and soil microorganisms. 'Saturation' occurs when all pore spaces of the soil are filled with water. Field capacity is defined as the soil water content after saturated soil has drained freely. At field capacity, the 'micropores' are generally water-filled while the so-called 'macropores' are mostly air-filled. Field capacity is often operationally defined as $-0.033\,MPa$ or $-0.010\,MPa\,\Psi_m$, although Ψ_m at field capacity varies somewhat with soil texture. 'Water-holding capacity' is another term sometimes used interchangeably with field capacity and sometimes used interchangeably with water content at saturation. Finally, water-filled pore space (WFPS), usually expressed as a percentage, is the ratio of volumetric soil water content to total porosity of the soil ($100 \times \theta_v/\varepsilon$, where $\varepsilon = cm^3$ pore space cm^{-3} soil). Soil at 100% WFPS is saturated. WFPS can be calculated from θ_g, bulk density, and particle density:

$$\% WFPS = [100 \times (\theta_g \times BD)]/[1 - (BD/PD)]$$

where BD is the bulk density (g dry soil cm^{-3}) and PD is the particle density ($2.65\,g$ dry soil cm^{-3} for most soils).

The WFPS parameter has the additional advantage of being largely comparable among soils of different texture. The parameters θ_g and θ_v cannot be easily compared among soils because a value, such as 0.2, may be relatively dry for a fine-textured or organic-rich soil and relatively wet for a coarse-textured soil. Soil matric potential accounts for the effects of soil texture directly, but Ψ_m is a more useful indicator of water availability to living organisms than it is a predictor of diffusion of substrates through gas and liquid phases (Skopp *et al.*, 1990). In contrast, WFPS is directly related to diffusivity. WFPS also accounts for variation in total porosity (ε) among soils as a result of compaction, although it does not indicate pore size distribution (Doran *et al.*, 1990). Where gaseous diffusion is known to be an important controlling factor, as is the case for production and consumption of NO, N_2O and CH_4, WFPS appears to be the most appropriate parameter to express soil water content.

Two mechanisms relate soil water content to regulation of nitrification, denitrification, CH_4 oxidation and methanogenesis. First, the supply of substrates to microbes (e.g. NH_4^+ to nitrifiers and acetate to methanogens) occurs via diffusion of these substrates in soil water films. Second, water in soil pores is the dominant control of gaseous diffusion in soil. Diffusion of a gas through water is about 10 000 times slower than through air. Soil water content therefore controls both the rate of O_2 and CH_4 diffusion into the soil, and the rate of NO, N_2O, CH_4 and CO_2 diffusion out of the soil. Oxygen and CH_4 must ultimately enter the liquid phase before being consumed by microorganisms living in water films, and this phase change may also be rate limiting in some cases (Skopp, 1985).

Linn & Doran (1984) noted that diffusion of organic carbon substrates in water films is probably rate limiting for microbial respiration in dry soils, whereas diffusion of O_2 is rate limiting for microbial respiration in wet soils. Similarly, chemoautotrophic nitrification could also be limited by diffusion of NH_4^+ in water films of dry soil and by diffusion of O_2 in wet soil (Papendick & Campbell, 1981). Skopp *et al.* (1990) used volumetric soil water content (θ_v) to predict diffusion of substrates in soil solution and used θ_v/ε (which equals WFPS) to predict diffusion of gases. The optimal soil water content was generally about 60% WFPS for aerobic processes, such as microbial respiration and nitrification, and was >80% WFPS for the anaerobic process of denitrification. The optimal value of 60% WFPS for aerobic processes represents the intersection of increasing availability of organic carbon and inorganic nitrogen and decreasing availability of O_2 with increasing soil water content. In model simulations, the optimal value of WFPS varied depending on empirical constants used to characterize diffusion in a given soil, but optimal values of about 60% WFPS were remarkably consistent for respiration in a variety of actual soils, indicating the robustness of this parameter (Doran et al., 1990).

The supply of O_2 is an important control on nitrification, denitrification, CH_4 oxidation and methanogenesis. Nitrification requires O_2, while denitrification is inhibited by O_2. The supply of O_2 also affects the relative proportions of end-products of both nitrification and denitrification (Firestone et al., 1979; Goreau et al., 1980). Nitrifiers produce more N_2O relative to NO_2^- under low O_2 conditions (Goreau et al., 1980). Hence, the effects of O_2 on N_2O production by nitrifiers is complicated; maximum N_2O production occurs when there is enough O_2 for some NH_4^+ oxidation, but when partial O_2 limitation results in production of N_2O as a significant end product. Similarly, the relative proportions of NO, N_2O and N_2 produced by denitrifiers is also affected by O_2 (Firestone et al., 1979; Firestone & Davidson, 1989). Denitrification occurs when partial pressures of O_2 are low, but completely anaerobic conditions can result in production of N_2 as the dominant end product of denitrification. Maximum N_2O production by denitrifiers occurs when partial pressures of O_2 are low enough to promote reduction of NO_3^-, but not so low as to promote reduction of N_2O.

Davidson (1993) has suggested that nitrification is a more important source of NO and N_2O than is denitrification when WFPS < 60%, and that the opposite is true when WFPS > 60%. It was also shown that the ratio of $N_2O : NO$ emitted from the soil was <1 when WFPS < 60%, and that this ratio was >1 when WFPS > 60%. It follows that nitrification produced more NO than N_2O, and denitrification produced more N_2O than NO. For many soils, field capacity occurs at about 60% WFPS, so that field capacity may be used in lieu of WFPS for inferring whether nitrification or denitrification is the dominant source and whether NO or N_2O is the dominant gas produced. Data from more sites are needed to support these generalizations, but, unfortunately, field capacity, bulk density and WFPS are often not reported. Analysis of a larger dataset (Davidson, 1993) required use of gravimetric soil water content, which, for the reasons described above, was not expected to be a good parameter

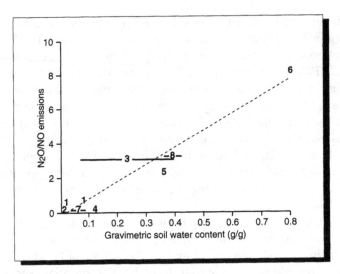

Fig. 10.3 The relation between gravimetric soil water content (θ_g) and the ratio of emissions of N_2O and NO in a variety of ecosystems. Where a range of soil water contents was reported, a horizontal bar indicates the range and the symbol marks either the midpoint or the arithmetic average of the reported soil water contents. The ratio of emissions was calculated from either the reported mean or median flux of NO and N_2O from each site. Symbols identify data from the following studies: 1, Mexican drought-deciduous forest during dry and wet seasons (Davidson *et al.*, 1993b); 2, Venezuelan savanna on sandy soil (Hao *et al.*, 1988; Johansson *et al.*, 1988); 3, Venezuelan savanna on sandy clay loam soil (Sanhueza *et al.*, 1990); 4 and 5, Brazilian humid tropical forest on sandy soil and clayey soil, respectively (Bakwin *et al.*, 1990; Matson *et al.*, 1990); 6, Costa Rican rainforest (Keller & Reiners, 1991); 7, managed pasture in Texas, USA (Hutchinson *et al.*, 1993); and 8, perennial grassland of Kansas, USA (Coyne, Dunkin and Firestone, University of California, Berkeley, unpublished data). ($\cdots$) Indicates the least squares simple linear regression. The regression equation is:

ratio of $N_2O/NO = 10.1(\theta_g) - 0.25$; $R_2 = 0.93$.

for comparing the effects of soil water content across sites of varying soil texture. Nevertheless, the ratio of $N_2O:NO$ was correlated to gravimetric water content (Fig. 10.3). Although the empirical relationship shown in Fig. 10.3 is intriguing, a more mechanistically based model would be more satisfying and might be more useful for modelling purposes. Because WFPS can be related to diffusivity, it would be the preferred parameter for assessing the effects of soil wetness on trace gas emissions.

Rates of CH_4 uptake have also been shown to be related to soil diffusivity (Born *et al.*, 1990). For upland soils the source of CH_4 is the atmosphere. Diffusion of CH_4 into the soil is therefore perhaps the prime control on CH_4 consumption (Schimel *et al.*, 1993), so CH_4 consumption should be enhanced in dry soils. This is counterbalanced by water potential effects that should reduce the activity of methanotrophs under extremely dry conditions. WFPS should be closely associated with CH_4 consumption activity, but this has not been examined.

None of the measures of soil water discussed above will be very useful in studying CH_4 production, as this process only occurs at substantial rates in fully saturated soils. The more useful measure of soil water in studies on CH_4 production is water table height. As the water table drops below the soil surface, CH_4 efflux rates drop rapidly (Harriss et al., 1982; Moore & Knowles, 1989). This is due to the combined effects of reduced CH_4 production as the anaerobic soil volume shrinks and increased CH_4 consumption in the aerobic surface soil. Water table height is easily measured simply by removing a soil core and measuring the depth to the water that fills the core hole.

Another variable related to soil water that has a major effect on CH_4 and N_2O production is redox potential. As methanogenesis is a strictly anaerobic process, it can only occur at low redox potentials. CH_4 production often begins at redox potentials below $+100 \, mV$, but usually does not become rapid until the redox drops below $0 \, mV$ (Yagi & Minami, 1990; Lindau et al., 1991). Production of N_2O generally occurs between $+200 \, mV$ and $+500 \, mV$, and reduction of N_2O generally occurs below $+250 \, mV$ (Smith et al., 1983). Redox is normally measured with a platinum electrode (Gambrell et al., 1975).

10.5.2 Acidity
Soil acidity affects nearly all microbial processes, including those discussed here. Nitrification, denitrification and CH_4 production can be significant in soils that are naturally acidic if other factors are favourable (Firestone, 1982; Robertson, 1982; Robertson & Tiedje, 1984, 1987, 1988; Martikainen, 1985; Stams et al., 1990; Keller & Reiners, 1991). The rates of these processes, however, are often reduced by acidification and increased by liming (Firestone, 1982; Nagele & Conrad, 1990a,b; Schimel et al., 1993).

In culture, chemoautotrophic nitrifying bacteria generally show a neutral or slightly alkaline pH optimum (Hankinson & Schmidt, 1984). The same is true for many denitrifying bacteria, but some denitrifiers are acid-adapted and have pH optima as low as 3.9 (Parkin et al., 1985). Whatever the pH optimum, however, the bulk soil acidity need not approximate the pH optimum for microbial processes to occur. Rather, the organisms need only survive and grow in microsites which are within their pH tolerance range; the pH of these microsites may be quite different than the bulk soil pH. Thus nitrifiers and denitrifiers may be active despite the fact that the bulk soil pH is below optimal.

Additional complexity to variation in acidity results from the interaction of the nitrification and dentrification rates with the balance of their end-products. Acidity or any other factor that slows the rate of any step in a multistep process (e.g. denitrification in Fig. 10.1) causes accumulation of intermediates, such as N_2O, in laboratory incubations (Betlach & Tiedje, 1981). This effect of acidity may explain why N_2O is a common final product of dentrification in acid forest soils (Melillo *et al.*, 1983), but this hypothesis has not been fully tested.

While variation in soil acidity should not be ignored, soil acidity exerts complex controls on microbial populations and enzyme activities that are not easily interpreted (Nagele & Conrad, 1990a). Variation in soil acidity can also be confounded with variation in nitrogen availability. Consideration of soil acidity may be crucial for studies that compare the effects of fertilizer, liming and fire regimes. It is our view, however, that indices of nitrogen availability and soil wetness, rather than soil pH, will prove to be the most useful parameters for understanding variation in NO, N_2O, and CH_4 fluxes at regional and global scales.

10.5.3 Temperature

The rates of most enzymatic processes increase exponentially with temperature provided that other factors are not limiting. Many instances of temperature dependence of production and consumption of NO, N_2O and CH_4 have been observed (King & Adamsen, 1992; Nesbit & Breitenbeck, 1992; Williams *et al.*, 1992; Schimel *et al.*, 1993). Although temperature can be a strong controller of microbially mediated production and consumption of trace gases, other factors, such as substrate and moisture availability can often be limiting such that the temperature effect is not expressed (Davidson *et al.*, 1993a; Schimel *et al.*, 1993).

Diel effects should be considered when extrapolating short-term flux measurements to daily or longer time scales. At a minimum, fluxes should be measured at the warmest and coolest times of the day at least once in a study to determine the magnitude of diel variation. The soil depth where temperature should be measured obviously depends on the depth to which diel variation in temperature occurs and where production and consumption of the trace gases occur. This depth may vary widely among ecosystems and among trace gases. Production of NO that escapes the soil probably occurs near the surface, whereas N_2O may emanate from much deeper in the soil. Production of CH_4 may occur near the water table where diel variation in temperature is small, but consumption of CH_4 may occur near the surface where diel variation in temperature is larger.

In temperate and boreal ecosystems, seasonal variation in temperature is confounded with other seasonal patterns of plant growth and mineralization of carbon and nitrogen substrates. Hence, relating seasonal variation in fluxes to a specific cause, such as temperature, moisture or substrate availability, must be done with caution.

10.6 Integrating the results of process investigations

Extrapolating the results of short-term, site-specific studies remains a challenge. Studies relating gas fluxes to physical, chemical and microbial soil parameters are often specific to the sites and study periods chosen. Understanding the mechanisms and the interactions of the controlling factors provides a way to integrate and expand the information from specific studies. This integration is achieved through the use of conceptual models, which, as they are developed, may be incorporated into numerical simulation models.

Appropriate questions regarding factors that affect fluxes can be effectively addressed with mechanistically based models. An example of this approach is provided by the 'leaky pipe' model presented by Firestone & Davidson (1989) (see Chapter 11, this volume). This conceptual model suggests three levels of regulation of soil emissions of NO and N_2O:

1 the factors that affect rates of nitrification and denitrification, metaphorically viewed as nitrogen flowing through two pipes;
2 the factors that affect the end-products of these processes, metaphorically viewed as the controls of the sizes of holes in the pipes through which NO and N_2O leak; and
3 the factors that affect diffusion of these gases in the soil and their possible consumption before reaching the soil surface.

If nitrification were identified as the dominant source of NO emitted from the soil of a particular site or ecosystem (using nitrapyrin or C_2H_2 as inhibitors as described above), then the obvious questions would become:

1 What affects the supply of NH_4^+ to nitrifiers (rate of mineralization, diffusion of NH_4^+, other fates for NH_4^+, nitrogen fertilization)?
2 What controls the ratio of NO to other end-products (soil wetness, soil acidity)? and
3 Are net NO emissions affected by consumption of NO by other processes?

The questions that would be posed if denitrification were the source are similar, but not identical to these. In particular, denitrification would be important in wetter soils, so diffusion of substrates in thin water films would probably not be limiting. Sources of NO_3^- and competing consumptive fates of NO_3^- would need to be addressed. Given the considerations of NO consumption in a denitrifying soil, production of NO by denitrifiers might be expected to occur near the surface of a wet soil, where it could readily diffuse to the atmosphere before being consumed.

As the mechanisms of gas production should apply to all soils, and

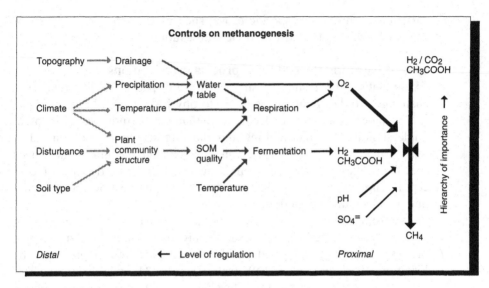

Fig. 10.4 Proximal and distal controls on methanogenesis (Schimel *et al.*, 1993).

these mechanisms can be identified and characterized, then the challenge becomes finding readily measured parameters that are mechanistically accurate and also are good, robust predictors of fluxes of trace gases. This involves another type of conceptual modelling that relates the immediate physiological (proximal) controls on processes to the progressively larger scale (distal) controls. This linking of different scale controls has been illustrated for nitrification and denitrification (Robertson, 1989), and for CH_4 production and consumption (Schimel *et al.*, 1993); Fig. 10.4 presents the example for methanogenesis. The primary controls on methanogensis in wetland soils are O_2 (the system must be anaerobic) and substrate availability. The two factors that are the primary controls on these are water table height and organic matter quality, which are in turn controlled by drainage, climate and plant community structure. It should therefore be possible to build a large-scale model of CH_4 production based on these larger-scale variables that incorporates the essential mechanistic detail.

To develop these larger-scale models of gas fluxes requires extensive investigations of the microbial processes. We have described methods to assess enzymatic capacities of soils, to measure rates of net and gross nitrogen mineralization and nitrification, and to express soil wetness appropriately. We believe that these are the most important and useful methods currently available to help describe variation in field studies of NO, N_2O and CH_4 from a mechanistic perspective. While none of these methods has emerged as a universal predictor of fluxes, these and other parameters have not been applied in a systematic manner to field studies of trace gases. If investigators of biogeochemistry begin applying the relatively few methods described here, larger and more robust data sets will result that may confirm the usefulness of some or all of these methods across a wide range of soils and ecosystems.

Acknowledgements

The authors thank G.P. Robertson and two anonymous reviewers for many useful comments on a draft of this manuscript. E.A.D. also thanks the A.W. Mellon Foundation for a grant to the Woods Hole Research Center for financial support while this manuscript was prepared.

References

Bakwin P.S., Wofsy S.C., Fan S.-M., Keller M., Trumbore S.E. & daCosta J.M. (1990) Emission of nitric oxide (NO) from tropical forest soils and exchange of NO between the forest canopy and atmospheric boundary layers. *Journal of Geophysical Research*, **95**, 16755–16764.

Bédard C. & Knowles R. (1989) Physiology, biochemistry, and specific inhibitors of CH_4, NH_4^+, and CO oxidation by methanotrophs and nitrifiers. *Microbiology Review*, **53**, 68–84.

Belser L.W. (1979) Population ecology of nitrifying bacteria. *Annual Review of Microbiology*, **33**, 309–333.

Belser L.W. & Mays W.L. (1980) Specific inhibition of nitrate oxidation by chlorate and its use in assessing nitrification in soils and sediments. *Applied Environmental Microbiology*, **39**, 505–510.

Bender M. & Conrad R. (1992) Kinetics of CH_4 oxidation in oxic soils exposed to ambient air or high CH_4 mixing ratios. *FEMS Microbiology and Ecology*, **101**, 261–270.

Berg P., Klemedtsson L. & Rosswall T. (1982) Inhibitory effect of low partial pressures of acetylene on nitrification. *Soil Biology and Biochemistry*, **14**, 301–303.

Betlach M.R. & Tiedje J.M. (1981) Kinetic explanation for accumulation of nitrite, nitric oxide, and nitrous oxide during bacterial denitrification. *Applied Environmental Microbiology*, **42**, 1074–1084.

Born M., Dörr H. & Levin I. (1990) Methane consumption in aerated soils of the temperate zone. *Tellus*, **42B**, 2–8.

Bowden W.B. (1986) Gaseous nitrogen emissions from undisturbed terrestrial ecosystems: An assessment of their impacts on local and global nitrogen budgets. *Biogeochemistry*, **2**, 244–279.

Burford J.R. & Bremner J.M. (1975) Relationships between the denitrification capacities of soils and total, water-soluble and readily decomposable soil organic matter. *Soil Biology and Biochemistry*, **7**, 389–394.

Burke R.A., Barber T.R. & Sackett W.M. (1988) Methane flux and stable hydrogen and carbon isotope composition of sedimentary methane from the Florida Everglades. *Global Biogeochemical Cycles*, **2**, 329–340.

Cawse P.A. & Cornfield A.H. (1972) Biological and chemical reduction of nitrate to nitrite in irradiated soils, and factors leading to eventual loss of nitrite. *Soil Biology and Biochemistry*, **4**, 497–511.

Chancy H.F. & Kamprath E.J. (1987) Effect of nitrapyrin rate on nitrification in soils having different organic matter contents. *Soil Science*, **144**, 29–35.

Chanton J.P. & Dacey J.W.H. (1991) Effects of vegetation on methane flux, reservoirs, and carbon isotopic composition. In: Sharkey T.D., Holland E.A. & Mooney H.A. (eds) *Trace Gas Emissions by Plants*, pp. 65–89. Academic Press, San Diego.

Chanton J.P., Pauly G.G., Martens C.S., Blair N.E. & Dacey J.W.H (1988) Carbon isotopic composition of methane in Florida Everglades soils and fractionation during its transport to the troposphere. *Global Biogechemical Cycles*, **2**, 245–252.

Conrad R. (1984) Capacity of aerobic microorganisms to utilize and grow on atmospheric trace gases (H_2, CO, CH_4). In: Klug M.J. & Reddy C.A. (eds) *Current Perspectives in Microbial Ecology*, pp. 461–467. American Society for Microbiology, Washington, DC.

Conrad R. (1989) Control of methane production in terrestrial ecosystems. In: Andreae M.O. & Schimel D.S. (eds) *Exchange of Trace Gases between Terrestrial Ecosystems and the Atmosphere*, pp. 39–58. Wiley, New York.

Davidson E.A. (1991) Fluxes of nitrous oxide and nitric oxide from terrestrial ecosystems. In: Rogers J.E. & Whitman W.B. (eds) *Microbial Production and Consumption of Greenhouse Gases: Methane, Nitrogen Oxides, and Halomethanes*, pp. 219–235. American Society for Microbiology, Washington, DC.

Davidson E.A. (1992) Sources of nitric oxide and nitrous oxide following wetting of dry soil. *Soil Science Society of America Journal*, **56**, 95–102.

Davidson E.A. (1993) Soil water content and the ratio of nitrous oxide to nitric oxide emitted from soil. In: Oremland R.S. (ed) *The Biogeochemistry of Global Change: Radiatively Active Trace Gases*, pp. 369–386. Chapman & Hall, New York.

Davidson E.A., Galloway L.F. & Strand M.K. (1987) Comparison of techniques for assessing available carbon in soil. *Communications in Soil Science and Plant Analysis*, **18**, 45–64.

Davidson E.A., Hart S.C., Shanks C.A. & Firestone M.K. (1991a) Measuring gross nitrogen mineralization, immobilization, and nitrification by ^{15}N isotopic pool dilution in intact soil cores. *Journal of Soil Science*, **42**, 335–349.

Davidson E.A., Herman D.J., Schuster A. & Firestone M.K. (1993a) Cattle grazing and oak trees as factors affecting soil emissions of nitric oxide from an annual grassland. In: Harper L.A., Mosier A.R. & Duxbury J.M. *et al.* (eds) *Agricultural Ecosystem Effects on Trace Gases and Global Climate Change*, pp. 109–119. ASA Special Publication No. 55, Agronomy Society of America, Madison, Wisconsin.

Davidson E.A., Matson P.A., Vitousek P.M. *et al.* (1993b) Processes regulating soil emissions of NO and N_2O in a seasonally dry forest. *Ecology*, **74**, 130–139.

Davidson E.A., Stark J.M. & Firestone M.K. (1990) Microbial production and consumption of nitrate in an annual grassland. *Ecology*, **71**, 1968–1975.

Davidson E.A., Swank W.T. & Perry T.O. (1986) Distinguishing between nitrification and denitrification as sources of gaseous-N production in soil. *Applied Environmental Microbiology*, **52**, 1280–1292.

Davidson E.A., Vitousek P.M., Matson P.A., Riley R., Garcia-Mendez G. & Maass J.M. (1991b) Soil emissions of nitric oxide in a seasonally dry tropical forest of Mexico. *Journal of Geophysical Research*, **96**, 15439–15445.

Doran J.W., Mielke L.N. & Power J.F. (1990) Microbial activity as regulated by soil water-filled pore space. In: *Transactions of the 14th International Congress of Soil Science. Symposium III-3; Ecology of Soil Microorganism in the Microhabital Environments.* August 12–18, 1990, Kyoto, Japan.

Eno C.F. (1960) Nitrate production in the field by incubating the soil in polyethylene bags. *Soil Science Society of America Proceedings*, **24**, 277–279.

Firestone M.K. (1982) Biological denitrification. In: Stevenson F.J. (eds) *Nitrogen in Agricultural Soils*, pp. 289–326. Agronomy Society of America, Madison, Wisconsin.

Firestone M.K. & Davidson E.A. (1989) Microbiological basis of NO and N_2O production and consumption in soil. In: Andreae M.O. & Schimel D.S. (eds) *Exchange of Trace Gases between Terrestrial Ecosystems and the Atmosphere*, pp. 7–21. Wiley, New York.

Firestone M.K., Smith M.S., Firestone R.B. & Tiedje J.M. (1979) The influence of nitrate, nitrite, and oxygen on the composition of the gaseous products of denitrification in soil. *Soil Science Society of America Journal*, **43**, 1140–1144.

Galchenko V.F., Lein A. & Ivanov M. (1989) Biological sinks of methane. In: Andreae M.O. & Schimel D.S. (eds) *Exchange of Trace Gases between Terrestrial Ecosystems and the Atmosphere*, pp. 59–71. Wiley, New York.

Gambrell R.P., Gilliam J.W. & Weed S.B. (1975) Denitrification in subsoils of the North Carolina coastal plain as affected by soil drainage. *Journal of Environmental Quality*, **4**, 311–316.

Goreau T.J., Kaplan W.A., Wofsy S.C., McElroy M.B., Valios F.W. & Watson S.W. (1980) Production of NO_2^- and N_2O by nitrifying bacteria at reduced concentrations of oxygen. *Applied Environmental Microbiology*, **40**, 526–532.

Groffman P.M. & Tiedje J.M. (1989) Denitrification in north temperate forest soils: relationship between denitrification and environmental parameters at the landscape scale. *Soil Biology and Biochemistry*, **21**, 621–626.

Haider K., Mosier A.R. & Heinemeyer O. (1983) Side effects of acetylene on the conversion of nitrate in soil. *Zeitschrift fuer Pflanzenernaehrung und Bodenkunde*, **146**, 623–633.

Hankinson T.R. & Schmidt E.L. (1984) Examination of an acid forest soil for ammonia- and nitrite-oxidizing autotrophic bacteria. *Canadian Journal of Microbiology*, **30**, 1125–1132.

Hao W.M., Scharffe D., Crutzen P.J. & Sanhueza E. (1988) Production of N_2O, CH_4, and CO_2 from soils in the tropical savanna during the dry season. *Journal of Atmospheric Chemistry*, **7**, 93–105.

Harriss R.C., Sebacher D.I. & Day F.P. Jr (1982) Methane flux in the Great Dismal Swamp. *Nature*, **297**, 673–674.

Hart S.C., Stark J.M., Davidson E.A. & Firestone M.K. (1994) Nitrogen mineralization immobilization, and nitrification. In: Weaver R.W., Angle, J.S. & Bottomley P.S. (eds) *Methods of Soil Analyses Part 2 – Microbiological and Biochemical Properties*. Soil Science Society of America, Madison Wisconcin.

Haynes R.J. (1986) Nitrification. In: Haynes R.J. (ed) *Mineral Nitrogen in the Plant–Soil System*, pp. 127–165. Academic Press, New York.

Hutchinson G.L. & Davidson E.A. (1993) Processes for production and consumption of gaseous nitrogen oxides in soil. In: Harper L.A., Mosier A.R. & Duxbury J.M. (eds) *Agricultural Ecosystem Effects on Trace Gases and Global Climate Change*, pp. 79–93. ASA Special Publication No. 55, Agronomy Society of America, Madison, Wisconsin.

Hutchinson G.L., Livingston G.P. & Brams E.A. (1993) Nitric and nitrous oxide evolution from managed subtropical grassland. In: Oremland R. (ed) *The Biochemistry of Global Change: Radiatively Active Trace Gases*, pp. 290–316. Chapman and Hall, New York.

Hynes R.K. & Knowles R. (1982) Effect of acetylene on autotrophic and heterotrophic nitrification. *Canadian Journal of Microbiology*, **28**, 334–340.

Johansson C. & Sanhueza E. (1988) Emission of NO from savanna soils during rainy season. *Journal of Geophysical Research*, **93**, 14193–14198.

Johansson C., Rodhe H. & Sanhueza E. (1988) Emission of NO in a tropical savanna and a cloud forest during the dry season. *Journal of Geophysical Research*, **93**, 7180–7192.

Jones R.D., Morita R.Y. & Griffiths R.P. (1984) Method for estimating in situ chemolithotrophic ammonium oxidation using carbon monoxide oxidation. *Marine Ecology Progress Series*, **17**, 259–269.

Keller M., Kaplan W.A., Wofsy S.C. & DaCosta J.M. (1988) Emissions of N_2O from tropical forest soils: Response to fertilization with NH_4^+, NO_3^-, and PO_4^{3-}. *Journal of Geophysical Research*, **93**, 1600–1604.

Keller M. & Reiners W.A. (1991) A seasonal study of the emissions of nitrous oxide and nitric oxide from soil in a wet forest area in Costa Rica: examining the effects of land use change. *EOS Transactions of the American Geophysics Union*, **72**, 110.

Killham K. (1986) Heterotrophic nitrification. In: Prosser J.I. (ed) *Nitrification*, pp. 117–126. Special Publication of the Society for General Microbiology, vol. 20 IRL Press, Oxford.

King G.M. (1990) Dynamics and controls of methane oxidation in a Danish wetland sediment. *FEMS Microbiology and Ecology*, **74**, 309–324.

King G.M. & Adamsen A.P.S. (1992) Effects of temperature on methane consumption in a forest soil and in pure cultures of the methanotroph *Methylomonas rubra*. *Applied Environmental Microbiology*, **58**, 2758–2763.

Klemmedtsson L., Svensson B.H. & Rosswall T. (1988) A method of selective inhibition to distinguish between nitrification and denitrification as sources of nitrous oxide in soil. *Biology and Fertility of Soils*, **6**, 112–119.

Krzycki J.A., Kenealy W.R., DeNiro M.J. & Zeikus J.G. (1987) Stable carbon isotope fractionation by *Methanosarcina barkeri* during methanogenesis from acetate, methanol, or carbon dioxide–hydrogen. *Applied Environmental Microbiology*, **53**, 2597–2599.

Lansdown J.M., Quay P.D. & King S.L. (1991) CH_4 production via CO_2 reduction in a temperate bog: a source of ^{13}C depleted CH_4. In: *Abstracts of the 10th International Symposium on Environmental Biogeochemistry*, San Francisco, California.

Large P.J. (1983) Methylotrophy and Methanogenesis. *American Society of Microbiology*, Washington, DC.

Lidstrom M.E. (1992) The aerobic methylotrophic bacteria. In: Balows A., Trüper H.G., Dworkin M., Harder W. & Schleifer K.-H. (eds) *The Procaryotes*, 2nd edn. Springer-Verlag, New York.

Lindau C.W., Bollich P.K., Delaune R.D., Patrick W.H. Jr & Law V.J. (1991) Effect of urea fertilizer and environmental factors on CH_4 emissions from a Louisiana, USA rice field. *Plant Soil*, **136**, 195–203.

Linn D.M. & Doran J.W. (1984) Effect of water-filled pore space on carbon dioxide and nitrous oxide production in tilled and nontilled soils. *Soil Science Society of America Journal*, **48**, 1267–1272.

Livingston G., Vitousek P.M. & Matson P.A. (1988) Nitrous oxide fluxes and nitrogen transformations across a landscape gradient in Amazonia. *Journal of Geophysical Research*, **93**, 1593–1599.

Matson P.A. & Vitousek P.M. (1990) Ecosystem approach to a global nitrous oxide budget. *Bioscience*, **40**, 667–672.

Matson P.A., Vitousek P.M., Livingston G.P. & Swanberg N.A. (1990) Sources of variation in nitrous oxide flux from amazonian ecosystems. *Journal of Geophysical Research*, **95**, 16789–16798.

Martens C.S., Blair N.E., Green C.D. & Des Marais D.J. (1986) Seasonal variations in the stable carbon isotopic signature of biogenic methane in a coastal sediment. *Science*, **233**, 1300–1303.

Martens C.S., Kelley C.A., Ussler W. III *et al.* (1991) Isotopic characterization of methane from Yukon–Kuskokwim tundra and minerotrophic fens in northern Quebec. In: *Abstracts of the 10th International Symposium on Envrionmental Biogeochemistry*, San Francisco, California.

Martikainen P.J. (1985) Nitrous oxide emission associated with autotrophic ammonium oxidation in acid coniferous forest soil. *Applied Environmental Microbiology*, **50**, 1519–1525.

Megraw S.R. & Knowles R. (1990) Effect of picolinic acid (2-pyridine carboxylic acid) on the oxidation of methane and ammonia in soil and in liquid culture. *Soil Biology and Biochemistry*, **22**, 635–641.

Melillo J.M., Aber J.D., Steudler P.A. & Schimel J.P. (1983) Denitrification potentials in a successional sequence of northern hardwood forest stands. *Ecology Bulletin (Stockholm)*, **35**, 217–228.

Moore T.R. & Knowles R. (1989) The influence of water table levels on methane and carbon dioxide emissions from peatland soils. *Canadian Journal of Soil Science*, **69**, 33–38.

Mosier A., Schimel D., Valentine D., Bronson K. & Parton W. (1991) Methane and nitrous oxide fluxes in native, fertilized and cultivated grasslands. *Nature*, **350**, 330–332.

Nagele W. & Conrad R. (1990a) Influence of pH on the release of NO and N_2O from fertilized and unfertilized soil. *Biology and Fertility of Soils*, **10**, 139–144.

Nagele W. & Conrad R. (1990b) Influence of soil pH on the nitrate-reducing microbial populations and their potential to reduce nitrate to No and N_2O. *FEMS Microbiology and Ecology*, **74**, 49–58.

Nason G.E. & Myrold D.D. (1991) ^{15}N in soil research: appropriate application of rate estimation procedures. *Agricultural Ecosystems and Environments*, **34**, 427–441.

Nelson D.W. (1982) Gaseous losses of nitrogen other than through denitrification. In: Stevenson F.J. (ed) *Nitrogen in Agricultural Soils*, pp. 327–364. American Society of Agronomy, Madison, Wisconsin.

Nesbit S.P. & Breitenbeck G.A. (1992) A laboratory study of factors influencing methane uptake by soils. *Agriculture, Ecosystems and Environment*, **41**, 39–54.

Oremland R.S. & Capone D.G. (1988) Use of 'specific' inhibitors in biogeochemistry and microbial ecology. *Advances in Microbiology and Ecology*, **10**, 285–383.

Oremland R.S. & Culbertson C.W. (1992a) Importance of methane-oxidizing bacteria in the methane budget as revealed by the use of a specific inhibitor. *Nature*, **356**, 421–423.

Oremland R.S. & Culbertson C.W. (1992b) Evaluation of methyl fluoride and dimethyl ether as inhibitors of aerobic methane oxidation. *Applied Environmental Microbiology*, **58**, 2983–2992.

Papen H., von Berg R., Hinkel I., Thoene B. & Rennenberg H. (1989) Heterotrophic nitrification by *Alcaligenes faecalis*: NO_2^-, NO_3^-, N_2O, and NO production in exponentially growing cultures. *Applications of Environmental Microbiology*, **55**, 2068–2072.

Papendick R.I. & Campbell G.S. (1981) Theory and measurement of water potential. In: Parr J.F., Gardner W.R. & Elliot L.F. (eds) *Water Potential Relations in Soil Microbiology*, pp. 1–22. Soil Science Society of America Special Publications Number 9, Madison, Wisconsin, USA.

Parkin T.B., Sexstone A.J. & Tiedje J.M. (1985) Adaptation of denitrifying populations to low soil pH. *Applications of Environmental Microbiology*, **49**, 1053–1056.

Poth M. & Focht D.D. (1985) ^{15}N kinetic analysis of N_2O production by *Nitrosomonas europaea*: an examination of nitrifier denitrification. *Applications of Environmental Microbiology*, **49**, 1134–1141.

Remde A., Slemr F. & Conrad R. (1989) Microbial production and uptake of nitric oxide in soil. *FEMS Microbiology and Ecology*, **62**, 221–230.

Robertson G.P. (1982) Nitrification in forested ecosystems. *Philosophical Transactions – Royal Society of London*, **B296**, 445–457.

Robertson G.P. (1989) Nitrification and denitrification in humid tropical ecosystems: potential controls on nitrogen retention. In: Proctor J. (ed) *Mineral Nutrients in Tropical Forest and Savannah Ecosystems*, pp. 55–69. British Ecological Society Special Publication Number 9, Blackwell Scientific Publications, Oxford.

Robertson G.P. & Tiedje J.M. (1984) Denitrification and nitrous oxide production in successional and old-growth Michigan forest. *Soil Science Society of America Journal*, **48**, 383–389.

Robertson G.P. & Tiedje J.M. (1987) Nitrous oxide sources in aerobic soils: nitrification, denitrification, and other biological processes. *Soil Biology and Biochemistry*, **19**, 187–193.

Robertson G.P. & Tiedje J.M. (1988) Deforestation alters denitrification in a lowland tropical rain forest. *Nature*, **336**, 756–759.

Rogers G.A. & Ashworth J. (1982) Bacteriostatic action of nitrification inhibitors. *Canadian Journal of Microbiology*, **28**, 1093–1100.

Rudaz A.O., Davidson E.A. & Firestone M.K. (1991) Sources of nitrous oxide production following wetting of dry soil. *FEMS Microbiology and Ecology*, **85**, 117–124.

Ryden J.C., Lund L.J. & Focht D.D. (1979) Direct measurement of denitrification loss from soils. I. Laboratory evaluation of acetylene inhibition of nitrous oxide reduction. *Science Society of America Journal*, **43**, 104–110.

Salvas P.L. & Taylor B.F. (1984) Effect of pyridine compounds on ammonia oxidation by autotrophic nitrifying bacteria and *Methylosinus trichosporium* OB3b. *Current Microbiology*, **10**, 53–56.

Sanhueza E., Hao W.M., Scharffe D., Donoso L. & Crutzen P.J. (1990) N_2O and NO emissions from soils of the northern part of the Guayana Shield, Venezuela. *Journal of Geophysical Research*, **95**, 22481–22488.

Schimel J.P., Holland E.A. & Valentine D. (1993) Controls on methane flux from terrestrial ecosystems. In: Harper L.A., Mosier A.R. & Duxbury J.M. (eds) *Agricultural Ecosystem Effects on Trace Gases and Global Climate Change*, pp. 167–182. ASA Special Publication No. 55, Agronomy Society of America, Madison, Wisconsin, USA.

Singh J.S. & Grupta S.R. (1977) Plant decomposition and soil respiration in terrestrial ecosystems. *Botany Review*, **43**, 449–528.

Skopp J. (1985) Oxygen uptake and transport in soils: analysis of the air–water interfacial area. *Soil Science Society of America Journal*, **49**, 1327–1331.

Skopp J., Jawson M.D. & Doran D.W. (1990) Steady-state aerobic microbial activity as a function of soil water content. *Soil Science Society of America Journal*, **54**, 1619–1625.

Smith C.J., Wright W.F. & Patrick W.H. Jr (1983) The effect of soil redox potential and pH on the reduction and production of nitrous oxide. *Journal of Environmental Quality*, **12**, 186–188.

Smith M.S. & Parsons L.L. (1985) Persistence of denitrifying enzyme activity in dried soils. *Applications of Environmental Microbiology*, **49**, 316–320.

Stams A.J.M., Flameling E.M. & Marnette E.C.L. (1990) The importance of autotrophic versus heterotrophic oxidation of atmospheric ammonium in forest ecosystems with acid soil. *FEMS Microbiology and Ecology*, **74**, 337–344.

Steudler P.A., Bowden R.D., Melillo J.M. & Aber J.D. (1989) Influence of nitrogen fertilization on methane uptake in temperate forest soils. *Nature*, **341**, 314–316.

Svensson B.H. (1984) Different temperature optima for methane formation when enrichments from acid peat are supplemented with acetate or hydrogen. *Applications of Environmental Microbiology*, **48**, 389–394.

Terry R.E. & Duxbury J.M. (1985) Acetylene decomposition in soils. *Soil Science Society of America Journal*, **49**, 90–94.

Tiedje J.M. (1982) Denitrification. In: Page A.L. (ed) *Methods of Soil Analysis, Part 2*, 2nd edn, pp. 1011–1026. American Society of Agronomy, Madison, Wisconsin, USA.

Tiedje J.M. (1988) Ecology of denitrification and dissimilatory nitrate reduction to ammonium. In: Zehnder J.B. (ed) *Biology of anaerobic microorganisms*, pp. 179–244. John Wiley and Sons, New York.

Tiedje J.M., Simpkins S. & Groffman P.M. (1989) Perspectives on measurement of denitrification in the field including recommended protocols for acetylene based methods. In: Clarholm M. & Bergstrom L. (eds) *Ecology of Arable Land*, pp. 217–240. Kluwer Academic Publishers.

Topp E. & Hanson R.S. (1991) Metabolism of radiatively important trace gases by methane-oxidizing bacteria. In: Rogers J.E. & Whitman W.B. (eds) *Microbial Production and Consumption of Greenhouse Gases: Methane, Nitrogen Oxides, and Halomethanes*. American Society for Microbiology, Washington.

Tortoso A.C. & Hutchinson G.L. (1990) Contributions of autotrophic and heterotrophic nitrifiers to soil NO and N_2O emissions. *Applications of Environmental Microbiology*, **56**, 1799–1805.

Valentine D., Holland E.A. & Schimel D.S. (1993) Ecosystem and physiological controls over methane production in northern wetlands. *Journal Geophysical Research*, **99**, 1563–1571.

Weast R.C. (1976) *Handbook of Chemistry and Physics*, 57th edn, page D-178. CRC Press, Cleveland.

Wessel W.W. & Tietema A. (1992) Calculating gross N transformation rates of ^{15}N pool dilution experiments with acid forest litter: analytical and numerical approaches. *Soil Biology and Biochemistry*, **24**, 931–942.

Whalen S.C. & Reeburgh W.S. (1990) Consumption of atmospheric methane by tundra soils. *Nature*, **346**, 160–162.

Whitman W.B., Bowen T.L. & Boone D.R. (1992) The methanogenic bacteria. In: Barlows A., Trüper H.G., Dworkin M., Harder W. & Schleifer K.-H. (eds) *The Prokaryotes*, 2nd edn. Springer-Verlag, New York.

Williams E.J. & Fehsenfeld F.C. (1991) Measurement of soil nitrogen oxide emissions at three North American ecosystems. *Journal of Geophysical Research*, **96**, 1033–1042.

Williams E.J., Hutchinson G.L. & Fehsenfeld F.C. (1992) NO_x and N_2O emissions from soil. *Global Biogeochemical Cycles*, **6**, 351–388.

Yagi K. & Minami K. (1990) Effect of organic matter application on methane emission from some Japanese paddy fields. *Soil Science and Plant Nutrition*, **36**, 599–610.

Yarrington M.R. & Wynn-Williams D.D. (1985) Methanogenesis and the anaerobic microbiology of a wet moss community at Signy Island. In: Siegfried W.R., Condy P.R. & Laws R.M. (eds) *Antarctic Nutrient Cycles and Food Webs*, pp. 229–233. Springer-Verlag, New York.

Yavitt J.B., Downey D.M., Lang G.E. & Sextone A.J. (1990) Methane consumption in two temperate forest soils. *Biogeochemistry*, **9**, 39–52.

Yavitt J.B., Lang G.E. & Downey D.M. (1988) Potential methane production and methane oxidation rates in peatland ecosystems of the Appalachian mountains, United States. *Global Biogeochemistry Cycles*, **2**, 253–268.

Yoshinari T., Hynes R. & Knowles R. (1977) Acetylene inhibition of nitrous oxide reduction and measurement of denitrification and nitrogen fixation in soil. *Soil Biology and Biochemistry*, **9**, 177–183.

CHAPTER 11

Process modelling and spatial extrapolation

D.S. SCHIMEL & C.S. POTTER

11.1 Introduction

The inclusion of modelling in a methods handbook reflects the importance of a variety of types of calculations in the analysis of atmospheric trace gases. Trace gas production and emission occurs on a microscopic scale, being mediated by physiological processes of plants and microorganisms and by physical properties of the soil. Interest in the consequences of trace gas emissions, however, is commonly related to either ecosystem element budgets or regional and global atmospheric processes. There is thus an inherent mismatch between the scale at which trace gas emissions occur (<1 cm) and are measured (10 cm to 10 km) and the scales at which they influence the Earth system (10 km to global). Similarly, trace gas emissions fluctuate over time, often rapidly, yet must be integrated up to annual or longer cycles. The logistics of sampling trace gas fluxes are such that the sampling problems in space and time must be solved by interpolation in order to accomplish the scientific objectives of most trace gas studies. The nature of heterogeneity of terrestrial systems is such that contouring, as applied in marine biogeochemistry, is of limited use and interpolation must take advantage of spatially extensive data sets of readily measured proxy data, and the calculation of fluxes using models. The interpolation problem in time and space is central to the analysis of many problems in trace gas biogeochemistry, and it is on the use of models in interpolation that this chapter will focus.

Modelling has several roles in the development and application of rigorous algorithms for spatial and temporal extrapolation:

1 The development and testing of models aids in the identification of parameters whose values (a) significantly affect process rates; and (b) have significant variability in the spatial or temporal domain of interest. Thus spatial integration of fluxes may be improved by reference to a related but more readily mapped variable. Model analysis can aid in identification of such proxy variables.

2 Models can be used to calculate fluxes from data sets describing the spatial or temporal variability of proxy variables and boundary conditions. Process models may be applied in this fashion even when non-linear interactions preclude the use of simple functions relating a proxy to a flux of interest.

3 Validated models allow the analysis of system behaviour under scenarios of possible past or future conditions (extrapolation in time).

The intent of this chapter is to present modelling as a tool to improve the interpretation of measurements in the context of the questions which drive trace gas research. Model development can interact with field studies in mutually beneficial ways. Results from experimental observations generate hypotheses and functional responses that motivate model representations of trace gas control and emission mechanisms. Models, in turn, can aid in the identification of gaps in field measurement data sets, guide new hypothesis formulation, provide theoretical tests of experimental hypotheses and facilitate scaling of trace gas production estimates.

Model validation is a crucial, but often misunderstood topic. Models are deliberate simplifications of 'reality' and as such cannot be expected always to make correct predications. More important than model validation is model evaluation, i.e. the determination of under what circumstances and with what reliability a model will function. There are inherent limits to predictability in non-linear models that impose certain constraints on model validation. Issues of validation are difficult both conceptually and in practice; a full discussion is outside the scope of this chapter. Nevertheless, a number of validation–evaluation issues are addressed.

In this chapter, several extant modelling paradigms are reviewed, their application in extrapolation discussed and, just as important, data requirements of different modelling approaches are discussed. While the majority of the book addresses the direct measurement of trace gases and trace gas exchange, we will indicate the types of measurements required to model fluxes, and to use models in spatial and temporal extrapolation. While this chapter will not detail specific techniques, we will cite key references as appropriate.

11.2 General principles
Trace gas exchange with the atmosphere occurs through a set of coupled processes. These are:

1 Production of substrate for processing by trace gas-producing organisms or pathways. Examples of such substrates include acetate for CH_4, pyruvate for isoprene and nitrate for N_2O.

2 Conversion of substrate to the atmospheric trace gas of interest; e.g. nitrate to N_2O, or conversion of an atmospheric trace species to an inorganic or organic plant or soil phase, e.g. plant scavenging of atmospheric NH_3, soil oxidation of CH_4.

3 Diffusion of the trace species into or out of the free atmosphere and into or out of the site of biological activity.

Key considerations are discussed related to each of the above processes separately. Rather similar modelling approaches for different trace gases are often used, despite the wide range of organisms and metabolic pathways involved. This commonality is due in part to the enzymatic regulation of all of trace gas metabolic processes with accompanying similarities in responses to temperature, precursor pool size and oxygen status. Differences arise between trace gas processes that are coupled to autotrophic rather than heterotrophic metabolism, and between aerobic and anaerobic processes. Below the 'building block' components of trace gas models are discussed, with examples of actual models to contrast strengths and limitations of the various approaches.

11.3 Building blocks

11.3.1 Substrate (gas precursor) production
While much research on trace gas emissions has focused on temperature and moisture as controls over trace gas fluxes, substrate is often limiting. For example, Matson et al. (1989) showed strong relationships between nitrogen mineralization and N_2O production in tropical ecosystems. Parton et al. (1988) and Mosier et al. (1992) similarly showed soil nitrogen availability to be a major control over N_2O emissions in temperate grasslands. NO emissions have also been related to nitrate concentrations (Williams et al., 1992). Valentine et al. (1994) showed acetate availability to limit methanogenesis in northern wetlands, supporting earlier work by Svensson & Rosswall (1984). Ojima et al. (1993) and Born et al. (1990) argue that a principal control over soil CH_4 uptake (where atmospheric CH_4 is the substrate) is atmospheric CH_4 concentration. Similarly, plants may either emit or scavenge NH_3 depending upon atmospheric concentration (Langford & Fehsenfeld, 1992; Langford et al., 1992). Lerdau (1991) argued for and has since shown (personal communication) relationships between leaf terpene concentrations and terpene emissions. Isoprene emissions are influenced by incident light because, in part, of the limitation of isoprene synthesis by the supply rate of new photosynthate (pyruvate).

In general, then, trace gas substrates are produced through one of three processes: (a) decomposition (and fermentation); (b) nitrogen mineralization; or (c) photosynthesis (and possibly photorespiration). Numerous models are used to represent all of these processes, a review of these models is beyond the scope of this chapter. Empirical models of decomposition may be found in Meentemayer (1984), Raich & Schlesinger (1992) and Townsend et al. (1992). Empirical models for nitrogen mineralization are widely used as well (Stanford & Smith, 1972). The reader is referred to Van Veen & Paul (1981), Bossata & Agren (1985) and Parton et al. (1987) for discussions of ecosystem simulations of nitrogen mineralization and decomposition. A wide range of models are used to simulate photosynthesis; notable contributions are von Caemmerer & Farquhar (1985) and Collatz *et al.* (1991).

Calculation of substrate availability for trace gas production from first principles requires application of models which couple primary production and decomposition. For example, calculation of CH_4 production requires knowledge of the amount and chemical composition of plant residues produced, along with pH, temperature and variability in water table level (Roulet *et al.*, 1992; Valentine *et al.*, 1994). Because the substrate for methanogenesis is produced by anaerobic decomposition, production and litter organic chemistry must be known. In the case of isoprene produced in foliage, rates of photosynthesis must be modelled. Modelling rates of photosynthesis requires knowledge of plant nitrogen status, which is determined by rates of nitrogen mineralization in soils. Nitrogen mineralization occurs during the microbial decomposition of plant residues and soil organic matter in soils. While 'stand-alone' models of trace gas production may be parameterized for individual research sites or for instantaneous fluxes, calculation of time or regionally integrated fluxes requires the use of ecosystem models linking plant and soil processes.

The extensive literature on applied soil physics offers trace gas modellers a comprehensive theoretical treatment of generalized processes involved in moisture and temperature fluxes. The theory of Philip & de Vries (1957) and de Vries (1958) forms the basis for many of the mathematical models of moisture and heat flows in porous media. One updated version of the theory calls for use of matric head (rather than moisture content) as a dependent variable in a system of partial differential equations that are solved by the finite element method (Milly, 1982). The reader is referred to standard references such as Hillel (1982) and Marshall & Holmes (1988) for examples of applications and validations of Darcy's and Fourier's laws in soil water and heat transfer models.

11.3.2 Trace gas production
The step most studied in trace gas studies is the production or consumption of the gaseous species of interest from its immediate precursor. Examples of this step include:

Nitrification:	$NH_4 \rightarrow$ (approx. 99%)$NO_3 +$ (approx. 0.1–1%)N_2O
Denitrification	$NO_3 \rightarrow N_2O \rightarrow N_2$
Methanogenesis	$CO_2 + H_2$ or acetate $\rightarrow CH_4$
No production	NO_3 (?) $\rightarrow NO$
Methanotrophy	$CH_4 \rightarrow CO_2$ or methanotroph biomass
Isoprene synthesis	pyruvate ... $\rightarrow$ ... isoprene (C_5H_8)

Table 11.1 Generalized equations for modelling trace gas production (dG/dt)

Model type	Formulation	References
Zero order	k	Focht, 1974; Starr & Parlange, 1976; Kanwar *et al.*, 1980
First order	$k \cdot S$	Cho, 1971; Misra *et al.*, 1974a; Kirda *et al.*, 1974
Michaelis–Menten	$k_{max} \cdot (S/K + S)$	McConnaughey & Bouldin, 1985
Double-Monod	$k_{max} \cdot (S_1/K_1 + S_1)(S_2/K_2 + S_2)$	McConnaughey & Bouldin, 1985; Grant, 1991
Pirt	$u/u_{max} + m(E_i/E)$	Leffelar & Wessel, 1988

k_{max}, reaction rate constant (maximum); S_i, concentration of substrate i; K_i, half-saturation constant for substrate i; u and u_{max}, actual and maximum microbial growth rates on substrate, respectively; m, maintenance coefficient with respect to substrate; E_i and E, concentrations of individual and total electron acceptor, respectively.

Modelling of microbial population dynamics can help explain observations of base metabolic activity associated with gas production processes. Common assumptions in microbial-level trace gas modelling studies include:

1   all microbes in the population are active and their growth rates are density independent;

2   microbial growth is not directly inhibited by growth of other species;

3   populations are homogeneously distributed and immobile in the model soil; and

4   substrates and oxidants react to produce trace gases within the same organism.

An array of physical, chemical and biological factors influence gas production in soils. Mathematical models ranging from simple zero-order reaction terms to equations based on competitive substrate-limitation have been tested. Generalized forms of such models are compared in Table 11.1.

Results from experimental systems in which substrate concentrations are relatively high tend to fit zero-order formulations, whereas systems in which substrate-limitations can be experimentally demonstrated are better represented using Michaelis–Menten or double-Monod kinetics. In the double-Monod scheme, the S_2 term can represent any electron acceptor for substrate (S_1) reduction reactions. For example, this equation can describe mineral nitrogen reduction to (intermediate) trace gas phase as a function of nitrogen substrate concentration (S_1) and carbon availability (S_2).

Half-saturation constants (K_i) for heterogeneous microbial populations reactions normally come from measurements of continuous cultures in soil extracts. Such kinetics parameters can be determined from the double reciprocal (Lineweaver–Burk) linearization of experimental data for reaction levels at various substrate concentrations (Robinson, 1985). The slope of the resulting line is equal to K_i/k_{max}, while the intercept is equal to $1/k_{max}$. It is assumed that k_{max} is a function of microbial population type and size, whereas K_i is independent of population size.

The Pirt (1965) equation can be used to model trace gas production or the consumption of electron acceptors. Results are calculated as the difference between the total amount of substrate consumed and the amount used for microbial cell synthesis. Several studies have attempted to separate and validate steady-state microbial growth and maintenance terms (Smith, 1979; Knapp et al., 1983; Smith et al., 1986).

Existing models illustrate the diverse application of these approaches. Using the Pirt equation, Leffelar & Wessel (1988) demonstrated that CO_2 production can be modelled as the difference between the total rate of labile carbon consumption and biomass synthesized during growth of soil bacteria, as shown in Eqn. 11.1

$$dCO_2/dt = dC/dt - dB/dt \qquad (11.1)$$

where CO_2 is production rate for carbon dioxide, C is the rate of soluble carbon consumption and B is the biomass synthesized.

Leffelar & Wessel (1988) maintain that incorporation of terms to estimate the effects of soil temperature, pH and moisture would not contribute to a better understanding of the processes reflected in relative growth rate and electron acceptor consumption equations of their model. This type of model is intended for use in a single soil under controlled laboratory conditions and, as such, field-level applications are not a prime consideration.

In moving toward field studies, a useful point of comparison is in the DNDC model of Li et al. (1992), which uses a similar approach based on growth and activity of denitrifier and nitrifier populations in a model of N_2O production (Eqn. 11.2):

$$dN_2O/dt = (u_{N_2O}/Y_{N_2O} + M_{N_2O} \cdot N_{N_2O}) \cdot B(t) \cdot pH_{N_2O} \cdot TE \qquad (11.2)$$

where u_{N_2O} is the relative growth rate of the denitrifier population, Y_{N_2O} is the maximum growth yield on NO_3 as a substrate, M_{N_2O} is the maintenance coefficient for growth on N_2O, N_{N_2O} is the amount of nitrogen available as NO_3, $B(t)$ is the denitrifier biomass at time t, pH_{N_2O} is a function relating pH to denitrification and TE is the temperature effect on denitrifier growth.

In this case, the denitrification submodel is activated by a rainfall event that leads to saturated soil moisture conditions, which explains the lack of water content parameter in the equation. Incorporation of a lag time factor related to microbial response to soil wetting is important in simulating the timing of trace gas release.

The transient microsite modelling approach described by McConnaughey & Bouldin (1985) is noteworthy in that it combines processes that influence trace gas production at the soil aggregate level. This model accounted for formation of anaerobic microsites, temporal changes in substrate availability and enzyme activity, and transient anaerobiosis. Simplification was made in the definition of an 'operational' (i.e. time dependent fraction of the aerobic electron transfer) k_{max} for Michaelis–Menten nitrate reduction calculations.

Whereas certain gas production processes can be quantitatively described and simulated by systems of competing reaction–diffusion processes, there have been few efforts to perform independent verification of model parameters such as kinetic reaction terms and reducing reaction rates. Advances have been made in modelling transient soil processes that capture anaerobic microsite formation and sequential reduction reactions, but these expressions generally require the modeller to specify numerous reaction terms, including those for pore-solute velocity scalars, substrate diffusion coefficients and reaction rate constants. More detailed information of the range of these variables from comparative field and laboratory experiments is a key to further refinement of highly mechanistic trace gas production models.

Trace gas consumption has been modelled somewhat differently from production. Recently, Born *et al.* (1990) and Ojima *et al.* (1991) calculated regional and global CH_4 consumption fluxes in aerobic soils. Ojima *et al.* (1991) used the model shown in Eqn. 11.3:

$$\text{annual } CH_4 \text{ uptake} = k \cdot \Delta CH_4 \tag{11.3}$$

where ΔCH_4 is the gradient between atmospheric and soil CH_4 concentrations, and k is an activity coefficient. Typical soil CH_4 concentrations were assumed to be about 0.4 ppmv, and k was estimated from a number of empirical studies. Clearly, k could also be calculated from the factors known to influence CH_4 consumption in soils, such as soil porosity, nitrogen content and moisture (Mosier *et al.*, 1991).

11.3.3 Trace gas transport

Transport of trace gases interacts strongly with production and consumption to influence total exchange. Transport is important for at least three reasons:

1 the residence time of trace gases produced in the soil or water column influences the time interval during which they are vulnerable to further chemical transformation;

2 the invasion rate of atmospheric trace gases is a significant control over the amount consumed by plant or soil processes;

3 when transport is inhibited, significant quantities of trace gases may be stored in soils or ground water, reducing the amount or changing the timing of emissions, and in some cases inhibiting further production through feedback.

Trace gas transport is generally modelled using Fick's first law model of diffusion (Eqn. 11.4).

$$J_s = -\phi \cdot D_s \cdot dC/dz \qquad (11.4)$$

where J_s is diffusive flux, ϕ is porosity, D_s is diffusion coefficient in the medium, and dC/dz is the concentration gradient over the distance z (Liss & Slater, 1974; Barber et al., 1988; Nobel, 1991). In this model, key parameters include the gradient in trace gas concentrations (driven by production rates), atmospheric concentration and the gas-specific diffusion coefficient.

In soils, diffusion is restricted by several factors that increase the complexity of modelling beyond that of a solute in solution:

1 soil structure creates tortuosity and increases the diffusional path length between the site of production and consumption and the atmosphere relative to the straight-line depth, a factor related to soil texture and structure;

2 soil moisture influences the connectedness of soil pores and the number of phase transitions (gas-solute) between the site of production and consumption and the atmosphere;

3 soil particles may adsorb the gas of interest. In water alone, resistance to diffusion is controlled largely by the solubility of the gas in water, and by its diffusion coefficient.

Consequently, several simplifying assumptions are commonly made in deriving diffusion transport equations for trace gases (Letey et al., 1980):

1 species concentration is distributed between three phases in the soil – adsorbed, dissolved and gaseous;

2 volumetric water content, air-filled and total porosity are constant throughout the profile;

3 adsorption and gas phase concentrations are proportional to liquid concentration (following Henry's law);

4 under unsaturated soil conditions, liquid diffusion is negligible compared to vapour diffusion.

In some cases, such as dissolved gases in sediments or peats, processes other than diffusion can become important. These processes include transport in vegetation (Dacey & Klug, 1979; Chanton et al., 1989), convection and advection of the liquid, which may transport gases within the soil/sediment column (Bowden, 1984) and ebullition, or the transport of gases through water in bubbles (Keller, 1990). Evidence indicates that ebullition is controlled by wind speed, influencing turbulence and by pressure head changes, as water tables drop or rise. While Keller (1990) has presented a regression model for ebullition of CH_4 in ponds, no other models exist for this process.

Based on existing field observations, we can identify several essential model components needed to address the complex set of mechanisms influencing the emissions of gases such as CH_4. Of the three processes controlling CH_4 emissions (production, oxidation and transport), only production reactions are fairly well understood. Further work is required in oxidation and transport modelling theory. Description of aerobic oxidation reactions occurring in areas where O_2 and CH_4 intermix must be thoroughly tested using Michaelis–Menten kinetics (Conrad, 1984). If O_2 is limiting oxidation in anaerobic–aerobic transition zones, a Monod equation may be more appropriate. In relatively dry agricultural and uplands soils, O_2 diffuses freely with CH_4 from the atmosphere to sites of oxidation, so that CH_4 diffusion rates alone may be adequate predictors. As understanding of physical controls increases, comparative ecosystem experiments will be needed to determine whether k_{max} and K_m change with increasing CH_4 substrate concentrations, which would be indicative of differentially adapted microbe populations.

One modelling approach to the overall CH_4 question is to combine substrate diffusion equations with Michaelis–Menten oxidation reactions integrated over aerobic zone soil layers (for example, see Kuivila *et al.*, 1988). Model calibration will require extensive measurements of oxygen concentrations and flux rates. Evaluation of the model may be conducted by measuring CH_4 flux into aerobic zones, and correcting for fluxes into overlying anaerobic layers to give the amount consumed via oxidation in the substrate. Alternatively, one could measure fluxes with and without addition of oxidation inhibitor (CH_3F for CH_4 oxidation; Oremland & Culberson, 1992). Comprehensive field level modelling approaches should consider water table dynamics, nitrogen and sulphur inhibition reactions, and pH, temperature and moisture stress effects on enzymatic reaction rates.

11.4 From microcosm to field level simulations

While a number of models have estimated trace gas flux rates from soils based on microbial metabolism (see Table 11.1), few have included information concerning the spatial and temporal variability of field-level controls such as soil temperature, moisture, texture and organic matter availability. On the other hand, ecosystem trace gas models contain comparatively little detail about substrate dynamics and gas transport processes.

By way of specific examples, Parton *et al.* (1987, 1988; Century model) used a largely empirical formulation in which N_2O production was estimated from soil moisture, soil temperature and soil NO_3 concentration. In this model, the relationships between N_2O production and environmental variables were developed from an extensive grasslands database, including both field and laboratory studies. The model, however, did not simulate the underlying enzyme kinetic and microbial growth processes. An important assumption in models like Century is that microbial biomass levels are always sufficient to achieve maximal rates permitted by prevailing temperature, moisture and NO_3. Primarily data driver requirements are average monthly or weekly climate (minimum/maximum temperature, precipitation amount) attributes, soil texture and management characterization.

In contrast, the DNDC model (Li *et al.*, 1992) adds several components, including pH and microbial growth controls on denitrification, in a rainfall event-driven approach to prediction of soil-wetting processes. This model was designed to simulate agroecosystem-level N_2O and N_2 fluxes with explicit consideration of enzyme kinetic and microbial biomass dynamics. It requires specification of daily precipitation event inputs for calculations in a one-dimensional soil heat flux and moisture flow submodel. The DNDC decomposition submodel is not unlike that of Century in that substrates are divided into labile and resistant pools with reactions leading to the formation of soil humus. The denitrification submodel, however, is relatively explicit in the manner it deals with nitrogen species reduction reactions, trace gas precursors and production using double-Monod kinetics (refer to Eqn. 11.2 above). As such, the model attempts to simulate what are conventionally thought of as soil microsite, electron donor–acceptor processes, at the field level.

In addition to the differences in the level of detail given to microbial growth and electron transfer reactions, the contrast between these modelling approaches (Century and DNDC) raises a central issue in moving from microcosm to field level simulations: i.e. the use of episodic vs. average climate drivers for trace gas emission predictions. The question implied herein is whether changes in soil moisture and heat conditions that occur at the hourly or daily time step can be captured adequately in a temporally aggregated parameter for longer (monthly or annual) time step ecosystem studies. This is essentially a scaling question that we take up again later in this chapter. It is nevertheless evident from microcosm-level work that trace gas responses to wetting/drying events and lags in microbial responses to moisture changes are critical, but not well understood, processes that have been addressed primarily at a relatively fine temporal resolution. One potential solution rests in nested, hierarchical modelling approaches that aggregate functions properly across space/time scales.

To summarize the status of field-level simulations, the next generation of comprehensive ecosystem models should include at least three modular components:

1 scalable soil thermal and hydraulic flow simulators using empirical or stochastic climate drivers and process based plant–soil system controls on latent heat flux calculations;

2 substrate production submodels that include decomposition, mineralization, fermentation reactions and root interactions for substrates from biochemically distinct plant and soil organomineral sources;

3 trace gas production and diffusion transport equations with verification of aggregated parameters from microcosm-level modelling experiments.

11.5 Regional to global scale modelling

Modellers face significant challenges in generalizing available field data on rate and timing of climate controls, soil properties, and management/disturbance effects on landscape, regional and global trace gas budgets. Most large-scale knowledge about terrestrial trace gas emissions is currently based on multiplication of field flux estimates by the area of functionally classified land area to extrapolate microscale flux estimates (Matthews & Fung, 1987; Matson *et al.*, 1989; Matson & Vitousek, 1990), correlations with climate–soil attributes (Bouwman *et al.*, 1993; Raich & Schlesinger, 1992), or inverse modelling from atmospheric concentrations and tracers (Wigley, 1991; Fung *et al.*, 1983). All of these modelling methods offer valuable insights into controls on global trace gas dynamics. They do not, however, provide direct process-level ecological understanding of controls on fluxes. Previous global emission budget studies for CH_4 have also been based on 'measure and multiply' techniques (Matthews & Fung, 1987; Cicerone & Oremland, 1988; Aselmann & Crutzen, 1989; Taylor *et al.*, 1991) and N_2O (Bouwman *et al.*, 1993).

Modelling for application of soil microbiological and ecosystem level controls over relatively coarse spatial scales is an alternative method (to microscale measurement and extrapolation: 'measure and multiply') for extension of the process knowledge represented by the equations in Table 11.1. As an example, King *et al.* (1989) applied carbon cycling models for tundra and coniferous forest ecosystems for regional CO_2 exchange analysis by integration with Monte Carlo simulation of climate drivers. Four major components of this method were identified as:

1 regional boundary delineation;

2 development of the local ecosystem model;

3 driver descriptions as spatially distributed random variables; and

4 a procedure for calculating the expected value of the regional model. Century and TEM, two current ecosystem models, have been applied in large-scale analyses using geographic information systems to store and manage mapped information on properties regulating ecosystem behaviour.

Among a selection of notable ecosystem models developed previously – including Phoenix (McGill *et al.*, 1981); modified Jabowa (Pastor & Post, 1986); Century (Parton *et al.*, 1987; 1988); Forest-BGC (Running & Coughlan, 1988); OBM (Esser, 1990); GEM (Rastetter *et al.*, 1991); and TEM (Raich *et al.*, 1991; McGuire *et al.*, 1992) – three (Century, OMB and TEM) have been designed specifically for extension of biogeochemical cycling patterns to regional and global scales. The OBM is a georeferenced global carbon cycling model which runs on a 2.5° lat./lon. grid. The model couples ecosystem processes such as NPP and litter decomposition to coarse-scale environmental factors (climate, soil type, land use and atmospheric dynamics) rather than to regional vegetation units. Soil processes are not represented in detail (e.g. to level of Century), and trace gas production estimates other than CO_2 have not been considered.

The TEM has been applied as a continental scale model (0.5° lat./lon. grid). Potential NPP was estimated for South American life zones. Compared to OBM, TEM is somewhat more mechanistic in design, although calibration relies predominantly on ecosystem research site data sets that are taken to represent vegetation major units. Regional trace gas budget estimates have not been provided in TEM publications to date. It is notable that, because both the OBM and TEM are calibrated from site-specific ecosystem databases, very few independent sources have been used for validation of flux predictions.

Century was designed for regional analyses (county to continental scale), and was originally developed for grasslands. A global version was recently described (Schimel *et al.*, 1994). Currently, versions exist for grasslands, savannas, alpine tundra, forests and croplands. The model is applied regionally by coupling to commercial geographic information systems (GIS) software. Similarly to TEM and OBM, the model requires a base map of land cover classes. Unlike TEM and OBM, within biome types, Century can calculate changes in the relative abundance of photo-synthetic pathways (C_3 vs. C_4) and growth forms (woody vs. herbaceous in savannas). The model computes nitrogen trace gas fluxes, partitioned between NH_3 and N_2O as part of its base calculation of nitrogen mass balance. While the calculations of above-ground NPP and soil carbon storage have been extensively validated, few data sets exist to test the computed nitrogen trace gas fluxes.

To improve knowledge of atmospheric trace gas dynamics significantly, models must simulate the global terrestrial biosphere. Such global simulations require that models be linked to large-scale geographic databases describing soils, vegetation, land use and climate. Mechanistic representations of ecosystem biogeochemical cycles must integrate short and long-term soil processes at a level comparable to Century, a process requiring appropriate information on soil texture, vegetation composition, nutrient budgets and climate. Schimel *et al.* (1990) and Ojima *et al.* (1991) described attempts to couple Century to climate drivers from global atmospheric general circulation models (GCM), and emphasized the importance of matching biotic and climatic temporal scales for model integration. The problem of extrapolating biogeochemical models is reviewed generally in Schimel *et al.* (1991). Selected digital global data sets to support future modelling of this nature include monthly temperature and precipitation (Shea, 1986; Leemans & Cramer, 1990; Legates & Willmott, 1990), Advanced very high resolution radiometer-Normalized Difference Vegetation Index. (Tucker *et al.*, 1986; Fung *et al.*, 1987), solar radiation (Bishop & Rossow, 1991), vegetation (Matthews, 1983), ecosystem (Olson, 1983) or life zone (Holdridge, 1967; Leemans, 1990) classification, cultivation intensity (Matthews, 1983), soil type, texture and slope (Zobler, 1986), soil carbon and nitrogen contents (Post *et al.*, 1985) and fractional inundation (Matthews & Fung, 1987).

11.6 Spatial extrapolation approaches

The purpose of this section is to present a conceptual overview of methods of spatial extrapolation, and not to discuss specific algorithms and relationships, many of which have been reviewed above. Several procedures exist for spatial extrapolation of fluxes. All exist in a number of variants, and may be employed with either empirical databased algorithms or process simulation models. The first approach is often referred to as 'measure and multiply'. In this approach, the simplest and most commonly used, measurements are made in a range of surface types (vegetation, soil, ecosystem or biome) and fluxes are multiplied by areas. Hence, a total flux over i biomes (or vegetation types, soils series, ecosystem types or whatever) is computed as:

$$F = \Sigma(A_i \cdot F_i) \tag{11.5}$$

where F is the total flux, A_i is the area of biome i, and F_i is the flux from biome i. Matthews & Fung's (1987) and Aselmann & Crutzen's (1988) estimates of global CH_4 fluxes are examples of such an approach, as is Matson & Vitousek's (1990) estimate of tropical N_2O flux. In these applications, the mapped biome or other unit is used as a surrogate for the actual independent variables, and no explicit use is made of correlations with forcing functions (such as temperature, soil nitrogen or moisture).

A slightly more complex approach is typically employed using GIS technology where multiple effects may be overlain. If the relationships between a flux rate or ecosystem attribute and a series of independent variables are known, the maps of the independent variable may be overlain, discretized and cells defined that have specified properties. A flux or ecosystem attribute is then associated with each cell, and the cells summed. This approach, where multiple fields of independent variables (e.g. climate, soil, etc.) are overlain to form discrete classes with defined combinations of values of the independent variable, is often referred to as 'painting by numbers' because many, often irregularly shaped, cells are filled with a value of a computed flux. The technique is expressed in Eqn. 11.6

$$F = \Sigma A_i[f(x, y, z)] \tag{11.6}$$

where A_i is the area of cell i, and $f(x, y, z)$ indicates that the flux is a function of the values of the independent variables x, y and z defined for that cell. An example of 'painting by numbers' is Jenny et al.'s (1949) analysis of soil carbon storage in the US Great Plains as the result of the spatially orthogonal patterns of rainfall (east–west) and temperature (north–south). This type of analysis was extended to consider, in addition, soil texture as an input (Burke et al., 1990), and trace gas emissions as an output (Schimel et al., 1990). In these latter two cases, a simulation model (Century) was employed, with one simulation per cell, and with each cell having defined and discrete climate and soil attributes. A variant on this approach is the use of a regular grid or 'raster' format where each cell has defined climate, soil and ecosystem attributes, rather than using irregular contours of independent variables to define irregular regions of common properties. There are many papers using raster approaches, since this format is similar to the format employed in atmospheric models and so is compatible and familiar. Raster calculations are vastly simpler and faster than calculations on an irregular grid (vector format). Examples include McGuire et al. (1992) and Friedlingstein et al. (1992).

In more complex calculations, where the independent variables are known or interpolated to a continuous or nearly continuous format, the total flux may be calculated by integration. For example, the effect of varying temperature as a continuous function may be integrated (Eqns 11.7 and 11.8):

$$F = \int f(x, y, z), \tag{11.7}$$

$$F = \Sigma A_i \int [f(x, y, z)]i. \tag{11.8}$$

In the former case (Eqn. 11.7), the relationship is integrated over the entire region. In the latter case (Eqn. 11.8), a continuous function (e.g. climate) is integrated over each of i discrete classes (e.g. vegetation types). In formulations like the two integral approaches above, the integration may take into account the mean and frequency distribution of variables within a region (in which case x, y and z become probability functions). A statistical solution can be derived in which trace gas flux is calculated dynamically from the distribution function of the independent variables (e.g. King *et al.*, 1989). This approach is important in calculating fluxes with a non-linear model where the exact spatial overlay of the independent variables is known imperfectly or not at all, but the distributions are known.

In all of the approaches which rely upon a knowledge of a functional relationship, the degree to which data may be aggregated for a specified error is dependent upon the functional form of the relationship. In cases of extreme non-linearity, significant spatial disaggregation may be required, or careful application of a statistical approach. In all applications of techniques for extrapolation, careful consideration of error and error propagation is required (see following section).

In summary, mathematical techniques for extrapolation of fluxes depend strongly upon knowledge of the relationships between independent variables and processes, and upon knowledge of the geography of the independent variables, or controls, over fluxes. In cases where the geography is not well known, knowledge of the distributions can help improve solutions. There are many other issues relevant to extrapolation in space and time, such as techniques for contouring or interpolation, data retrieval from remote sensing, and scale dependencies. All of these techniques are useful, however, only when some relationship is known. Thus, extrapolation using any technique (except, arguably, measure and multiply) requires a mathematical framework describing relationships. That mathematical framework, as described above, may vary considerably as a function of scale (e.g. Groffman, 1991).

11.7 Functional responses and aggregation errors

Current ecosystem models perform adequately at predicting biogeochemical rates when detailed site-specific data are available. Most current ecosystem and trace gas models were originally developed for site-specific applications or are based on conceptual approaches first developed at that level. The greatest challenges to extrapolation generally arise from the inadequacy of current databases describing process controls (Schimel *et al.*, 1991). The broader application of conventional biogeochemical models can be limited if they are sensitive to values of site-specific data

Table 11.2 Factors controlling trace gas fluxes at various spatial scales. (After Groffman, 1991.)

Scale of model	Trace gas controls
Soil microorganism	Carbon and nitrogen substrate, oxygen, temperature, moisture
Ecosystem	Litter carbon/nitrogen/lignin, soil water-filled pore space
Biome	Vegetation type, soil texture, land use
Global	Climate

describing controls over processes (King *et al.*, 1989). Serious error propagation problems can arise when calibrations made at fine scales are applied directly to describe behaviour over large areas (O'Neill, 1979). Relationships developed for processes understood on a centimetre to metres scale must be transformable to relationships at scales of tens to hundreds of kilometres in order to yield useful information about, for example, effects of trace gas fluxes and concomitant climate change on ecosystem structure and function. Groffman (1991) has provided a useful classification of factors affecting denitrification activity at various scales of investigation (Table 11.2), illustrating scale transformation of process models. He argues that explicit representation of microbial processes is progressively lost when moving from the smallest to the largest scale, but can be replaced by empirical correlates of activity.

Process modelling of soil trace gas fluxes at coarse spatial (meso- to macro-scale) and temporal (daily to monthly time step) resolutions may call for an approach in which the overall dynamics of fine resolution processes are represented at a more aggregated level of system behaviour (e.g. among those shown in Table 11.2). For example, trace gas models which require specification of pore-solute velocity scalars, substrate diffusion coefficients and reaction rate constants are difficult to calibrate at macro-scales, due primarily to the lack of appropriate soil property information except at very well-studied research sites (Zobler, 1986). Therefore, in the scaling process, one important research objective is to assess the degree of accuracy that is lost in describing any given process at one of the higher levels in Table 11.2.

An initial qualitative assessment of the contribution of a specific process to aggregation errors can be carried out by plotting the fine-scale relationship of flux scalar vs. empirical controlling variable when the functional form of the relationship is known (Rastetter *et al.*, 1992); for example NO_3 diffusion coefficient vs. soil moisture. When such plots are strongly non-linear, large aggregation errors in coarse scale analyses may occur. Combination of functional response forms with information about the range and distributions of controlling variables can lead to a quantitative assessment of extrapolation errors. An example of aggregation error analysis is described in Burke *et al.* (1991), which describes spatial aggregation errors associated with soil texture effects. In the following sections, we provide illustrations for important trace gas controllers.

11.7.1 Substrate availability

Aggregation errors associated with substrate regulation of trace gas production should be minimal over the range of concentrations in which zero order kinetics operate. Using denitrification as a case in point, previous zero order kinetic models (Misra *et al.*, 1974b; Starr & Parlange, 1976; Kanwar *et al.*, 1980) have been calibrated using nitrogen substrate concentrations in the range of $5-100 \, \text{mg} \, \text{l}^{-1}$. Average forest soil solution nitrate levels, on the other hand, rarely exceed $0.3 \, \text{mg} \, \text{l}^{-1}$ (Vitousek *et al.*, 1981), occasionally reaching levels within a range of $7-30 \, \text{mg} \, \text{l}^{-1}$, which leads to the conclusion that zero order kinetic conditions are not typical of natural, undisturbed soils. Parton *et al.* (1988) showed that inorganic nitrogen availability frequently limited N_2O production in grasslands, and that first order kinetics prevailed. In the field, however, heterogeneity in soil structure attributes makes average substrate concentrations less meaningful. The occurrence of large pore spaces in predominately coarse-textured soils may result in dramatic substrate gradients along a soil profile, so that both zero and first order kinetics may apply in a single soil. Measured substrate concentration averages in fine-textured soils are likely to more closely reflect pore space conditions (Kanwar *et al.*, 1980).

Assuming first order kinetics for most intact soils, the hyperbolic substrate regulation relationship (Fig. 11.1a) used by McConnaughey & Bouldin (1985) and Leffelar & Wessel (1988) may be associated with important aggregation errors during scaling. The degree of non-linearity over the range of substrate concentrations will depend on the half-saturation constant (K). Typical K values ($\text{kg} \, \text{m}^{-3}$) are 1.7 and 8.6 for carbon and nitrogen electron acceptors, respectively (Leffelar & Wessel, 1988).

11.7.2 Soil moisture status

The water-filled pore space (WFPS) index has been used to examine the relationship of carbon (CO_2) and nitrogen trace gas fluxes to soil moisture and oxygen status (Linn & Doran, 1984; Doran *et al.*, 1990; Davidson, 1991). These authors report a strongly concave function over the range of $20-100\%$ WFPS (Fig. 11.1b). At fine spatial resolution, the range and distribution of soil textures (and hence microsite structures) will have an

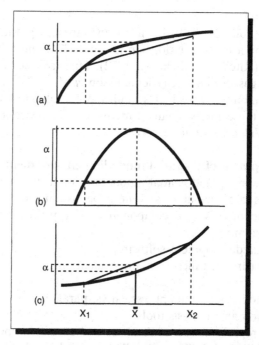

Fig. 11.1 Relationship of flux scalar (*y* axis) vs. controlling variable (*x* axis) for effects on soil trace gas emissions. (a) Substrate availability (Michaelis–Menten); (b) soil water-filled pore space (after Doran, 1990); and (c) soil temperature (exponential Q_{10} function). Potential aggregation errors (α) are represented by flux scalar differences between the variable response curve and the linear approximation at an arithmatic mean (*X*) of a measured range of controlling variable values (X_1–X_2).

impact on WFPS. We anticipate that net soil moisture inputs (calculated as the difference between precipitation and evapotranspiration) are likely to contribute strongly to aggregation errors in meso- to macro-scale analyses.

11.7.3 Soil temperature

A general $Q_{10} = 2$ relationship (Fig. 11.1c) has often been used to describe the relationship of trace gas fluxes to temperature (Stanford *et al.*, 1975). Support for consistent model application of the $Q_{10} = 2$ across biomes and temperature regimes comes from the work of Holland *et al.* (submitted). These authors found no significant difference in Q_{10} functional response values affecting soil respiration in a variety of climate and soil types. Aggregation errors associated with temperature effects on trace gas fluxes are likely to be less important than those associated with more episodic variables, such as substrate concentration and soil moisture.

11.7.4 Soil pH

The effect of soil pH (4–9) on nitrogen trace gas production has been represented by a linear relationship (Focht, 1974) and by slightly hyperbolic curves (Ritchie & Nicholas, 1972). High soil acidity can severely restrict trace gas production (Klemedtsson *et al.*, 1978). CH_4 production is strongly inhibited by acid pHs, and may be inhibited at very alkaline pHs, though these rarely occur in methanogenic environments (Conrad, 1984, 1989; Valentine *et al.*, 1994).

11.8 Development of a spatial modelling environment

Although few integrated spatial model development environments exist, key elements are available and may be assembled to provide most required functions. Major components of a process-based ecological modelling system include:

1 model code development software;
2 databases (driver, calibration and evaluation);
3 GIS;
4 computational environment – serial vs. parallel;
5 visualization and analysis tools.

A schematic representation of database and ecosystem model integration is shown in Fig. 11.2. The modelling procedure should begin with a conceptual design that recognizes the parameter limitations imposed by availability and resolution of initial state and driving variables in geographic databases. For example, in the development of a process model for macro-scale trace gas emissions, soil porosity attributes are likely to be derived from the Food and Agriculture Organization map of the world at $1° \times 1°$ lat./lon. resolution (Zobler, 1986).

Conversion of conceptual to mathematical models can be greatly facilitated by commercially available software packages for the personal computer (Costanza, 1987; Costanza & Maxwell, 1991). These applications are based on principles of systems dynamics (Forrester, 1961). Icons representing stock, flow and converter variables may be assembled visually and defined mathematically to create models at any scale. Modern software development environments, available for personal computers and workstations, greatly facilitate development of efficient, documented code for Fortran or C implementations of models. As the complexity and effort spent on code development increases, the importance of producing documented code that can be modified as new knowledge becomes available increases.

To adapt a single point model to spatial modelling, driving and initiaization variables may be entered as raster or vector map arrays in a GIS. In raster maps, data are described in a regular grid of values. In vector databases, more akin to traditional maps, irregularly shaped polygons are

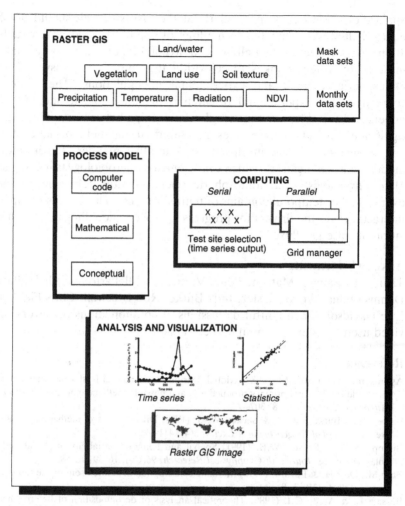

Fig. 11.2 A framework for spatial database and ecosystem model integration. GIS may be combined with routines to generate probability densities of input variables for model forcing functions. Parallel processing of gridded spatial models speeds computation times.

defined for each mapping unit. As a stand-alone system, a GIS handles digitization, transformation and storage of data sets. Integration of dynamic model variables with the GIS adds the capability for output map layer overlay, weighting or averaging by reference (e.g. biome type, life zone or soil type) databases, and rapid display of output in map format.

11.9 Concluding remarks
Impressive advances are being made in the area of high performance computing, visualization, data handling tools, GIS and geostatistical

applications. Modelling, particularly at the process level, should not be equated, however, with the technology of computing and data visualization. As we see it, modelling is a hypothesis-driven activity that is most useful when it takes place within the context of ecosystem and atmospheric research and experimental manipulations. The value of coupled model–experimental research is in being able to identify and understand variables, or functional relationships between variables, that significantly affect process rates measured in the field or laboratory, and to test those relationships in an interative manner. Furthermore, modelling is the means by which soil scientists, microbiologists, ecologists, plant physiologists and atmospheric scientists ultimately communicate over spatial or temporal domains of mutual interest. These cross-discipline interactions are the drivers for progress in process modelling and extrapolation of trace gas fluxes.

Acknowledgements
Thanks to Pamela Matson, Peter Vitousek, Bill Parton, Beth Holland, Dennis Ojima, Arvin Mosier, Indy Burke, Rob Braswell, Chris Field and Eric Davidson for insightful discussions. Two anonymous reviewers provided useful comments on an earlier version.

References
Aselmann I. & Crutzen P.J. (1989) Global distribution of natural freshwater wetlands and rice paddies, their net primary productivity, and possible methane emissions. *Journal of Atmospheric Chemistry*, **8**, 307–358.
Barber T.R., Burke R.A. Jr & Sackett W.M. (1988) Diffusive flux of methane from warm wetlands. *Global Biogeochemical Cycles*, **2**(4), 411–425.
Bishop J.K.B. & Rossow W.B. (1991) Spatial and temporal variability of global surface solar irradiance. *Journal of Geophysical Research*, **96**(C9), 16839–16858.
Born M., Dorr H. & Levin I. (1990) Methane consumption in aerated soils of the temperate zones. *Tellus*, **42B**, 2–8.
Bossata E. & Agren G.I. (1985) Theoretical analysis of decomposition of heterogeneous substrates. *Soil Biology and Biochemistry*, **17**, 601–610.
Bouwman A.F., Fung I., Matthews E. & John J. (1993) Global analysis of the potential for N_2O production in natural soils. *Global Biogeochemical Cycles*, **7**(3), 557–597.
Bowden W.B. (1984) Nitrogen and phosphorus in the sediments of a tidal freshwater marsh in Massachusetts (USA). *Estuaries*, **7**, 108–118.
Burke I.C., Schimel D.S., Yonker C.M., Parton W.J., Joyce L.A. & Lauenroth W.K. (1990) Regional modeling of grassland biogeochemistry using GIS. *Landscape Ecology*, **4**(1), 45–54.
Burke I.C., Kittel T.G.F., Lauenroth W.K., Snook P., Yonker C.M. & Parton W.J. (1991) Regional analysis of the Central Great Plains. *BioScience*, **41**, 685–692.
Chanton J.P., Martens C.S. & Kelley C.A. (1989) Gas transport from methane-saturated tidal freshwater and wetland sediments. *Limnology and Oceanography*, **34**, 807–819.
Cho C.M. (1971) Convective transport of ammonium with nitrification in soil. *Canadian Journal of Soil Science*, **51**, 339–350.
Cicerone R. & Oremland R. (1988) Biogeochemical aspects of atmospheric methane. *Global Biogeochemical Cycles*, **2**, 299–327.

Collatz G.L., Ball J.T., Grivet C. & Berry J.A. (1991) Physiological and environmental regulation of stomatal conductance, photosynthesis and transpiration: a model that includes a laminar boundary layer. *Agricultural and Forest Meteorology*, **54**, 107–136.

Conrad R. (1984) Capacity of aerobic microorganisms to utilize and grow on atmospheric trace gases (H_2, CO CH_4). In: Klug M.J. & Reddy C.A. (eds) *Current Perspectives in Microbial Ecology*, pp. 461–467. American Society for Microbiology, Washington, DC.

Conrad R. (1989) Control of methane production in terrestrial ecosystems. In: Andreae M.O. & Schimel D.S. (eds) *Exchange of Trace Gases between Terrestrial Ecosystems and the Atmosphere*, pp. 39–58. Wiley, New York.

Costanza R. (1987) Simulation modeling on the Macintosh using STELLA. *BioScience*, **37**, 129–132.

Costanza R. & Maxwell T. (1991) Spatial ecosystem modeling using parallel processors. *Ecological Modelling*, **58**, 159–183.

Dacey J.W. & Klug M.J. (1979) Methane efflux from lake sediments through water lilies. *Science*, **203**, 1253–1254.

Davidson E.A. (1991) Fluxes of nitrous and nitric oxide from terrestrial ecosystems. In: Rogers J.E. & Whitman W.B. (eds) *Microbial Production and Consumption of Greenhouse Gases: Methane, Nitrogen Oxides and Halomethanes*, pp. 219–235. American Society for Microbiology, Washington, DC.

de Vries D.A. (1958) Simultaneous transfer of heat and moisture in porous media. *Transactions, American Geophysical Union*, **39**, 909–916.

Doran J.W., Mielke L.N. & Power J.F. (1990) Microbial activity as regulated by soil water filled pore space. In: *Transactions of the 14th International Congress of Soil Science*, pp. 94–99. Kyoto, Japan.

Esser G. (1990) Modelling global terrestrial sources and sinks of CO_2 with special reference to soil organic matter. In: Bouwman A.F. (ed) *Soils and the Greenhouse Effects*, pp. 247–262. John Wiley and Sons, New York.

Focht D.D. (1974) The effect of temperature, pH, and aeration on the production of nitrous oxide and gaseous nitrogen – A zero order kinetic model. *Soil Science*, **118**, 173–179.

Forrester J.W. (1961) *Industrial Dynamics*. MIT Press, Cambridge, Massachusetts.

Friedlingstein P., Delire C., Müller J.F. & Gérard J.C. (1992) The climate induced variation of the continental biosphere: A model simulation of the last glacial maximum. *Geophysical Research Letters*, **19**(9), 897–900.

Fung I., Prentice K., Matthews E., Lerner J. & Russell G. (1983) A 3-D tracer model study of atmospheric CO_2: Response to seasonal exchanges with the terrestrial biosphere. *Journal of Geophysical Research*, **88**, 1282–1294.

Fung I.Y., Tucker C.J. & Prentice K.C. (1987) Application of advanced very high resolution radiometer vegetation index to study of atmosphere–biosphere exchange of CO_2. *Journal of Geophysical Research*, **92**, 2999–3015.

Grant R.F. (1991) A technique for estimating denitrification rates at different soil temperatures, water contents, and nitrate concentrations. *Soil Science*, **152**(1), 41–52.

Groffman P.M. (1991) Ecology of nitrification and denitrification in soil evaluated at scales relevant to atmospheric chemistry. In: Rogers J.E. & Whitman W.B. (eds) *Microbial Production and Consumption of Greenhouse Gases: Methane, Nitrogen Oxides and Halomethanes*, pp. 201–217. American Society for Microbiology, Washington, DC.

Hillel D. (1982) *Introduction to Soil Physics*. Academic Press, New York.

Hinds A.A. & Lowe L.E. (1980) Distribution of carbon, nitrogen, sulphur and phosphorus in particle-size separates from gleysolic soils. *Canadian Journal of Soil Science*, **60**, 783–786.

Holdridge L.R. (1967) *Life Zone Ecology*. Tropical Science Center. San Jose, Costa Rica.

Holland E.A., Townsend A.K. & Vitousek P.M. Temperature and substrate controls over soil respiration in tropical soils. *Global Change Biology*, (submitted).

Jenny H., Gessel S.P. & Bingham F.T. (1949) Comparative study of decomposition rates of organic matter in temperate and tropical regions. *Soil Science*, **68**, 419–432.

Kanwar R.S., Baker J.L., Johnson H.P. & Kirkham D. (1980) Nitrate movement with zero-order denitrification in a soil profile. *Soil Science Society of America Journal*, **44**, 898–902.

Keller M.M. (1990) *Biological sources and sinks of methane in tropical habitats and tropical atmospheric chemistry*. PhD thesis, Princeton University, Princeton, New Jersey.

King A.W., O'Neill R.V. & DeAngelis D.L. (1989) Using ecosystem models to predict regional CO_2 exchange between the atmosphere and the terrestrial biosphere. *Global Biogeochemical Cycles*, **3**(4), 337–361.

Kirda C., Starr J.L., Mirsa C., Biggar J.W. & Nielson D.R. (1974) Nitrification and denitrification during miscible displacement in unsaturated soil. *Soil Science Society of America Proceedings*, **38**, 772–776.

Klemedtsson L., Svensson B.H., Lindberg T. & Rosswall T. (1978) The use of acetylene inhibition of nitrous oxide reductase in quantifying denitrification in soils. *Swedish Journal of Agricultural Research*, **7**, 179–185.

Knapp E.B., Elliott L.F. & Campbell G.S. (1983) Carbon, nitrogen, and microbial biomass interrelationships during decomposition of wheat straw: a mechanistic simulation model. *Soil Biology and Biochemistry*, **15**(4), 455–461.

Kuivila K.M., Murray J.W., Devol A.H., Lidstrom M.E. & Reimers C.E. (1988) Methane cycling in the sediments of Lake Washington. *Limnology and Oceanography*, **33**(4,1), 571–581.

Langford A.D. & Fehsenfeld F.C. (1992) The role of natural vegetation of a source or sink for atmospheric ammonia: A case study. *Science*, **255**, 581–583.

Langford A.O., Fehsenfeld F.C., Zachariassen J. & Schimel D.S. (1992) Gaseous ammonia fluxes and background concentrations in terrestrial ecosystems of the United States. *Global Biogeochemical Cycles*, **6**(4), 459–483.

Leemans R. (1990) *Global Holdridge Life Zone Classifications. Digital Data*. International Institute of Applied Systems Analysis, Laxenburg, Austria.

Leemans R. & Cramer W.P. (1990) The IIASA database for mean monthly values of temperature, precipitation and cloudiness of a global terrestrial grid. Wp-41. *International Institute of Applied Systems Analysis*. Laxenburg Working Paper, Austria.

Leffelar P.A. & Wessel W.W. (1988) Denitrification in a homogeneous closed system: Experiment and simulation. *Soil Science*, **146**, 335–349.

Legates D.R. & Willmott C.J. (1990) Mean seasonal and spatial variability in gauge-corrected global precipitation. *International Journal of Climatology*, **10**, 111–127.

Lerdau M.T. (1991) Plant function and biogenic terpene emission. In: Sharkey T., Holland E. & Mooney H. (eds) *Trace Gas Emissions by Plants*, pp. 121–134. Academic Press, San Diego.

Letey J., Jury W.A., Hadas A. & Valoras N. (1980) Gas diffusion as a factor in laboratory incubation studies on denitrification. *Journal of Environmental Quality*, **9**(2), 223–227.

Li C., Frolking S.E. & Frolking T.A. (1992) A model of nitrous oxide evolution from soil driven by rainfall events: I. Model structure and sensitivity. *Journal of Geophysical Research*, **97**(D9), 9759–9776.

Linn D.M. & Doran J.W. (1984) Effect of water-filled pore space on carbon dioxide and nitrous oxide production in tilled and nontilled soils. *Soil Society of America Journal*, **48**, 1267–1272.

Liss P.S. & Slater P.G. (1974) Flux of gases across the air–sea interface. *Nature*, **247**, 181–184.

Marshall T.J. & Holmes J.W. (1988) *Soil Physics*. Cambridge University Press, Cambridge.

Matthews E. (1983) Global vegetation and land use: New high-resolution databases for climate studies. *Journal of Climate and Applied Meteorology*, **22**, 474–487.

Matthews E. & Fung I. (1987) Methane emission from natural wetlands: Global distribution, area and environmental characteristics of sources. *Global Biogeochemical Cycles*, **1**(1), 61–86.

Matson P.A. & Vitousek P.M. (1990) An ecosystem approach to the development of a global nitrous oxide budget. *BioScience*, **40**, 677–672.

Matson P.A., Vitousek P.M. & Schimel D.S. (1989) Regional extrapolation of trace gas flux based on soils and ecosystems. In: Andreae M.O. & Schimel D.S. (eds) *Exchange of Trace Gases Between Ecosystems and the Atmosphere*, pp. 97–108. John Wiley and Sons, New York.

McConnaughey P.K. & Bouldin D.R. (1985) Transient microsite models of denitrification: I. Model development. *Soil Science Society of America Journal*, **49**, 886–891.

McGill W.B., Hunt H.W., Woodmansee R.G. & Reuss J.O. (1981) A model of the dynamics of carbon and nitrogen in grassland soils. *Ecological Bulletin (Stockholm)*, **33**, 49–116.

McGuire A.D., Melillo J.M., Joyce L.A. *et al.* (1992) Interactions between carbon and nitrogen dynamics in estimating net primary production for potential vegetation in North America. *Global Biogeochemical Cycles*, **6**(2), 101–124.

Meentemeyer V. (1984) The geography of organic decomposition rates. *Annual Association American Geographers*, **74**, 551–560.

Milly P.C.D. (1982) Moisture and heat transport in hysteretic, inhomogeneous porous media: A matric head-based formulation and a numeric model. *Water Resources Research*, **18**, 489–498.

Misra C., Nielsen D.R. & Biggar J.W. (1974a) Nitrogen transformations in soil during leaching: I. Theoretical considerations. *Soil Science Society Proceedings*, **38**, 289–293.

Misra C., Nielsen D.R. & Biggar J.W. (1974b) Nitrogen transformations in soil during leaching: III. Nitrate reduction in soil columns. *Soil Science Society Proceedings*, **38**, 300–304.

Mosier A.R., Schimel D.S., Valentine D., Bronson K. & Parton W. (1991) Methane and nitrous oxide fluxes in native, fertilized and cultivated grasslands. *Nature*, **350**, 330–332.

Nobel P.S. (1991) *Physiochemical and Environmental Plant Physiology*. Academic Press, San Diego, California.

Ojima D.S., Kittel T.G.F., Rosswall T. & Walker B.H. (1991) Critical issues for understanding global change effects on terrestrial ecosystems. *Ecological Applications*, **1**(3), 316–325.

Ojima D.S., Valentine D.W., Mosier A.R., Parton W.J. & Schimel D.S. (1993) Effect of land use change on methane oxidation in temperate forest and grassland soils. *Chemosphere*, **26**(1–4), 675–685.

Olson J. (1983) *Carbon in Live Vegetation of the Major World Ecosystems*. DOE/NBB-0037, Oak Ridge National Laboratory, Tennessee.

O'Neill R.V. (1979) Transmutations across hierarchical levels. In: Innis G.S. & O'Neill R.V. (eds) *Systems Analysis of Ecosystems*, pp. 59–78. International Cooperative Publishing House, Fairland, Maryland.

Oremland R.S. & Culberson C.W. (1992) Importance of methane-oxidizing bacteria in the methane budget as revealed by the use of a specific inhibitor. *Nature*, **356**, 412–423.

Parton W.J., Mosier A.R. & Schimel D.S. (1988) Dynamics of C, N, P, and S in grassland soils: a model. *Biogeochemistry*, **5**, 109–131.

Parton W.J., Schimel D.S., Cole C.V. & Ojima D.S. (1987) Analysis of factors controlling soil organic matter levels in Great Plains grasslands. *Soil Science Society of America Journal*, **51**(5), 1173–1179.

Pastor J. & Post W.M. (1986) Influence of climate, soil moisture and succession on forest carbon and nitrogen cycles. *Biogeochemistry*, **2**, 3–27.

Philip J.R. & de Vries D.A. (1957) Moisture movement in porous materials under temperature gradients. *Transactions of the American Geophysics Union*, **38**, 222–232.

Pirt S.J. (1965) The maintenance energy of bacteria in growing cultures. *Proceedings of the Royal Society, London*, Series B, **163**, 224–231.

Post W.M., Pastor J., Zinke P.J. & Stangenberger A.G. (1985) Global patterns of soil nitrogen storage. *Nature*, **317**, 613–616.

Raich J.W., Rastetter E.B., Melillo J.M. *et al.* (1991) Potential net primary production in South America: application of a global model. *Ecological Applications*, **1**(4), 399–429.

Raich J.W. & Schlesinger W.H. (1992) The global carbon dioxide flux in soil respiration and its relationship to climate. *Tellus*, **44B**, 81–99.

Rastetter E.B., King A.W., Cosby B.J., Hornberger G.M., O'Neill R.V. & Hobbie J.E. (1992) Aggregating fine-scale ecological knowledge to model coarser-scale attributes of ecosystems. *Ecological Applications*, **2**(1), 55–70.

Rastetter E.B., Ryan M.G., Shaver G.R. *et al.* (1991) A general biogeochemical model describing the responses of the C and N cycles in terrestrial ecosystems to changes in CO_2, climate, and N deposition. *Tree Physiology*, **9**, 101–126.

Ritchie G.A.F. & Nicholas D.J.D. (1972) Identification of the sources of nitrous oxide produced by oxidative and reductive processes in Nitrosomonas europea. *Biochemical Journal*, **126**, 1181–1191.

Robinson J.A. (1985) Determining microbial kinetic parameters using non-linear analysis: Advances and limitations in microbial ecology. *Advances Microbial Ecology*, **8**, 61–114.

Roulet N., Moore T., Bubier J. & LaFleur P. (1992) Northern fens: methane flux and climatic change. *Tellus*, **44B**, 100–105.

Running S.W. & Coughlan J.C. (1988) A general model of forest ecosystem processes for regional applications. I. Hydrologic balance, canopy gas exchange and primary processes. *Ecological Modelling*, **42**, 125–154.

Schimel D.S., Braswell B.H., Holland E.A. *et al.* (1994) Climatic edaphic, and biotic controls over carbon and turnover of carbon is soils. *Global Biogeochemical Cycles*, **8**(3), 279–293.

Schimel D.S., Kittel T.G.F. & Parton W.J. (1991) Terrestrial biogeochemical cycles: global interactions with the atmosphere and hydrology. *Tellus*, **43AB**, 188–203.

Schimel D.S., Parton W.J., Kittel T.G.F., Ojima D.S. & Cole C.V. (1990) Grassland biogeochemistry: links to atmospheric processes. *Climatic Change*, **15**, **17**, 13–25.

Shea D.J. (1986) *Climatological Atlas 1950–1979*. Surface air temperature, precipitation, sea-level pressure and sea surface temperature (45°S–90°N). Atmospheric Analysis and Prediction Division, National Center for Atmospheric Research, NCAR Technical Note 269+STR.

Smith J.L., McNeal B.L., Cheng H.H. & Campbell G.S. (1986) Calculation of microbial maintenance rates and net nitrogen mineralization in soil at steady state. *Soil Science Society of America Journal*, **50**, 332–338.

Smith O.L. (1979) An analytical model of the decomposition of soil organic matter. *Soil Biology and Biochemistry*, **11**, 585–606.

Stanford G. & Smith S.J. (1972) Nitrogen mineralization potentials of soils. *Soil Science Society of America Proceedings*, **36**, 465–472.

Stanford G., Dzienia S. & Vander Pol R.A. (1975) Effect of temperature on denitrification rate in soils. *Soil Science Society of America Proceedings*, **39**, 867–870.

Starr J.L. & Parlange J.Y. (1976) Relation between the kinetics of nitrogen transformation and biomass distribution in a soil column during continuous leaching. *Soil Science Society of America Journal*, **40**, 458–460.

Svensson B.H. & Rosswall T. (1984) *In situ* methane production from acid peat in plant communities with different moisture regimes in a subarctic mire. *Oikos*, **43**, 341–350.

Taylor J.A., Brasseaur G.P., Zimmerman P.R. & Cicerone R.J. (1991) A study of the sources and sinks of methane and methyl chloroform using a global three-dimensional Lagrangian tropospheric tracer transport model. *Journal of Geophysical Research*, **96**(D2), 3013–3044.

Townsend A.R., Vitousek P.M. & Holland E.A. (1992) Tropical soils dominate the short-term carbon cycle feedbacks to atmospheric carbon dioxide. *Climatic Change*, **22**, 293–303.

Tucker C.J., Fung I.Y., Keeling C.D. & Gammon R.H. (1986) Relationship between atmospheric CO_2 variations and a satellite-derived vegetation index. *Nature*, **319**, 195–199.

Valentine D.W., Holland E.A. & Schimel D.S. (1994) Ecosystem and physiological controls over methane production in northern wetlands. *Journal of Geophysical Research*, **99(D1)**, 1563–1571.

Van Veen J.A. & Paul E.A. (1981) Organic carbon dynamics in grassland soils. I. Background information and computer simulation. *Canadian Journal of Soil Science*, **61**(2), 185–201.

Vitousek P.M., Reiners W.A., Melillo J.M., Grier C.C. & Gosz J.R. (1981) Nitrogen cycling and loss following forest perturbation: The components of response. In: Barrett G.W. & Rosenberg R. (eds) *Stress Effects on Natural Ecosystems*, pp. 115–127. John Wiley and Sons, New York.

von Caemmerer S. & Farquhar G.D. (1985) Kinetics and activation of Rubisco and some preliminary modeling of RuP_2 pool sizes. In: Vill J., Grishina G. & Laisk A. (eds) *Proceedings of the 1983 Conference at Tallinn*, pp. 46–58. Estonian Academy of Sciences, Tallinn.

Wigley T.M.L. (1991) A simple inverse carbon cycle model. *Global Biogeochemical Cycles*, **5**(4), 373–382.

Williams E.J., Guenther A. & Fehsenfeld F.C. (1992) An inventory of nitric oxide emissions from soils in the United States. *Journal of Geophysical Research*, **97**(D9), 7511–7519.

Zobler, L. (1986) *A World Soil File for Global Climate Modeling*. NASA Technical Memorandum 87802.

Index

Page numbers in *italic* indicate figures and **bold** indicate tables

ABLE-3A *see* Arctic Boundary Layer Experiment
acid bubblers 220
acidity of soil 376
  and microbial processes 346–7
advective transport 17
air–water interface 70
  gas exchange at *53–4*
aircraft techniques for surface flux measurements 131, 157–9
ammonia
  fluxes 2, 223–5, 273
    air–water interface 53
    enclosure-based 25, 223, 225
    foliage reabsorption *19*
  isotopic studies
    isotopic abundance 292, **293**
    sample purification 302
  measurement **167**
    acid bubblers 220
    ambient concentrations 217–25
      fate in atmosphere 217–19
      techniques 219–23, **242**
    chemical methods 170
    citric acid denuder **242**
    DIAL 222
    diffusion denuders 220
    filter packs 220–1, **242**
    Fourier transform infrared spectroscopy 280–1
    laser-induced fluorescence **242**
    molybdenum oxide denuder *224*, **242**
    thermal denuders 221–2
    tungsten oxide denuder *224*, **242**
    vegetative transport 19
analytical methods 164–257
  ammonia 217–23
    techniques 167, 219–25, **242**
  calibration 234–6
    analytical interferences 236
    diffusion/effusion sources 236
    permeation tube 235
    standard reference gas mixture 234–5

carbon dioxide exchange 181–2
carbon monoxide 183–7, 225–8
  GC and ECD 186–7
  GC and HgO reduction 186
  methanization and GC-FID 185
  NDIR gas filter correlation 184–5
  solar spectroscopy 183
  techniques 166, 183–7, **243**
  tunable diode laser spectroscopy 183–4
hydrocarbons 187–9
microbial *see* microbial processes
nitrogen gases 189–93, 207–25
  nitric oxide, NO_x, NO_y 189–90, 209–15
  nitrous oxide 190–3
optical 171, *172*
ozone 232–4, **248**
  fate in atmosphere 232
  techniques 232–4, **248**
reduced sulphur gas 193–8
reference gases and materials 187
standardization 165, 168, 234–5
sulphur compounds 193–8, 228–32
  techniques 229–32, **244–7**
see also individual trace gases and analytical methods
Arctic Boundary Layer Experiment 6–7
atmospheric boundary layer 127–39, 133, 135
  basic considerations 127, *128*–30
  turbulence characteristics 15, 130–9
atmospheric fate
  ammonia 217–19
  carbon monoxide 225
  nitrogen oxides 207, *208–9*
  sulphur compounds 228–9
atmospheric pressure ionization/mass spectrometry 285

bacteria
  denitrifying *see* denitrification
  nitrifying 328–9
Beer–Lambert law 171
boundary layers
  atmospheric 127–39
    basic considerations 127, **128**–30
    turbulence characteristics 130, *131*–2,

385